Ibrahim A. Mirsal

Soil Pollution
Origin, Monitoring & Remediation

Ibrahim A. Mirsal

Soil Pollution

Origin, Monitoring & Remediation

Second edition
With 197 Figures and 34 Tables

Author

Prof. Dr. Ibrahim A. Mirsal
Oberroßbacherstr. 53
35685 Dillenburg, Germany
E-mail: IbrahimMirsal@web.de

ISBN: 978-3-642-08968-8   e-ISBN: 978-3-540-70777-6

1st Edition 2004

© 2008 Springer-Verlag Berlin Heidelberg
Softcover reprint of the hardcover 2nd edition 2008

This work is subject to copyright. All rights are reserved, whether the whole or part of the material is concerned, specifically the rights of translation, reprinting, reuse of illustrations, recitations, broadcasting, reproduction on microfilm or in any other way, and storage in data banks. Duplication of this publication or parts thereof is permitted only under the provisions of the German Copyright Law of September 9, 1965, in its current version, and permission for use must always be obtained from Springer. Violations are liable to prosecution under the German Copyright Law.

The use of general descriptive names, registered names, trademarks, etc. in this publication does not imply, even in the absence of a specific statement, that such names are exempt from the relevant protective laws and regulations and therefore free for general use.

Cover design: deblik, Berlin

Printed on acid-free paper   30/2132/AO – 5 4 3 2 1 0

springer.com

To Gabi, Miriam and Jasmin

Preface to the Second Edition

Despite having been published about two years ago for the first time, the continuous demand for this book encouraged me to prepare this revised and enlarged edition. Many parts of the text have been rewritten, type errors traced and corrected, and the bibliography largely modified to include many of the references published about the subject of soil pollution in the previous ten years.

I should like to express my thanks to the staff of Springer-Verlag, Heidelberg, for their cooperative efforts in preparing this edition. I also would like to thank Mr. Michael Sidwell (B.A.) for the extreme but characteristic care with which he read and revised the proofs.

I hope that, in this new edition, the book may continue to serve the needs of students and professionals alike interested in the subject of soil pollution.

Ibrahim A. Mirsal

Preface to the First Edition

Whoever has enjoyed following the legendary duel between the Egyptian Pharaoh and his magicians (Alchemists) on one side, and Moses and his brother Aaron on the other, as is vividly narrated in the Bible, must have realised that people (at least those living at, or near the eternal battlefields of the Middle East) have always had knowledge about the terrible consequences of soil pollution by chemicals. This knowledge must have existed long before Moses and his Pharaoh. Nobody knows when people became aware of this, yet it must have been born in very early times, reaching back to the dawn of human conscious.

As history teaches, human knowledge explodes in logarithmic dimensions, and times have come where pollution has attained alarming levels, with chemicals in industrial and military use becoming a threat to life everywhere on this Earth. At the end of the 20th century, states commenced projects for pollution control and remediation, major universities altered their programs to include environmental studies at central positions within their curricula, and nations, putting their differences aside, came together to sign the Chemical Weapons Convention (CWC) and other treaties allowing pollution control on a worldwide scale.

The present book is designed as a contribution to the understanding of the origins, mechanisms and consequences of the environmental setbacks brought about by soil pollution. It is based on university lectures held by the author over the last twenty years at the University of Marburg, Germany; the University of the Philippines, Manila, Philippines; and Bogaziçi University, Istanbul, Turkey.

This book is useful for students of Earth Sciences, Environmental Sciences and Agriculture, as well as for professionals of all these fields who are seeking an integration of all the isolated parts they have learnt about soil pollution. It may also be of help for members of environmental NGO's and environment community officers.

I wish to acknowledge with gratitude my indebtedness to the members of my family for their support and patience during the preparation of this book. My thanks are also due to many of my colleagues and students in Manila and Istanbul. I am particularly indebted to the library staff of Bogaziçi University, Istanbul, who always offered me their assistance and support in searching for the required literature.

Finally, I wish to thank all individuals and departments who helped to produce this work.

Ibrahim A. Mirsal

Contents

Part I Soil – Its Nature and Origin 1

1 The Origin of Soil 3
1.1 Physical or Mechanical Weathering 3
1.2 Chemical Weathering – The Gate to Pedogenesis 3
1.3 Weathering by Biological Agents 5
    1.3.1 The Pedogenic Cycle – A Cycle within the
           Global Sedimentary Cycle 6
    1.3.2 Transport Routes and Material Transfer within the Soil Body 6
1.4 Factors Controlling Soil Formation 10
1.5 Morphology of Soil 11

2 Soil Constituents 15
2.1 The Mineral Solid Phase 15
    2.1.1 The Orthosilicates 16
    2.1.2 Chain Silicates or Inosilicates 17
    2.1.3 Sheet Silicates or Phyllosilicates 18
    2.1.4 Framework Silicates or Tectosilicates 23
2.2 Organic Matter and Soil Organisms 23
    2.2.1 Soil Organisms 25
    2.2.2 Dead Organic Matter 28
2.3 The Liquid Phase – Soil Water 42
    2.3.1 Composition of Soil Waters 44
2.4 The Gaseous Phase – Soil Air, Origin, Composition and Properties 44

3 Soil Properties 47
3.1 Physical Properties 47
    3.1.1 Colour 47
    3.1.2 Texture 47
    3.1.3 Structure 47
    3.1.4 Consistence 48
    3.1.5 Porosity 49
3.2 Chemical Properties 50
    3.2.1 Soil Acidity (pH) 50
    3.2.2 Ion Exchange 51

	3.2.3	Cation Exchange Capacity (CEC)	52
	3.2.4	The Interaction of Organic Soil Matter with Mineral Components	53
	3.2.5	Oxidation-Reduction Status	55

4 Soil Types and Classification ... 57
4.1 The Soil Taxonomy System – Criteria of Classification ... 57
- 4.1.1 Morphological Criteria of Classification in the Soil Taxonomy System (Diagnostic Horizons) ... 57
- 4.1.2 Description of the Environmental Criteria of Classification ... 60
- 4.1.3 Description of the Chemical Criteria of Classification ... 62
- 4.1.4 Categories of the Taxonomy System Based on the Criteria of Classification ... 63
- 4.1.5 The Soil Orders of Taxonomy ... 63

4.2 The FAO-UNESCO Soil Classification System ... 88
- 4.2.1 Description of the Reference Soil Groups of the World Reference Base for Soil Resources (WRB) ... 88
- 4.2.2 Relation Between the WRB System and the USDA Taxonomy System ... 92

4.3 Other Systems of Classification ... 92
Examples of National Systems ... 93

5 Soil Degradation ... 95
5.1 Soil Degradation and Soil Quality ... 95
- 5.1.1 Biological Indicators of Soil Quality – Soil Respiration Rates ... 95
- 5.1.2 Physical Indicators of Soil Quality ... 96
- 5.1.3 Chemical Indicators of Soil Quality ... 98

5.2 Physical Soil Degradation ... 100
- 5.2.1 Soil Erosion ... 100
- 5.2.2 Soil Compaction ... 105
- 5.2.3 Soil Crusting and Sealing ... 108

5.3 Chemical Soil Degradation ... 110
- 5.3.1 Acidification ... 110
- 5.3.2 Salinization and Sodification ... 112

Part II Soil Pollution ... 115

6 Major Types of Soil Pollutants ... 117
6.1 Heavy Metals and Their Salts ... 117
- 6.1.1 Heavy Metals and the Soil System ... 121
- 6.1.2 Transport of Heavy Metals within the Soil System ... 121
- 6.1.3 Bioavailability of Heavy Metals ... 122
- 6.1.4 Biochemical Effects of Heavy Metals ... 123
- 6.1.5 Major Environmental Accidents Involving Pollution by Heavy Metals ... 124

6.2	Other Inorganic Pollutants	127
6.3	Radionuclides	128
	6.3.1 Speciation and Behaviour of Radionuclides in the Soil System	129
	6.3.2 Uptake of Radionuclides by Plants	131
6.4	Nuclear Debris from Weapon Tests and Belligerent Activities	131
6.5	Nuclear Debris from Major Nuclear Accidents	133

7 Sources of Soil Pollution … 137

7.1	Pollutants of Agrochemical Sources	137
	7.1.1 Insecticides	139
	7.1.2 Herbicides	144
	7.1.3 Fungicides	146
	7.1.4 Fuel Spills in Farms	147
7.2	Soil Pollutants of Urban Sources	147
	7.2.1 Power Generation Emissions	148
	7.2.2 Soil Pollution through Transport Activities	148
	7.2.3 Soil Pollution by Waste and Sewage Sludge	151
7.3	Soil Pollution through Chemical Warfare	153
	7.3.1 Pollutants, Toxic Chemicals, and Chemical Weapons	155
	7.3.2 Soil Pollution by Military Activities During the Cold War	162
7.4	Soil Pollution through Biological Warfare (BW)	165
	7.4.1 Bacteria	166
	7.4.2 Viruses	168
	7.4.3 Rickettsiae	171
	7.4.4 Chlamydia	171
	7.4.5 Fungi	172
	7.4.6 Toxins	172

8 Pollution Mechanisms and Soil-Pollutants Interaction … 175

8.1	Physical Processes and Mechanisms of Pollution	175
	8.1.1 Adsorptive Retention	177
	8.1.2 Nonadsorptive Retention	187
8.2	Contaminants Transport	189
	8.2.1 Microscopic Dispersion: Molecular Diffusion	190
	8.2.2 Macroscopic Dispersion	192
8.3	Behaviour of Non-Aqueous Phase Liquids (NAPLs) in Soils	194
	8.3.1 NAPLs Lighter than Water (LNAPLs)	195
	8.3.2 NAPLs Denser than Water (DNAPLs)	196

9 Pollutants' Alteration, Transformation, and Initiation of Chemical Changes within the Soil … 199

9.1	Processes Related to Chemical Mobility	199
	9.1.1 Immiscible Phase Separation	199
	9.1.2 Acid-Base Equilibrium	200
	9.1.3 Dissolution-Precipitation Reactions	202

9.2	Chemical Transformation Processes	202
	9.2.1 Hydrolysis	206
9.3	Biodegradation and Biologically Supported Transformations	206
9.4	Enzymatic Transformations: A Primer on Enzymes, Their Types and Mode of Action	207
	9.4.1 The Hydrolases	209
	9.4.2 The Transferases	211
	9.4.3 The Oxidoreductases	216
	9.4.4 The Lyases	218
	9.4.5 The Ligases	218
9.5	Transformations Assisted by Bacterial Action	219
	9.5.1 Sulphur Bacteria	220
	9.5.2 Nitrifying Bacteria	221
	9.5.3 Iron Oxidising Bacteria	222
	9.5.4 Methane Oxidising Bacteria	222
	9.5.5 Hydrogen Bacteria	222

Part III Monitoring of Soil Pollution ... 223

10 Monitoring and Monitoring Plans ... 225
10.1 Site Characterisation ... 226
10.2 Data Acquisition ... 227
    10.2.1 Sampling-Planning and Realisation ... 228
    10.2.2 Sampling Procedures ... 229
10.3 Field and Laboratory Investigations ... 234
    10.3.1 Investigation of Solid Matter ... 234
    10.3.2 Investigation of Soil Solution ... 235
10.4 Monitoring of Groundwater Flows ... 237
    10.4.1 The Different Zones of Groundwater ... 237
    10.4.2 Monitoring Flow Directions ... 238
    10.4.3 Monitoring Hydraulic Heads ... 239
    10.4.4 Measuring Hydraulic Heads in the Vadose Zone ... 240

11 Biological Monitoring ... 241
11.1 Planning and Implementation of Biological Monitoring ... 242
11.2 Foliage Sampling and Investigation ... 243
11.3 Chemical Investigation of Foliage ... 243
11.4 Sampling and Investigation of Litterfall ... 243

Part IV Modelling of Soil Pollution ... 245

12 Models and Their Construction ... 247
12.1 Types of Models ... 248
    12.1.1 Space Analogue Models ... 248
    12.1.2 Mathematical Modelling of Fluid Flows in Soil ... 252

Part V Soil Remediation 263

13 Planning and Realisation of Soil Remediation 265
13.1 Categories of Pollutants 265
13.2 Scale of Pollution 266
13.3 Risk Level 267
13.4 Remediation Technologies 267
    13.4.1 Chemical and Physical Remedial Techniques 269
    13.4.2 Biological Treatment (Bioremediation) 273
    13.4.3 Solidification/Stabilisation Methods 279
    13.4.4 Thermal Treatment 280

References 283

Index 301

Part I
Soil – Its Nature and Origin

Soil is essentially a natural body of mineral and organic constituents produced by solid material recycling, during a myriad of complex processes of solid crust modifications, which are closely related to the hydrologic cycle. It is the interface at which all forces acting on the Earth's crust meet to produce a medium of unconsolidated material, acting as an environment for further changes and developments, keeping pace with the evolution of the global Earth system as a whole. It offers shelter and habitat for a countless number of organisms, provides an incubation and living medium for plants, while playing its role perfectly in the universal cycle of material flow between the four main geospheres (atmosphere, lithosphere, hydrosphere and biosphere). For this reason, some authors consider soil as a separate geosphere and give it the name *pedosphere*. The pedosphere is formed by, and is eternally evolving through, weathering forces. Despite their complicated nature and intricately structured processes, they are found to belong to a few basic types, the nature of which and importance for the evolution of our Earth environments will be the subject of close inspection in the first few chapters of the present work.

Chapter 1

The Origin of Soil

1.1
Physical or Mechanical Weathering

Physical and biological agents, such as wind, running water, temperature changes, and living organisms, perpetually modify the Earth's crust, changing its upper surface into products that are more closely in equilibrium with the atmosphere, the hydrosphere, and the biosphere. Earth scientists sum up all processes through which these alterations take place under the collective term *weathering*. One speaks of mechanical weathering in the case that the dominant forces are mainly mechanical, such as the eroding action of running water, the abrading action of stream load or the physical action of wind and severe temperature fluctuations. Similarly, one speaks of biological weathering when the forces producing changes are directly or indirectly related to living organisms. Of these, we can mention several examples, such as the action of burrowing animals, the penetrating forces of plant roots, and the destructive action of algae, bacteria, and their acid-producing symbiotic community of the lichens, or simply the destructive action of man, who continuously disturbs the Earth's crust through various activities.

Processes of disintegration, during which mantle rocks are broken down to form particles of smaller size, without considerable change in chemical or mineralogical composition are known as *physical weathering processes*. Changes of this type prevail under extreme climatic conditions as in deserts or arctic regions. They are also prevailing in areas of mountainous relief. The most prominent agents of physical weathering are:

- differential stress caused by unloading of deep-seated rocks on emerging to the suface;
- differential thermal expansion under extreme climatic conditions;
- expansion of interstitial water volume by freezing, that leads to rupturing along crystal boundaries.

Other mechanical agents enhance the effect of mechanical weathering. These may include processes such as gravity, abrasion by glacial ice or wind blown particles.

1.2
Chemical Weathering – The Gate to Pedogenesis

Chemical weathering is the fundamental natural process underlying the group of processes known collectively as *pedogenesis*, or *soil forming processes*. It depends principally on the presence of water and is largely initiated by the preceding physical weathering, since disintegration of the solid material leads to activation of the solid phase and eventually to more favourable energetic conditions for subsequent chemical alterations. The effect of chemical weathering is the most decisive in the geologic cycle, whereby dramatic changes may completely obliterate the parent rock and vast geomorphologic changes may occur. Under these conditions, pedogenesis consists of the following categories of processes (Ellis and Mellor 1995).

a *Oxidation.* In the vadose (aerated) zone (see Fig. 10.13 in Sect. 10.4), where most processes leading to soil formation take place, the availability of oxygen, water, and dissolved gases leads to a dominance of oxidation reactions, leaving their marks on the formed soil horizons, represented by the characteristic colours of the resulting products. The typical yellow, brown, or red colour of soil in some warm areas (e.g. Mediterranean terra rossa) is due largely to the oxidation of ferrous iron in the minerals pyroxene, amphibole, and olivine into ferric iron. Besides Fe^{2+}, Mn^{2+} and S^{2+} are the most commonly affected elements by oxidation. They are normally oxidised to Mn^{4+} and S^{6+}. Other examples are V, Cr, Cu, As, Se, Mo, Pd, Sn, Sb, W, Pt, Hg, and U.

A good example of oxidation reactions during weathering is the oxidation of pyrite to form sulphuric acid, which attacks the rocks (see Eq. 1.1), developing solution pits and stains.

$$2FeS_2 + 2H_2O + 7O_2 \rightleftarrows 2FeSO_4 + 2H_2SO_4 \quad (1.1)$$

Oxidation functions most effectively in the vadose zone, yet in some places oxidation fluids can descend to depths far below the water table before their oxidising power is consumed (Rose et al. 1979).

b *Hydration and hydrolysis.* Hydration and hydrolysis are the most important processes encountered during soil formation by weathering. While in hydrolysis a proper chemical reaction between water and the mineral substance takes place to produce or consume a proton (H^+) or an electron $(OH)^-$, water in hydration forms an envelope around the cations to form a hydrate – a compound in which it is integrated within the crystalline structure of the substance. A typical example of hydration is the conversion of anhydrite ($CaSO_4$) into gypsum ($CaSO_4 \cdot 2H_2O$).

Clays that, together with organic substance and other colloidal matter, give soil its characteristic nature, are typical products of hydrolysis processes steering the change of aluminium or iron silicates into clay minerals and/or iron oxides. An example of this is the reaction of the mineral albite in the course of weathering (see Fig. 1.1) with weak acids to yield kaolinite (clay), silica and sodium ions (Na^+).

$$\underset{\text{Albite}}{2NaAlSi_3O_8} + 2H^+ + H_2O \rightleftarrows \underset{\text{Kaolinite}}{Al_2Si_2O_5(OH)_4} + 4SiO_2 + 2Na^+ \quad (1.2)$$

The protons involved in this reaction are generally provided by naturally occurring acids, such as carbonic acid or by the rather abundant humic acids. The released Na-cations will be sorbed on the surface of colloidal particles or released to the solution. SiO_2 precipitates as colloidal silica or quartz.

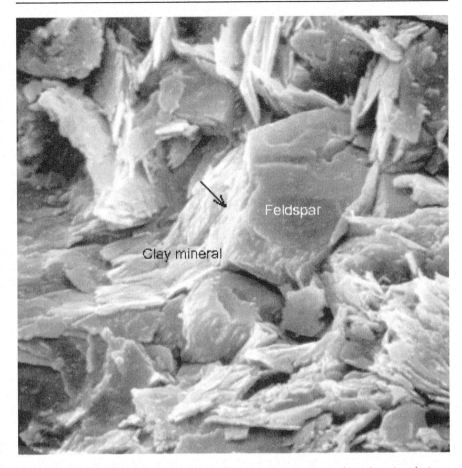

Fig. 1.1 Chemical weathering of Feldspars (*arrow*) by percolating waters to form clay minerals. Scanning electron photomicrograph of a sample collected near Marburg (Germany). 1 cm = 2 μm

1.3
Weathering by Biological Agents

Biological effect is a factor, which has never been absent in any soil forming process during weathering. It is always there, whether the dominant processes were mechanical or chemical; it always accompanies emergence and evolution of soil. One needs only to consider the mechanical forces exerted by intruding roots, or the enormous work of worms and rodents in mixing and disintegrating rock bodies in the upper surface environment, to realise how important this factor is for soil formation and its later evolution. The chemical dimension of biological weathering vary from simple dissolution reactions, occurring at the extensive acidic environment at root tips, to complex biochemical processes by which certain elements are extracted, concentrated or bound into complex by plants or by bacterial action. As an example, we may take the oxidation of iron and sulphur or the fixation of nitrogen by bacteria.

1.3.1
The Pedogenic Cycle – A Cycle within the Global Sedimentary Cycle

One of the most fundamental definitions of soil is that it is the relatively thin upper layer of the unconsolidated mantle of disintegrated and decomposed rock material, which overlies the consolidated bedrock. It is composed of organic material derived from the surface and mineral particles, which were contained in the parent material forming the bedrock. Soil matter, thus formed, seems in most cases to have been recycled in place, without being transported for long distances. A good example that can be seen in the field is given by the so-called *terra rossa* – red clay formed on top of limestone in some Mediterranean regions. In such cases, one can observe the conspicuous separating surface between the soil and underlying rocks with residual material, documenting its derivation from limestone by chemical erosion.

Generally the unconsolidated material (soil mantle) is zoned from above downward, forming the so-called *soil zones* (see below). This leads to the conclusion that pedogenic cycles are minor cycles within the main sedimentary cycle, modifying the Earth's crust (see Fig. 1.2). It is governed by the forces demonstrated by the typical stages leading to the formation of sedimentary rocks in all terrestrial environments.

1.3.2
Transport Routes and Material Transfer within the Soil Body

Once a soil body or a precursor of such a body is formed, material flows within the body, whether in suspension or in solution they will contribute to the maturing of this body into a well zoned soil profile as that shown in Fig. 1.3. Routes and patterns of flows in a landscape prism run actually in all horizontal and vertical directions according to the physical conditions prevailing in the soil body (grain size, porosities), or also concentration gradients in the soil water. A general pattern, however, was recognised and summarised by many authors, the most well known simpler pattern is that of Kozlovskiy (1972). He indentifed, according to direction of flow, three types of transport routes as shown in Fig. 1.3.

A short description of the different patterns of flow can be given as follows:

1. *Main migrational cycle* (Fig. 1.3a). This has a predominantly vertical movement and corresponds to the narrower concept of biogeochemical cycle in a more or less closed loop. It may, however, include the transport of fine material suspended in water from the upper parts of the soil body, to deposit them at the lower parts – *illuviation* – forming coatings or skins on the coarser grains (sometimes known as *clay skins*, see Fig. 1.4). Besides forming coats on coarser grains, clay skins, may in many cases form fine linings in the interstitial space between coarser grains (see Fig. 1.5). In the latter case, however, the clays in the interstitial space may be chemical (diagenetic) products formed in place along the boundaries between feldspar grains, due to increased solution activities following the pressure arising along these boundaries due to compaction (pressure solution).

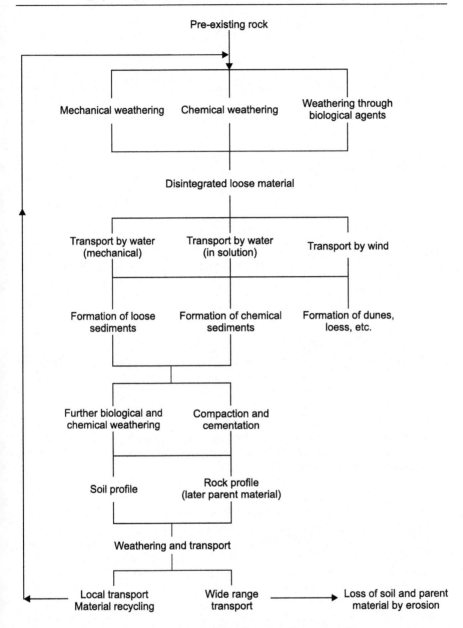

Fig. 1.2 The position of pedogenic processes within the main sedimentary cycle

Acid waters may, on their way downwards, leach cations from upper horizons to precipitate them as chemical sediments in the form of newly formed chemical sediments in the lower parts, as it is the case with the formation of carbonate rich horizons in some soils.

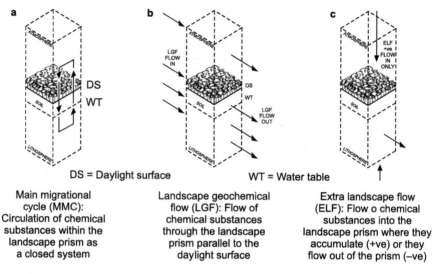

Fig. 1.3 Illustration of the three flow patterns according to Fortescue (1980)

In the course of material flow during an active pedogenic cycle, the vertical flow forms one of the main routes for the development of mature soil horizons, as well as the distribution of organic and mineral components among these horizons.

Examples of these are numerous, among which we may mention the following:

Flow of organic material down the profile of an active pedogenic cycle may result in the transfer of organic material – from the surface – to be deposited, either unchanged in the form of organic debris (roots, animal rests, etc.), or completely changed after being decomposed and homogenised by several organic reactions to form typical humic substances (see below).

2. *Landscape geochemical flow* (Fig. 1.3b). This involves a progressive transport of material parallel to soil surface (see Fig. 1.3b). It takes place within a prism of the landscape, including portions of the atmosphere, the pedosphere, and the lithosphere as shown in Fig. 1.3b. It also includes among others air (gas) migration. An example of chemically active air migrant in the LGF is carbon dioxide and other gases that would dissolve in soil water, causing a shift in its chemical constitution and in some cases precipitation of new chemical compounds (e.g. carbonates).

Complex organic compounds under conditions, prevailing during this type of flow (e.g. oxidation during air flow) may be decomposed allowing cations that form the bases for plant nutrients such as Ca^{2+}, Mg^{2+}, and K^+ to be released, forming salts (carbonates, sulphates, etc.) in a process fundamental to pedogenesis known as *mineralisation*. This process may occur at any time of the cycle, yet it is characteristic to changes taking place during landscape geochemical flows. Mineralisation processes may be reversed in subsequent changes to form simple organic compounds, as it will be shown later in Chap. 2 when polymerisation processes will be explained.

3. *Extra landscape flow (ELF)* (Fig. 1.3c). A third type of material flow in landscapes is the Extra Landscape Flow. Applying this to soils as a portion of the landscape prism,

Fig. 1.4 Formation of grain coats by illuviation (SEM)

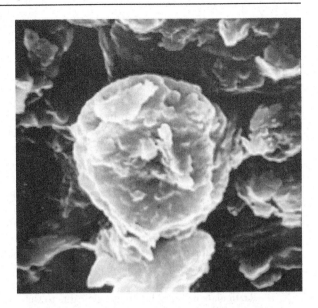

Fig. 1.5 Clay flakes (booklets) lining interstices between coarse feldspar grains (SEM). Sample collected near Marburg (Germany)

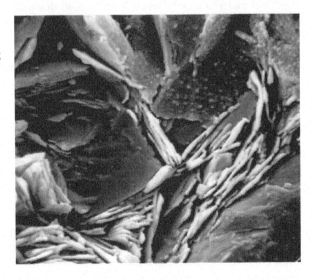

we may define it as the flow of chemical substances into the soil, where they would be accumulated (+ve flow), or out of it (–ve flow), where they would cause material loss from the site (see Fig. 1.3c). This type of flow characterises two main types of pedogenic processes, namely: *additions* and *losses*.

- *Additions* prevail during positive flows, i.e. when substances flow into the soil profile, where they would be accumulated – a process that may occur in vertical as well as in horizontal directions, either at the surface or in lower parts within the soil body. Soil additions may involve plant debris, animal bodies or their parts as well as organic material formed by decomposition of pre-existing molecules, whereby such detrital material could indicate – according to its degree of erosion or decomposition – whether it was locally derived or transported from distant regions other than the site of direct pedogenesis.

Following the addition of organic or inorganic material to the soil body, further chemical reactions continue driving the energetic conditions of the system towards chemical equilibrium, by transforming those elements of the system that might be metastable (under the prevailing environmental conditions) to stable ones. These transformations involve inorganic as well as organic soil material.

- *Losses*. In the case of negative flows, losses from the soil body characterise the net result of material flow. In this case, the site of pedolisation loses material to its surroundings both at the surface and in the lower parts of the landscape prism. Speaking about loss, would call to mind soil erosion in the first place. However, this conclusion is not completely right, even though it describes a great deal of the reality when loss is mainly defined on a physical or mechanical basis. Material loss during negative flows occurs mainly in four forms: gases, solutes, and particulate material, to which we may also add removed plants (Ellis and Mellor 1995).

Particulate material loss may occur due to water or wind erosion, while gaseous material may be lost from soil either due to degassing, by changing partial pressure gradients between a soil and its surroundings, or due to oxidation of organic compounds. Loss of nutritive material can follow after harvesting or removing plants from the soil. In deeper parts of the soil column, subsurface flow may carry dissolved material, leading to its loss in a portion of the soil. It might be later deposited outside the site of pedolisation. Thus leading to a net loss of soil material through extra landscape flow (ELF).

1.4
Factors Controlling Soil Formation

Climate. Weathering in general, and soil formation processes in particular, are dependent largely on climatic factors. These, not only control the main processes and directions in the main cycles of material flow, but also affect organic addition, and the rate of mineral transformation via crystal lattice break down. Ross (1989) found that, in such transformations, the rates of chemical reaction double for every 10 °C rise in temperature, and that the maximum rate of organic matter decay takes place in the temperature range from 25–35 °C. This may also follow from the observation that, on a global level, the rate of mineral transformation and organic matter decay increases from high to low latitudes.

Biota. Actually, the role of organisms cannot be discussed apart from the climatic control effect, since these are generally related to bio-geographical conditions. Aside from the mechanical work done by rodents and burrowing animals, the chemical reactions triggered off by bacteria and plant roots play, as mentioned before, a crucial role in the process of soil formation.

Parent material. Since the principal source of soil is the pre-existing rock or parent material, the main control on soil formation will be directly related to the susceptibility of this material to weathering processes, and the chemical and physical changes accompanying them. Physical properties (such as hardness, cleavage, porosity and grain size) form primary factors in determining whether water can percolate into a rock layer to initiate its disintegration into an unconsolidated material, or its decomposition into a different mineralogical constitution; the properties and characteristics of the resulting soil will also be directly related to parent material. Soils formed on parent material highly resistant to weathering will normally have relatively less thickness than those formed on easily weathered landscapes. They also contain more regolith or stony material than the latter.

1.5
Morphology of Soil

One of the characteristic properties of soil is the organisation of its constituents into layers related to present day surface. Each of these layers, which may easily be identified in the field through colour or texture, reflects subtle differences in chemical properties and composition, of which the most significant are pH, organic matter content, mineral assemblages, and metal concentrations, especially iron and manganese.

Soil layers, normally referred to as horizons, may range from a few centimetres to a metre or more in thickness. They are classified according to their position in profile, which is also closely related to their mineralogical constitution and grain size into few basic types (as shown in Fig. 1.6)

At the top of the profile, a layer of partially decomposed organic debris is referred to as the O-horizon (also A_0). It contains about 20–30% organic matter, derived from plant and animal litter. It is in this region that the principle process of soil formation, known as *humification*, i.e. complete change of organic debris into soil organic matter, takes place. The resulting material, humus, made up of a mixture of organic substances, is characterised by its dark colour and rather acidic nature. It is mainly produced by the work of consumers and decomposers among the microorganisms, living in the site of the soil formation.

In the best-developed soils, which render ideal profiles, three main horizons follow. They form a transition between the O-horizon and the base of the profile, made of the parent material, which is given the name *R-horizon*. This may be rock in situ, transported alluvial, glacial or wind blown overburden, or even soil of a past pedological cycle.

The three middle horizons, identified by the letters A, B, and C, are composed of sand, silt clay and other weathered by-products (Table 1.1 presents a description of grain size). They represent two main subsequent stages of soil formation, whereby the lowest one, C-horizon, represents the stage nearest to the parent material. It is made up of partially or poorly weathered bedrock having minimum content of organic matter and clay. The A- and B-horizons are viewed together as representing the real soil emerging from the complete weathering of the C-horizon; they are collectively known under the name *solum*.

The A-horizon, which is generally a dark coloured horizon, rich in organic matter, may in some cases have a structure made up of three identifiable subdivisions known as A_1, A_2 and A_3. Marked colour differences, resulting from leaching processes, make it possible to identify these subdivisions in the field. In fact, the resolution of the A-

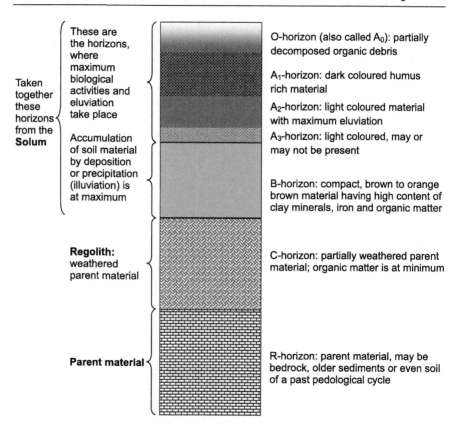

Fig. 1.6 Diagrammatic representation of a hypothetical soil profile

Table 1.1 Grain sizes of clastic sediments and related rock types

Grain size (mm)	Unconsolidated sediment	Lithified rock
>80	Boulders, cobbels	Conglomerate (Breccia if angular)
>2	Pebbels, gravel	
0.5 – 2	Coarse sand	Sandstone
0.02 – 0.5	Medium to fine sand	Sandstone
0.002 – 0.02	Silt	Siltstone (mudstone)
<0.002	Clay	Shale

horizon into a dark upper layer containing humus with mineral grains (A_1) and an underlying light coloured horizon with little organic matter, is due to leaching processes, initiated by water percolating downward through the rich organic material on the top of the A-horizon. On its course downwards, water carrying in solution organic acids and complexing agents, generated in the humus by bacterial action, performs a process of leaching known as *eluviation* – a word from Latin meaning "to wash out". Eluviation, enhanced by carbonic acid, resulting from the decay of humus, displaces bases (calcium, sodium, magnesium, potassium) from the exchange sites of clay minerals. These bases move down the soil profile as colloidal particles, dissolved ions or as free ions complexed with hydroxyl. Silica is also leached in the course of eluviation. It is largely dissolved as silicic acid or colloidal silica. Resistant primary mineral matter, however, remains behind in the upper soil (Rose et al. 1979).

Material dissolved in the A-horizon finds in some cases its way to the saturated zone of groundwater, yet the greatest part of it is normally re-deposited in the underlying layers forming the B-horizon. In this process, known as *illuviation* (from the Latin to wash in), colloidal material and metal oxides are deposited or precipitated in the B-horizon, resulting in an enrichment of its layers in clay and aluminium oxide. Iron oxides, if present, give the horizon its red or yellow brown colour.

Chapter 2
Soil Constituents

Generally speaking soil is a three-dimensional system, made of a solid, a liquid and a gaseous phase, each in an amount depending on the abundance of its constituents and their kinetic roles in the complex series of reactions, leading to soil formation. Figure 2.1 illustrates the composition by volume of an average soil.

2.1
The Mineral Solid Phase

Mineral matter in soil depends largely on the nature and composition of the parent rock. However, since about three fourths of the Earth's crust is made up of silicon and oxygen, we find that silicate minerals occupy a central position in any description of the mineral constituents of a given soil. All silicates are formed of a fundamental structural unit, comprising one silicon ion (Si^{4+}) and four oxygen ions O^{2-}, closely surrounding the silicon in a tetrahedral lattice, as shown by Fig. 2.2.

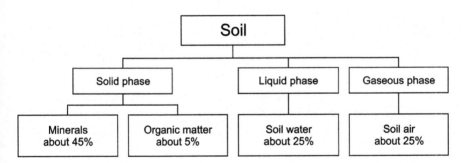

Fig. 2.1 Composition by volume of an average soil

Fig. 2.2 A silicate tetrahedron

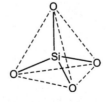

The tetrahedra may, based on their net and residual charges, come together in a multiform of combinatorial structures and three-dimensional arrangements to form various kinds and varieties of silicates, which can be categorised into the following fundamental groups:

- *Orthosilicates.* These are discrete units of individual or grouped tetrahedra, made of one, two, three or six tetrahedra per unit – a property, which, as we can see later, is used for further classification of the group.
- *Chain silicates.* In this group, individual tetrahedra catinate together, by sharing the four tetrahedrally co-ordinated oxygen atoms with the neighbouring silicon atoms, to form infinite chains of formula $(SiO_3)_n^{2-}$.
- *Sheet silicates.* In this group, three co-ordinated oxygen atoms at the corners of a tetrahedron are shared with adjacent silicon atoms, resulting in the formation of a sheet or a layer of tetrahedra connected together at the three basal corners.
- *Framework silicates.* These are formed, if all four oxygen atoms per SiO_4 tetrahedron are shared with adjacent tetrahedra in a framework structure. In the following, each of these four silicate groups or classes will be discussed in some detail.

2.1.1
The Orthosilicates

The orthosilicates contain two subcategories: nesosilicates and sorosilicates. In the first category, the SiO_4-tetrahedra occur as separate units, without shared oxygen atoms, linked by metallic cations. This structure (Fig. 2.3a) is not very common in minerals. However, some minerals like olivine, $(Mg,Fe,Mn)_2SiO_4$, which is an important constitu-

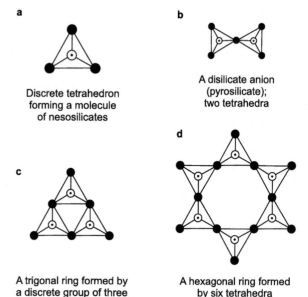

Fig. 2.3 Structure of the orthosilicates

a Discrete tetrahedron forming a molecule of nesosilicates

b A disilicate anion (pyrosilicate); two tetrahedra

c A trigonal ring formed by a discrete group of three tetrahedra (cyclosilicates)

d A hexagonal ring formed by six tetrahedra (cyclosilicates)

2.1 · The Mineral Solid Phase

ent of basalt, adopt it. Other minerals, made of single silicon tetrahedra are zircon $ZrSiO_4$, topaz $Al_2(FOH)_2SiO_4$, and the garnets, with the general formula:

$$M_3^{II}M_2^{III}(SiO_4)_3$$

where M^{II} can be Ca^{2+}, Mg^{2+} or Fe^{2+}, and M^{III} is Al^{3+}, Cr^{3+}, or Fe^{3+}. This group of minerals occurs in soils formed on igneous rocks due to their higher resistance to weathering. The sorosilicates, themselves, may be further classified into two groups: the *pyrosilicates* and the *cyclosilicates*. In the pyrosilicates (also called *disilicates*), discrete groups of two tetrahedra share one of the co-ordinated oxygen atoms to form the disilicate anion Si_2O_7 (see Fig. 2.3b). An example is the mineral hemimorphite that has the general formula $Zn_4(OH)_2Si_2O_7$; it sometimes occurs in soils formed on limestones. However, generally minerals having this structure are quite rare. In the cyclosilicates (the second category of the sorosilicates), three or six tetrahedra may share one or more of their co-ordinated oxygen atoms to form a trigonal ring, Si_3O_9 (Fig. 2.3c), or a hexagonal ring, Si_6O_{18} (Fig. 2.3d). Sorosilicates forming trigonal rings are represented by minerals like wollastonite, $Ca_3Si_3O_9$, or rhodonite, $Mn_3Si_3O_9$. Cyclosilicates having hexagonal ring structures are represented by minerals like beryl, $Be_3Al_2Si_6O_{18}$, or dioptase $Cu_6Si_6O_{18} \cdot 6H_2O$. Figure 2.4 summarises the classification of the orthosilicates.

2.1.2
Chain Silicates or Inosilicates

This group derives its structure from the self-association of metasilicate anions (SiO_3^{2-}) into infinite chains of formula $(SiO_3)_n^{2n}$. The simplest of these is the string-like chain characteristic of the pyroxenes. In this arrangement, the silicon atoms share two of

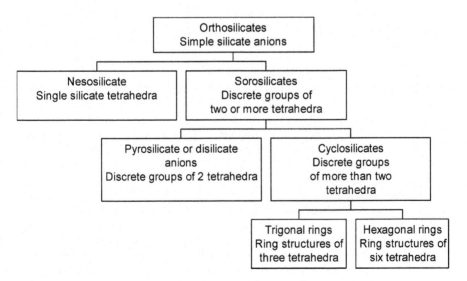

Fig. 2.4 Classification of the orthosilicates

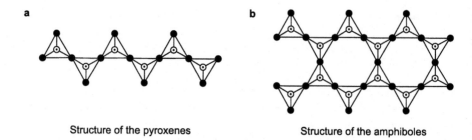

Fig. 2.5 Structure of the chain silicates

the four tetrahedrally co-ordinated oxygen atoms with adjacent silicon atoms (Fig. 2.5). Examples of the pyroxenes include enstatite, $MgSiO_3$ and diopside, $CaMg(SiO_3)_2$. If further sharing of the oxygen atoms by half of the silicon atoms occurs, a double chain or band structure is formed. This is the structure of the amphiboles. Amphiboles are more complicated, containing the basic $(Si_4O_{11})^6$ repeating unit as well as metal and hydroxide ions. Examples of the amphiboles are tremolite, $Ca_2Mg_5(OH)_2Si_8O_{22}$ and actinolite, $Ca_2(Mg,Fe)_5(OH)_2Si_8O_{22}$.

2.1.3
Sheet Silicates or Phyllosilicates

If a complete sharing of the three basal oxygen atoms in a silicon tetrahedron is established, a layer or a sheet structure would result, made of various associated tetrahedra, and having an empirical formula $(Si_2O_5)_n$, creating a completely new type of silicates, known as *the phyllosilicates* or *the sheet silicates*.

Connections between the central silicon atoms in the individual tetrahedra lead to the appearance of a network of hexagonal holes on the sheets, lending them a pronounced pseudo-hexagonal symmetry (Fig. 2.6).

The apical atoms of the tetrahedra contemplating neutrality, form ionic-covalent bonds with other metal cations. They commonly associate themselves into octahedral sheets of gibbsite $(Al_2OH_6)_n$, or brucite $[Mg_3(OH)_6]_n$. Gibbsite sheets are formed by edge-to-edge linking of two octahedra of Al equidistantly surrounded by six OH-groups (see Fig. 2.7). Edge-to-edge association of three octahedra of $Mg(OH)_6$, results in the formation of *brucite sheets*, which are also of octahedral structure, having an empirical formula: $[Mg_3(OH)_6]_n$.

The bonding of silica tetrahedral sheets (also known as *siloxane sheets*) with gibbsite or brucite octahedral sheets makes the basic structural units for the clay minerals, which are considered the most important group of mineral constituents of soils. These fundamental units (known as *layers*) are of two types; the first one (Fig. 2.8a), called here (for the sake of convenience) doublet structure while the other (also for convenience, called triplet structure) is made of a 2:1 lattice, comprising an octahedral brucite sheet sandwiched between two siloxane sheets (Fig. 2.8b).

2.1 · The Mineral Solid Phase

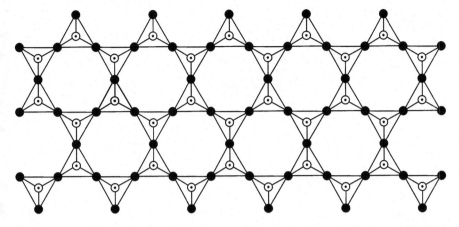

- Oxygen atom
- Oxygen atom with central silicon, showing in the middle

Fig. 2.6 Structure of the phyllosilicates. Observe the hexagonal network of holes

Fig. 2.7 Diagrammatic representation of an allumina octahedron

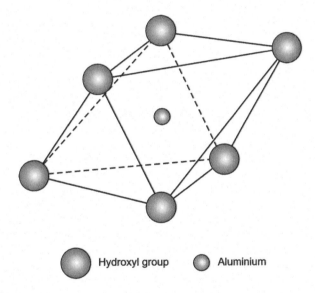

Hydroxyl group   Aluminium

Classification of the Clay Minerals

Before explaining the fundamentals of clay mineral's classification in detail, the terms used henceforth will be shortly recapitulated. There are three fundamental elements delineating the structure of clay minerals: *sheets*, *layers* and *stacks*. The sheets are structurally of two types (tetrahedral silica and octahedral Al-OH or Mg-OH sheets). Combinations of sheets make the layers, of these there are also two types – "doublets", made

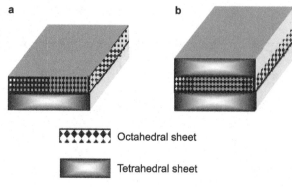

Fig. 2.8 a A doublet layer (1:1), an octahedral sheet linked to a tetrahedral (siloxane) sheet; **b** a triplet layer (2:1), an octahedral sheet sandwiched between two tetrahedral sheets

of a tetrahedral sheet linked to an octahedral one and "triplets" made of an octahedral sheet sandwiched between two tetrahedral ones. A stack is a combination of layers alternating in a vertical direction. Layers in a stack may all be of the same type or they may be different, in which case the clay mineral is called a *mixed layer-clay*. The alternation of layers in a stack may follow a regular rhythm or may be random. Figure 2.9 summarises these relations.

Clay minerals are classified according to the type of layer structure (doublets or triplets), the interlayer or basal spacing between the unit layers, and the interlayer components or species. Accordingly, the following groups are identified:

- *The kaolinite group.* This group is made of stacks of doublets (see above). The main member of the group is the mineral kaolinite, which has the composition $(OH)_8Al_4Si_4O_{10}$. Another member of this group is halloysite, having almost the same composition except for having two layers of water molecules within the stack. Figure 2.10 represents the structure of the group.
- *The montmorillonite group.* This group is made of triplets stacked in vertical direction (Fig. 2.11). It has the empirical formula $(OH)_4Al_4Si_8O_{10} \cdot nH_2O$. Mg^{2+}, Fe^{2+}, or any other divalent cations substitute generally for Al^{3+} in the octahedral sheet, producing a rest charge, which is normally balanced by accommodating Ca^{2+}, or Na^+, between the layers. The interlayer space in this group provides a shelter not only for neutralising ions, but also for organic material as well as for varying amounts of water, making the lattice expandable to accommodate variable numbers of water molecules. The interlayer ions may be displaced by other dissolved ions when their solutions enter the interlayer space. This property makes members of this group suitable for use as ion exchangers. Other members of the montmorillonite group include smectite, in which Fe^{2+} and Mg^{2+} substitute for Al^{3+} and nontronite with Fe^{3+} replacing Al^{3+}.
- *The illite (Hydromica) group.* The illite or hydromica group (also known as *muscovite group*) is represented by its main member illite. It has the general formula: $OH_4K_y(Al_4Fe_4Mg_4Mg_6)(Si_{8-y}Al_y)O_{20}$ and has a structure made of stacked triplets (Fig. 2.12). However, about three quarters of the tetrahedral positions in the siloxane sheets are occupied by Al^{3+} rather than Si^{4+}, leaving a net layer charge of about −2

2.1 · The Mineral Solid Phase

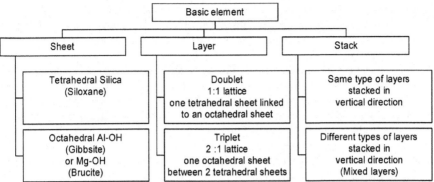

Fig. 2.9 Chart summarising the main basic structural elements of the clay minerals

Fig. 2.10 Structure of the kaolinite group

Fig. 2.11 Structure of the montmorillonite group

Fig. 2.12 Structure of the illite group

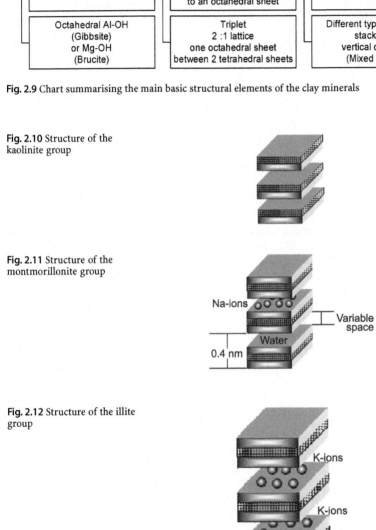

on each layer. This rest charge is balanced by accommodating K-ions in the interlayer space, making the interlayer bonding much stronger than the weaker ionic bonding in montmorillonite and rendering the illite lattice, unlike that of montmorillonite, non expandable.

- *The chlorite group.* The chlorite group can be structurally viewed as being derived from montmorillonite by inserting a sheet of $Mg(OH)_6$ (brucite) between each two adjacent montmorillonite layers (Fig. 2.13). This case, in which a single sheet functions by itself as a layer, is substantiated by the observation that montmorillonite is altered to chlorite in magnesium-rich seawater, that can provide the magnesium ions required to produce brucite sheets.
- *Mixed layer clays.* This group has a structure resulting from ordered or random stacking of the basic clay mineral groups in a vertical direction. It might be alternation between stacks of doublets with stacks of triplets (e.g. kaolinite-illite), or, as it is the case in chlorite, just a regular alternation of triplets with a single brucite or gibbsite sheets or even an alternation between stacks of chlorite and illite. The alternations may be regular, following a fixed rhythm like in chlorite or irregular like in some minor soil constituents

Factors Controlling the Formation and Alteration of Clay Minerals

The principal factors, controlling formation and alteration of clay minerals, are the chemical composition of the parent material and the physicochemical environment in which the process takes place. Kaolinite for instance, having only Al and Si as cations, will be formed in an environment where bases are continuously removed from the solution, giving rise to a residue with high Al/Si-ratios (kaolinite has the highest Al/Si-ratio among the clay minerals). Such an environment as here described, should be acidic and having a very active percolation of fluids to provide the effective removal of the bases. Therefore, kaolinite develops much more readily in relatively humid climates with free drainage and an enhanced percolation of groundwater.

Contrary to kaolinite, formation of members of the montmorillonite group (of low Al/Si-ratio about 0.5:1 and high content of iron, manganese, sodium, calcium, magnesium, and potassium), warrants the retaining of the bases. This can readily occur in a

Fig. 2.13 Structure of the chlorite group

Brucite sheet

neutral or a slightly alkaline environment of impeded drainage or high evaporation rates. Accordingly, montmorillonite is considered as a typical product of weathering in water logged terrain or semi-arid climates. The mere occurrence of montmorillonite in a landscape adds to the poor drainage character of the soil, due to its expandable lattice and its capacity to accommodate and retain water molecules.

2.1.4
Framework Silicates or Tectosilicates

In this category, complete sharing of all four oxygen atoms per SiO_4 tetrahedron is attained, giving rise to a framework structure. Minerals like quartz and the feldspars, which are very common in most soils, adopt this structure. Quartz is made up of Si-O tetrahedra linked through the O^{2-} ions, having the general formula $(SiO_2)_n$ or silicon dioxide. The name *quartz* stands actually for a mineral species under which about 300 varieties are included. The feldspars are far more important, as rock constituents, than any other group of minerals; in fact, they may be compared in igneous rocks to all other groups combined, since they constitute nearly 60% of such rocks and serve as the basis for their classification. They undergo *isomorphic substitution*, which is one of the fundamental properties of silicate minerals. In this process, an ion may be replaced by another ion in the silicate lattice, causing an imbalance in the electric charge of the crystal. To re-establish electric neutrality, extra ions or ionic species are incorporated or expelled from the crystal lattice. In feldspars a considerable part of the Si^{4+} ions are replaced by Al^{3+} ions with extra base cations like Ca^{2+}, Na^+ or K^+ to re-establish neutrality. The chief chemical types of feldspars are: $KAlSi_3O_8$ (orthoclase and microcline), $NaAlSi_3O_8$ (albite), and $CaAl_2Si_2O_8$ (anorthite). Natural crystals of microcline and of orthoclase contain 10 to 25% $NaAlSi_3O_8$ Albite and anorthite are completely miscible with each other and form mix-crystals (in all proportions), known as *plagioclase feldspars*, which are stable at any temperature. They normally contain 5 to 15% $KAlSi_3O_8$.

Accessory soil minerals. Apart from silicate minerals, non-silicate minerals of minor occurrence may play an important role in the development of soil properties. Among those most widely distributed in soils (see Fig. 2.14) are the oxides of iron and aluminium, which are often lumped under the collective term *sesquioxides*. Limonite ($Fe_2O_3 \cdot nH_2O$), hematite (Fe_2O_3), goethite ($Fe_2O_3 \cdot H_2O$), diaspore ($Al_2O_3 \cdot H_2O$), and gibbsite (Al_2O_3) best exemplify the sesquioxides.

Hydrous iron and aluminum oxides attain their highest concentrations in soils under humid tropical conditions. However, iron concentrations will principally depend on the Eh of the environment, such that, under oxidising conditions, iron that is not required for clay formation is precipitated as hydrated ferric oxide, while under reducing conditions ferrous iron is removed in solution. Other accessory minerals occurring in soils developed on recent volcanic deposits are anatase, TiO_2, and amorphous silica.

2.2
Organic Matter and Soil Organisms

A major agent determining the character of the entire soil profile is humus, which is generally concentrated in the upper horizons. Brady (1974) has defined humus as:

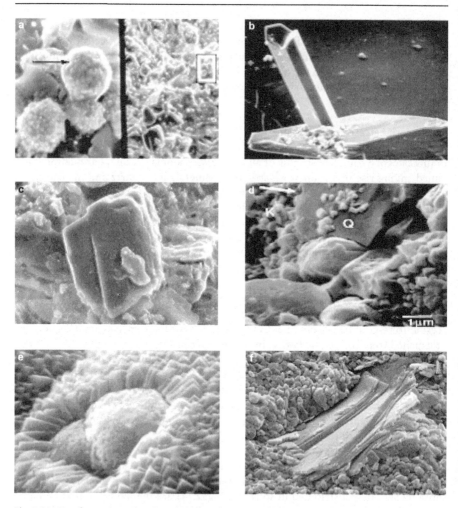

Fig. 2.14 a Raspberry shaped pyrite grains (*arrow*) close up from the right hand side half of the picture (*rectangle*); **b** gypsum crystals (swallow tail-grain); **c** corroded dolomite crystal; **d** quartz grains (*arrow*); **e** calcite crystals encrusting a grain; **f** detrital mica plates (scanning electron photomicrographs)

"A complex and rather resistant mixture of brown or dark brown amorphous and colloidal substances modified from the original plant tissues or synthesised by various soil organisms."

It is principally the product of decay of surfacial organic debris (litter), together with the decomposition products of roots within the uppermost soil horizons. In average soils, humus contains 4–6% organic substances, which, in turn, are made of 85% dead matter, 8.5% living roots and rootlets and about 6.5% soil organisms.

2.2.1
Soil Organisms

These are generally classified according to their biological activities in soils into producers, consumers, and decomposers. Decomposers form the basis of the nutritive chain among the three groups. They produce, by degradation of organic litter, the primary resources (CO_2, N_2, O_2, etc.) used by the producers to synthesise complex nutritive material, which in turn will be consumed by the group of consumers. All three groups work hand in hand to change and continuously develop the soil profile. Soil organisms may also be classified according to their size into microfauna (<200 μm), mesofauna (200–1 000 μm), and macrofauna (>1 000 μm). Table 2.1 illustrates the approximate distribution in volume % of soil organisms in the organic fraction of an average European soil.

Macrofauna

Macrofauna living on and in the soil includes large molluscs, beetles, large insect larvae, as well as vertebrates like moles, rabbits, foxes, and badgers. These normally bury deep in the soil and feed on other smaller organisms. Moles in particular consume a great deal of smaller soil dwellers.

Mesofauna

Soil dwellers of this category belong to four main groups – *nematodes, arthropods, annelids* and *molluscs* (see Fig. 2.15). Nematodes, the unsegmented roundworms (also called eelworms) are about 0.5 to 1 mm in length. They are considered to be the smallest soil fauna next to protozoa. In a soil block of 1 m^2 surface area and 30 cm depth, $10^6 - 2\times10^7$ individuals of these worms, with a total weight between 1 and 20 g may be present. They feed on plant debris, bacteria and in some cases on protozoans.

- *Arthropods*, as shown by Table 2.2, comprise of various categories. To these, we count acari (mites), colembola (springtails), myriapods (centipedes and millipedes), isopods (wood lice), beetles, insect larvae, and termites. Among these categories, the

Table 2.1 Distribution of soil organisms in average European soils

Class of organisms	Vol% in humus of an average European soil
Bacteria and actinomycetes	50
Fungi	25
Lumbriscid worms	14
Macrofauna	5
Mesofauna	2.5
Microfauna	3.5

Fig. 2.15 Some of the common mesofauna in European soils (based on Ellis and Mellor 1995)

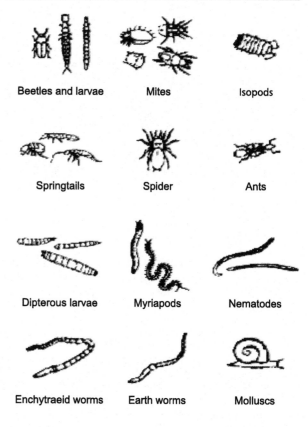

Table 2.2 Some data on small arthropods in soils

Group	Number of individuals	Total weight (g)	Characteristics
Mites	10^5 – 4×10^5	1 – 10	Common in acidic litter
Spring tails	5×10^4 – 4×10^5	0.6 – 10	Abundant with mites
Centipedes (chilopods)	50 – 300	0.4 – 2	Carnivorous
Millipedes (diplopods)	100 – 2000	0.05 – 1	Herbivorous
Isopods (wood lice)	50 – 200	0.5 – 1.5	Typical decomposers

Table 2.3 Number of individuals and total weight of microorganisms in a soil block of 1 m² surface area and 30 cm depth (after Klötzli 1993)

Group	Number (individuals m^{-2})	Total weight (g)
Bacteria	10^{12} – 10^{15}	50 – 500
Fungi	10^{10} – 10^{12}	100 – 1000
Algae	10^6 – 10^{10}	1 – 15

mites and springtails are the most abundant, especially in acidic litter, where they may form 80% of the soil organisms. Both mites and springtails feed on plant debris, bacteria and fungi. Ants and termites are most abundant in tropical soils, they are very active soil mixers. Earth pillars heaped by termites (termintaria) may reach several metres in height. Table 2.3 summarises some information on selected soil arthropods. The numbers are based on a soil block of 1 m² surface area and 30 cm depth.

- *Annelida* (ringed worms) is the name of the phylum to which the segmented earthworm, the leech, and the nereis belong. The Lumbricid worms (the earthworms) represent the first class of soil annelids. These occur in countless numbers in moist soils all over the world, emerging only at night and retreating under ground in the morning. Darwin made their activities the object of a careful study and concluded: *"it may be doubted if there are any other animals which have played such an important part in the history of the world as these lowly organised creatures."* Indeed, the quantity of earth burrowed, mixed and brought up from below and deposited on the surface by these worms has been estimated to be as high as 18 tons per acre per year, or, if spread out uniformly, about 3 cm in 10 years. The effects of worms on the soil include (beside bringing out deeper parts of the soil to the surface and exposing them to the air) improving the drainage through the intricate net of burrows and adding, in the form of excretory waste, a great deal of organic material to the soil. Their numbers in European soils are estimated to be between 80 and 800, with a total weight of 40 to 400 grams per square metre. Some giant earthworms are found in tropical regions, especially in Australia. *Megascolecides australis* may be 3 m in length; it lives in burrows with volcano-shaped openings. The *enchytraeid*, or potworms, are the second class of soil annelids. They are smaller than the earthworms (0.1–5.0 cm) and have a thread-like appearance. Potworms feed on algae, fungi, bacteria, and other soil organic materials. They may attain numbers as high as 200 000 individuals per square metre.
- *Mesofaunal molluscs* (2–20 mm), found in soils, include slugs and snails. Their numbers per square metre may range between 50 and 1 000. They feed on plants, fungi, and faecal remains.

Microorganisms (Microfauna and Microflora)

These are mainly represented by four groups: *bacteria, fungi, algae* and *protozoa*. All four groups occur in great numbers in soils (see Table 2.3).

- *Bacteria*. These are normally unicellular organisms with cellular sizes between 0.0001 and 0.02 mm, being the smallest organisms visible under the light microscope. They occur, due to their tremendous reproduction rates, in very high numbers (about 10^{12}–10^{15} m^{-2}) mostly in water films surrounding the soil particles. According to their morphology, bacteria are classified into three main classes:
  - *Eubacteria*. This is the most representative class of bacteria and can be further subdivided into spherical bacteria (*cocci*) and rod-shaped (*bacilli*) bacteria. Prominent examples of this group are the bacteria responsible for nitrogen fixation in the soil – *Nitrobacter*.

- *Chlamydobacteria (thread-shaped bacteri)*. In this class the bacterial cells are attached together by a filament (*filamenatum*) into a linear chain similar to bead strings. Iron bacteria, playing a very important role in weathering and diagenesis of sediments, belong to this class.
- *Actinomycetes*. In this class, rod-like cells are united to form stellar forms of bacteria, which in their vegetative stage of reproduction, resemble fungi, consisting of fine branching filaments (about 1 μm in diameter). These morphological peculiarities make it difficult to determine the real systematic position of these organisms, yet Actinomycetes are considered as an independent class of the Bacteriophyta.

Another method of classification of bacteria is their susceptibility to staining by the so-called *Gram solution* (after the Danish bacteriologist H. C. J. Gram, 1853–1938). Bacterial cells, stained by this solution, are called *gram-positive*; others are collectively termed *gram-negative*.

The effects of bacteria on the soil are numerous. They decompose a wide range of materials under various conditions; examples of these range from oxidation of Fe^{2+} and reduced sulphur compounds, under catalytic action of *Thiobacillus ferrooxidans*, to the formation of nitrogen-fixing nodules on the roots of leguminous plants by *Rhizobium* sp. Some bacteria are also capable of metabolising a wide range of chemicals, for instance *Pseudomonas*, a species capable of metabolising pesticides.

- *Fungi*. Mycophyta or Fungi are, like the bacteriophyta, in their overwhelming majority, parasitic; their carbon needs are dependent on nutritive material synthesised by other organisms. They are characterised by filamentous structures (hyphae), which are about 0.5–10 μm in diameter and which grow into a dense network called *mycelium*. Fungi live mostly in the surface layers of the soil, preferring acidic conditions, yet some of them live symbiotically in plant tissues. They may, under favourable (acidic) conditions, be responsible for the decomposition of up to 80% of the soil organic matter.
- *Algae*. These are photosynthetic organisms confined largely to the upper surface of the soil. They include Cyanophyceae (blue-green algae) and Chlorophyceae (green algae). Blue-green algae regulate the nitrogen cycle in the soil.
- *Protozoa*. A variety of protozoa, like rhizopoda, ciliates, and flagellates live in water films surrounding the soil particles. They control the numbers of bacteria and fungi on which they live. Table 2.4 shows the numbers and total weights of some protozoa in a soil block of 1 m^2 surface area and 30 cm depth.

2.2.2
Dead Organic Matter

Soil organic matter formed by metabolic action of the soil organisms, as well as by the break down of pre-existing organic material, can be classified into four main classes:

- Organic compounds free from nitrogen (other than lipids)
- Nitrogen compounds
- Lipids
- Complex substances including humic acids

2.2 · Organic Matter and Soil Organisms

Table 2.4 Some data on soil protozoa

Group	Number (individuals m^{-2})	Total weight (g)
Flagellates	5×10^{11}–10^{12}	10–100
Rhizopoda	10^{11} – 5×10^{11}	10–100
Ciliates	10^{6} – 10^{8}	10–100

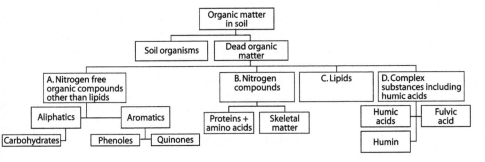

Fig. 2.16 Classification of soil organic matter

Each of these categories can be further subdivided as shown in the organisation chart in Fig. 2.16.

2.2.2.1
Organic Compounds Free from Nitrogen

Nitrogen Free Aliphatic Compounds
Carbohydrates are the most important representatives of this class of soil organic substances. Plants and soil organisms depend upon them for their structural material, which eventually ends up as important soil constituents. It is estimated that about 5–30% of the soil carbon exists as carbohydrates. They are so called because many of them (but by no means all) have a molecular formula which may be summarised by: $C_x(H_2O)_y$, where x and y can be equal or different numbers (see glyceraldehyde formula below).

Carbohydrates are *polyhydroxyaldehydes* or *polyhydroxyketones*. On hydrolysis, they again give only substances, which are either polyhydroxyketones or polyhydroxyaldehydes. The simplest molecule that fulfils this definition, besides displaying optical activity, which is one of the basic properties of carbohydrates, is glyceraldehyde (a polyhydroxyaldehyde). It has the formula:

$$CH_3(H_2O)_3 \text{ or } CH_2OH-CHOH-CHO$$

Classification of Carbohydrates
A key reaction of carbohydrates that may be used as a basis for a general classification of the group is hydrolysis. Carbohydrates on hydrolysis, as it was mentioned before, dissociate to simpler molecules having the same nature of being polyhydroxyketones or

polyhydroxyaldehydes. So if a carbohydrate does not hydrolyse to simpler molecules of the said nature, it would be called a *monosaccharide*; so, glyceraldehyde is a monosaccharide, as is glucose, fructose, ribose, etc.

The ending "-ose" is commonly used to indicate a carbohydrate. Structural formulae of some monosaccharides are given in Fig. 2.17.

A carbohydrate that hydrolyses to two monosaccharide molecules is by analogy called a *disaccharide*. Thus sucrose, lactose and maltose are disaccharides (see Fig. 2.18).

Similarly carbohydrates that hydrolyse to a large number of monosaccharides may be termed *polysaccharides*. An example is given here by starch, having the formula $(C_6H_{10}O_5)_n$. It hydrolyses to (n) monosaccharide molecules according to the equation shown in Fig. 2.19

Another way of classifying carbohydrates may be based on whether a carbohydrate has an aldehyde or a ketone group. In the former case, the carbohydrate may be called an *aldose*, while in the latter case the carbohydrate is called a *ketose*. The two groups resulting from this classification are corresponding isomeric groups having the same composition but different structures. Examples are:

- $CH_2OH–CHOH–CHO$ (glyceraldehyde)
- $CH_2OH–CO–CH_2OH$ (dihydroxyacetone)
- $CH_2OH(CHOH)_3CHOH–CHO$ (glucose)
- $CH_2OH(CHOH)_3CO–CH_2OH$ (fructose)

As a matter of fact monosaccharides form rings, the skeletons of which may be as shown for glucose in Fig. 2.20.

Such rings may be attached together by a process called *condensation*, taking place between the hydroxyl groups, giving rise to polysaccharides. Condensation may occur between similar molecules or by acting on other substances containing –OH groups. In this case the products are known as *glycosides*. These are often named after the sugars they contain e.g. *glucosides*, formed by reaction of glucose with methyl alcohol and hydrogen chloride.

The arrangement of the H's and OH's on the rings determines the chemical character of the substance formed, and that is why there might be several sugars of different chemical properties having the same formula of glucose, for instance α-glucose and β-glucose (see Fig. 2.21).

Nitrogen Free Aromatic Compounds

It is known that numerous low molecular weight *phenols* and *quinones* are liberated from dead vegetable matter. These may be utilised by fungi in building molecules with greater number of aromatic rings. It is also known that fungi (during the decomposition of lignin) produce extracellular phenol oxidases (enzymes, Sect. 9.4), which catalyse the introduction of hydroxyl groups into the phenol rings, giving rise to some aromatic soil constituents.

Phenols

Phenols form a class of aromatic compounds, the name of which is derived from "phen" – an old name of benzene. In this class one or more OH-groups are directly linked to

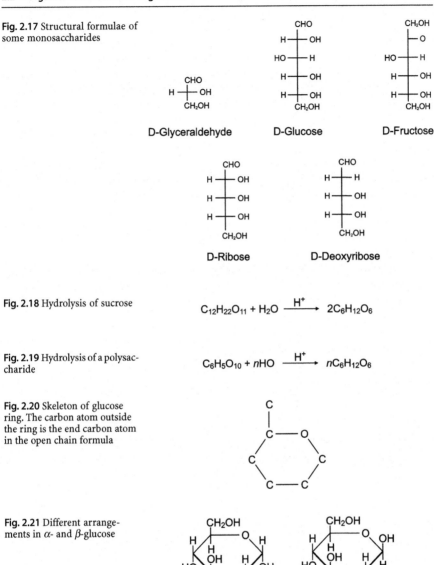

Fig. 2.17 Structural formulae of some monosaccharides

Fig. 2.18 Hydrolysis of sucrose

Fig. 2.19 Hydrolysis of a polysaccharide

Fig. 2.20 Skeleton of glucose ring. The carbon atom outside the ring is the end carbon atom in the open chain formula

Fig. 2.21 Different arrangements in α- and β-glucose

the benzene nucleus. Accordingly there may be mono-, di-, or tri-hydric phenols depending on the number of hydroxyl groups in the molecule (see Fig. 2.22).

Quinones – The Colour Dispensers
"Hail you O' land of Egypt, tell me O charming brunette where have you got this beautiful tan". This song of an Egyptian peasant praising the dark colour of his field re-

veals how the enigma of soil colour has always kindled the imagination of people, living on the banks of rivers where cultural centres first appeared. The name *Egypt* itself is derived from a Greek word meaning black or dark brown soil. It is perhaps not by accident that Egyptians have always made parallels between the colour of soil and the "*beautiful*" dark tan of *Henna* or *Egyptian privet*, which is used in the East, since ancient times, as a drug and cosmetic as well as a dye for leather and skin. The colour producing pigment in henna is a substance known as *lowsone*. It has the structure shown in Fig. 2.23, and is a member of a class of compounds known as the *quinones*.

Quinones, known since early times in history as dyes, are responsible for many characteristic colours in the plant kingdom. They often occur in the bark and roots of plants. Some fungi also owe their brilliant colours to the presence of quinones in their tissues. Such colourations are by no means limited to plants; quinones are also responsible for most of the dark colour pigmentation of animals. For example, the natural colouring of sea urchins is due largely to naphthaquinones (see formula, Fig. 2.24). Dark colourations of humus are, as it will be explained below, largely due to reactions of quinones and their derivatives in soil.

Quinones have a conjugated structure (see Figs. 2.23 and 2.24) and are reactive substances that often give hydroxy derivatives: *o*-quinone, benzoquinone, naphthaquinone, anthraquinone.

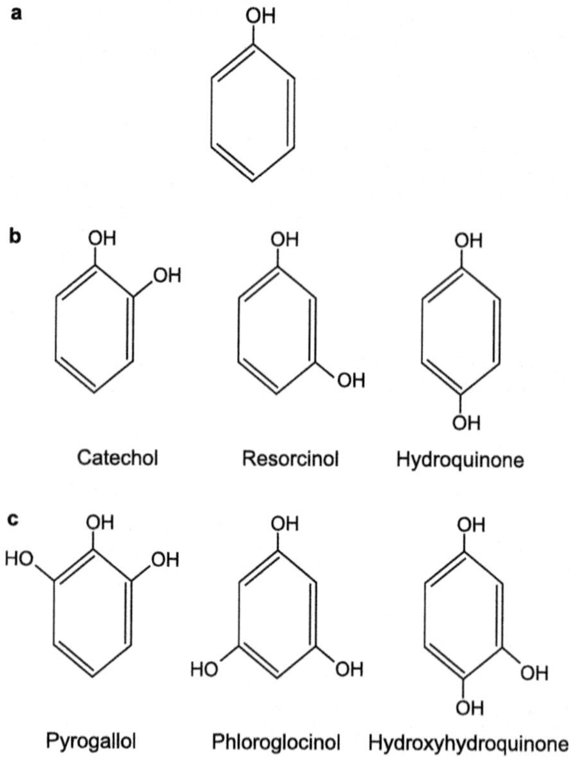

Fig. 2.22 Phenols classified according to the number of hydroxyl groups; **a** monohydric phenol (hydroxybenzene); **b** dihydric phenols: catechol (*left*), resorcinol (*middle*), hydroquinone (*right*); **c** trihydric phenols (trihydroxybenzenes): pyrogallol (*left*), phloroglucinol (*middle*), hydroxyhydroquinone (*right*)

Fig. 2.23 Structure of Lowsone

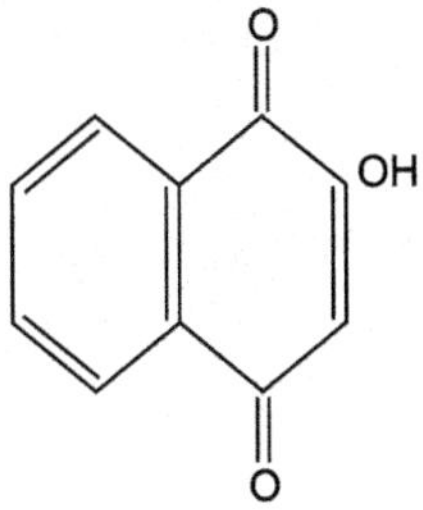

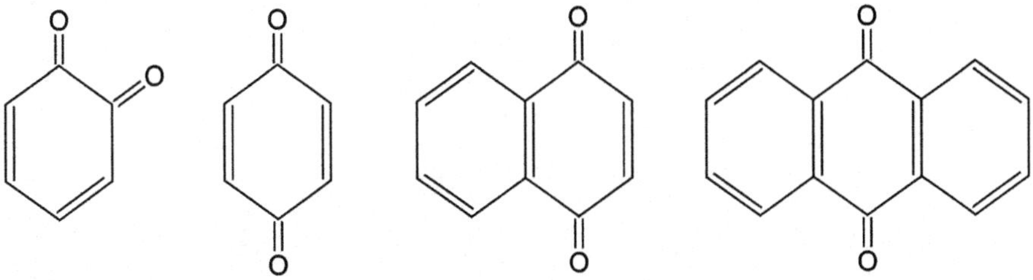

Fig. 2.24 Structure of different quinones

o-Quinone Benzoquinone Naphthaquinone Anthraquinone

2.2.2.2
Nitrogenous Organic Compounds

Proteins and Amino Acids
The most important nitrogenous organic compounds, found in soils, are proteins and amino acids. It is estimated that about 20–50% of organic nitrogen in soils exists as amino acids. These are compounds that contain carboxyl, –COOH and amino group, –NH_2. Thus many of them are neutral, but can react as acids or bases according to the prevailing conditions. Polymerisation of amino acids produce chain polymers known as *polypeptides*; very long chains of polypeptides are known as *proteins*. Both proteins and amino acids persist in the soil by being absorbed on the surfaces of clay minerals or being incorporated into other organic material. Proteins in soils ultimately break down, through bacterial or fungal action, into methane, CH_4; amines (compounds similar to ammonia, NH_3, in which a part or all the hydrogen has been replaced by organic groups); urea, NH_2–CO–NH_2; carbon dioxide or water. Figure 2.25 summarises the breakdown series of the proteins.

We should, however, bear in mind that not only decomposition takes place; in some cases, bacteria may resynthesize the simple molecules generated by protein decomposition.

In addition to the above-mentioned groups, nitrogenous organic compounds in soils may exist as *amino sugars*, nucleic acids or any other biological compounds.

Amino sugars are derived from sugars by substitution of one of the –OH groups by an amino group –NH_2. D-glucosamine (Fig. 2.26), formed from glucose, serves as an example.

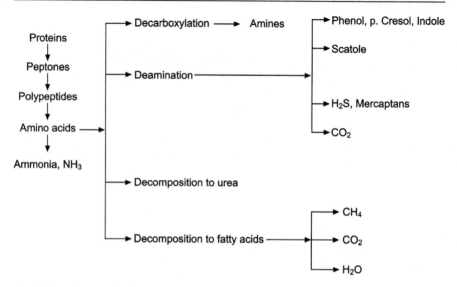

Fig. 2.25 Breakdown series of the proteins

Fig. 2.26 D-glucosamine (after Stevenson 1994)

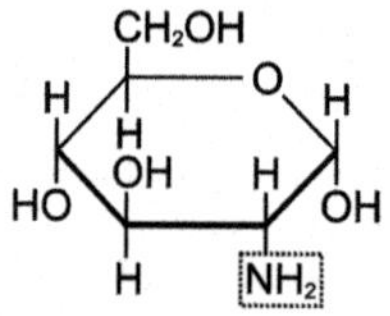

Nucleic acid is the name given to a substance found in all living cells. It may also occur in soil as a complex high molecular weight biopolymer, the monomers of which are known as *nucleotides*. Each of these nucleotides consists of three components: a nitrogenous heterocyclic base (either a purine or a pyrimidine), a pentose sugar and a phosphate group. According to the pentose sugar in the chain forming the single nucleotide, two types of nucleic acid are known. If the sugar was ribose (see Fig. 2.27) the acid would be called *ribonucleic acid* (RNA). In case of deoxyribose (see Fig. 2.27), the acid formed would be a *deoxyribonucleic acid* (DNA).

Skeletal Nitrogen Substances
These include cartilage of vertebrates and chitin of invertebrates. They are unlikely to decompose and persist in an almost unchanged state in the soil. Cartilage is largely made of a hardened mucous substance known as *mucoprotein* or *glycoprotein*. This is a covalently linked conjugate of a protein and a polysaccharide, whereby the latter forms about 4 to 30% of the whole compound. Another mucoprotein that may form a gel in soil at a low pH comes from mammalian urine, where it is most abundant.

Chitin (the name is derived from the same Greek word meaning *tunic*, referring to hardness) is a polymer made of a chain of an indefinite number of N-acetyl glucosamine groups (see Fig. 2.28). It is one of the main components in the cell walls of fungi, the exoskeletons of insects and skeletal parts of other arthropods.

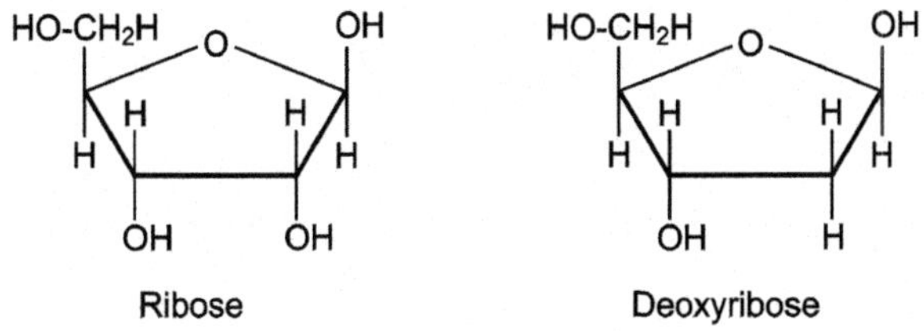

Fig. 2.27 Structures of ribose (left) and desoxyribose (right)

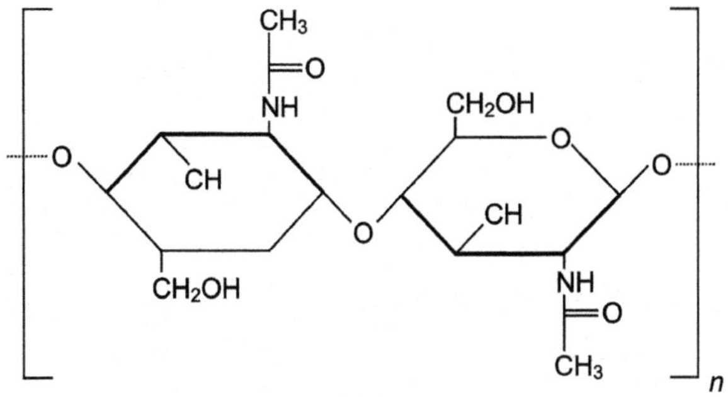

Ribose

Deoxyribose

Fig. 2.28 Structural model of chitin

2.2.2.3
Lipids

Lipids are esters of fatty acids with glycerol:

$$CH_2OH-CHOH-CH_2OH$$

Animals and plant waxes are esters of higher acids with other higher alcohols. Other lipids are compounds containing phosphorus or nitrogen or both. On hydrolysis, they yield the parent acids with glycerol, aliphatic alcohols, carbohydrates, nitrogenous bases, or sterols. Lipids are resistant to decomposition, but they are soluble in ordinary "fat" solvents such as ether and chloroform

2.2.2.4
Complex Substances Including Humic Acids

Humic substances are generally defined as natural organic substances, which are normally rich in nitrogen and may result from bacterial decay of plant and animal remains. They include a complex of poorly understood compounds, having different shades of brown and yellow colours.

It has been shown by many authors, that they comprise of three main fractions (*humic acids, fulvic acids and humin*), having molecules of 6–50 nm of no single structural formula and characterised by high molecular weights (1 500–30 000) and a large specific surface area. The three fractions are differentiated according to their solubility in water at different pH-values – a method that is normally used for their laboratory fractionation (see Fig. 2.29).

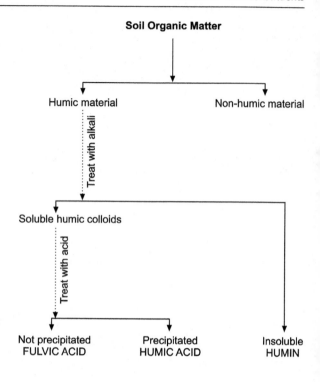

Fig. 2.29 Fractionation of humic substances (according to Ross 1989)

Humic Acid

This is the fraction soluble in water at a pH-value lower than 2, indicating that humic acid is only soluble under alkaline conditions (Stevenson 1994). It is of brown to grey black colours and may be extracted from soil using aqueous alkaline solvents. Humic acids were first reported by Achard in 1786, who extracted them from peat bogs near Berlin, Germany, and spent later a long time studying their properties and trying to decipher their origin and mode of formation. These studies led him to the conclusion that humic acids were derived from simple sugars. He then formulated this in the *sugar-amine condensation theory*, which was held, for a long time, as an explanation for the origin of humic material.

Amine-Sugar Condensation Theory

According to this theory, condensation reactions between reducing sugars (ketoses and aldoses) and amino compounds lead to the formation of an N–substituted glycosylmine. This is supposed to give, in the course of dehydration and fragmentation, the brown nitrogenous polymers forming the bulk of what is known as *humic acid*. This may be understood by considering the so-called *Maillard reaction* (the browning reaction), known from food chemistry, which was described by Luis Camille Maillard in the early 1920s. This reaction consists of, according to Lee and Nagy (1983) and Mauron (1981), three main stages:

1. *The initial stage (formation of glycoside).* In this stage condensation between the carbonyl group of an aldose and the free amino group of an amino acid results in

the formation of an N-substituted glycosylamine according to the equation shown in Fig. 2.30.

The initial stage is made up of two partial steps; in the first partial step, condensation between the carbonyl group of the aldose and the amino group in the amino acid, results in a condensation product (right upper part of Fig. 2.30). As expected when primary amines undergo condensation with aldehydes, the condensation product, in the second partial step, rapidly looses water to form a dehydration product, known generally as a *Schiff's base* – a reaction that is reversible and acid-base catalysed. The Schiff's base subsequently changes into the cyclic compound aldosylamine (in this case D-glucosylamine). During the initial stage no browning occurs.

2. *The medium stage (formation of ketosamine)*. Fragmentation of the products formed in the initial stage produces charged species known as *immonium ions*. These are the charged ends resulting from the fragmentation of peptide chains. The immonium ions are rearranged, after their formation, by isomerism (see Fig. 2.31) and subsequently form a compound called *ketosamine*. This process is known as *the Amadori rearrangement*.

3. *The final stage (formation of reactive substances that polymerise to produce dark brown humic acids)*. In the final stage the ketosamines may undergo dehydration to form reductones (compounds containing a hydroxyl group on each of the carbon atoms of a C=C double bond; so called because of their reducing power, e.g. ascorbic acid: vitamin C) and dehydroreductones. Ketosamines may undergo fragmentation as well, forming compounds such as hydroxyacetone (*acetol:* CH_3–CO–CH_2–OH), or pyruvaldehyde (Fig. 2.32) which subsequently undergoes degradation to form the so-called *Strecker* aldehyde.

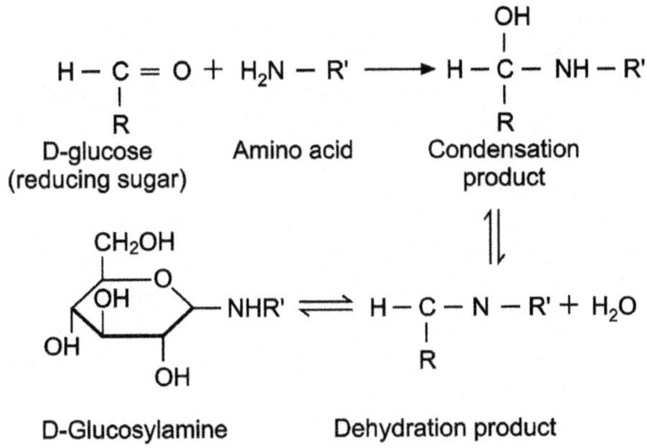

Fig. 2.30 Initial stage of the Maillard reaction

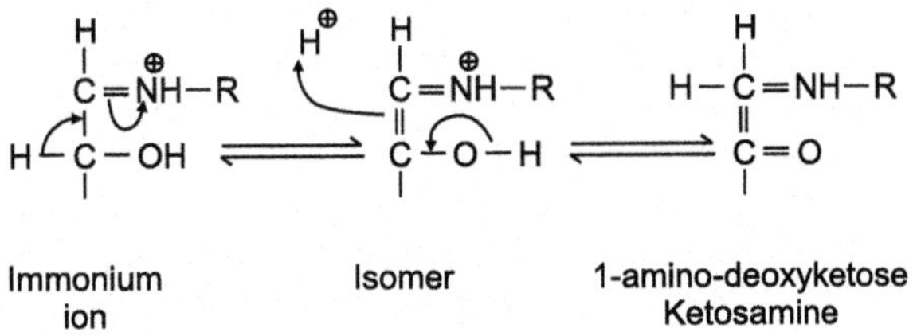

Fig. 2.31 Formation of the ketosamine (*Amadori rearrangement*)

Fig. 2.32 Formation of Strecker aldehyde

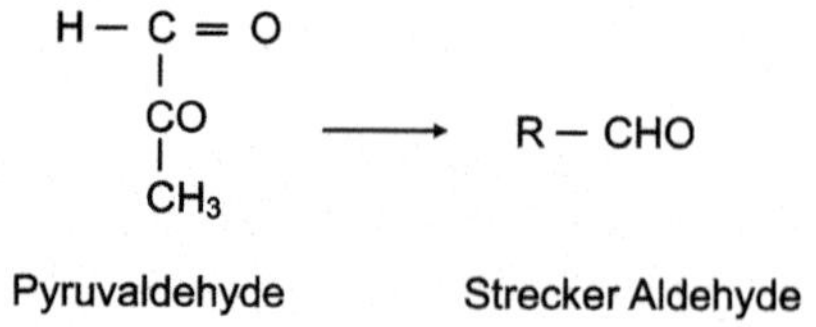

$$\begin{array}{c} H-C=O \\ | \\ CO \\ | \\ CH_3 \end{array} \longrightarrow R-CHO$$

Pyruvaldehyde   Strecker Aldehyde

These compounds are highly reactive. They eventually polymerise to form macromolecules of humic acids. A model of humic acid, according to Stevenson (1994), is shown in Fig. 2.33.

The model shows a macromolecule made of free and bound phenolic OH groups, quinone structures, nitrogen and oxygen as bridge units and COOH groups variously placed on aromatic rings.

It remains to say that the Maillard reaction, as known from food chemistry, is a temperature controlled reaction which works best at temperatures above 300 °C. Such temperatures are naturally not available at near surface sedimentary environments. That is why it is thought that microorganisms may be playing a key role in this process. Mineral catalysts such as clay minerals may also be supportive in overcoming the energetic barriers of the reaction. (Gonzales and Laird 2004)

However, due to the recognition that sugars were not the only possible precursors of humic acids, many authors challenged the amine-sugar condensation theory during the 1920s. In Germany, coal petrographers drew attention to the possibility that Lignin may be the parent material for humic acids, thus delivering the first arguments for what was later known as the *lignin theory*.

The Lignin Theory

Arguments delivered by coal petrographers that sugars in the natural environment were originally assimilated by microorganisms, and that they were not capable of persistence for a long time, paved the way for alternative notions considering lignin to be the parent material for humic acid (Waksman 1932). By the 1950s, the lignin origin theory seemed to be holding sway.

Lignin – a component in the cell walls of vascular plants – is a polymer with an aromatic structure, built up of phenyl propane monomers (see Fig. 2.34); it can be completely degraded by white rot fungi (Kirk and Farrell 1987). These fungi are capable of producing lignolytic enzymes that oxidatively cleave the phenylpropane monomers.

According to Waksman, fragmentation of lignin macromolecules is followed by elimination of methyl groups, oxidation and condensation with N-compounds, to give humic acids (see Fig. 2.35).

The lignin theory according to Waksman (1932) is supported by the following facts:

- The great majority of fungi and bacteria are capable of decomposing lignin and humic acids, albeit with considerable difficulty.
- Both polymers (lignin and humic acids) are soluble in alcohol and pyridine, hinting at a certain degree of kinship.
- Both are also soluble in alkali and precipitated by acids. They are both acidic in nature.
- Heating lignin with aqueous alkali gives methoxyl-containing humic acids.
- Humic acids and oxidised lignins have similar properties.

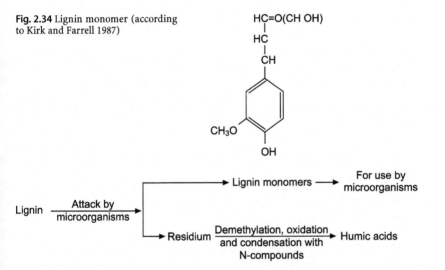

Fig. 2.33 A structural model of humic acid (after Stevenson 1994)

Fig. 2.34 Lignin monomer (according to Kirk and Farrell 1987)

Fig. 2.35 The lignin theory according to Waksman (1932)

As a matter of fact, the lignin theory, as formulated by Waksman, is considered obsolete today. Yet lignin still plays a key role in the subsequent theories, which are supported by most authors at present. Among such modern theories the *polyphenols theory* holds a central position.

The Polyphenols Theory
Polyphenols are chemical substances, principally found in plants, giving some flowers and vegetables their characteristic colours (see above). Chemically, they are phenolic compounds having more than one hydroxyl group. An example of these is Catechol (Fig. 2.36), which causes the dark colouration of some plants and vegetables when cut opened and exposed to air (apples, potatoes, etc.). The darkening occurs in this case due to the change of catechol into o-quinone by oxidation (see Fig. 2.36).

Quinones are capable of condensation with amino compounds (e.g. proteins and other macromolecules) to form strong dark coloured complexes of high molecular

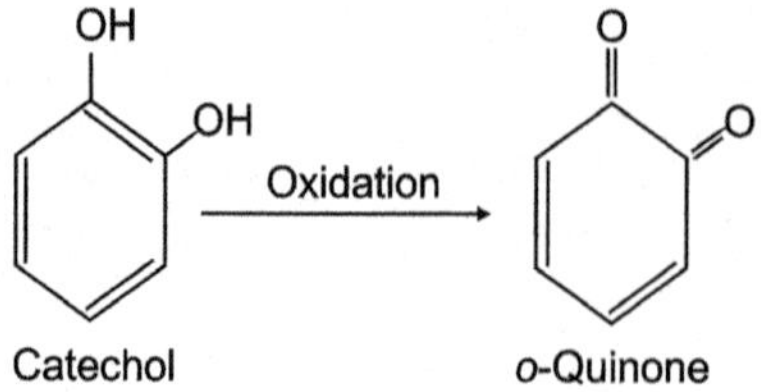

Fig. 2.36 Formation of *o*-quinone from a polyphenol (catechol) by oxidation

weights similar to humic acids. For this reason, it is easy to imagine a similar process through which lignin, cellulose or any complexes of polysaccharides, could be attacked by microorganisms to yield phenolic aldehydes and acids, which in turn would produce polyphenols. The latter would eventually be oxidised with the help of phenoloxidase enzymes (microorganisms) to produce quinones (Marshall et al. 2000). This, on condensation with amino compounds, can change into humic macromolecules. The whole process can be summarized in the schematic representation shown in Fig. 2.37 (Stevenson 1994).

Inspection of the different theories about the origin of humic acids shows clearly that the whole controversy lies in one point: that is at which stage in the alteration of plant and animal material does the reaction heads towards the formation of humic macromolecules and what chemical mechanisms steer the reaction. Does the reaction start at the stage of simple sugars? Does it start directly at the moment when lignin or cellulose is attacked by microorganisms? This remains a matter of speculation. Yet one point remains clear: humic acids are produced as a result of decomposition of organic materials, especially of plant origin – a matter substantiated by observation. It is also needless to say that the decomposition of plant remains is one of the central processes that keep the bio-geochemical cycle going. Plant tissues are sinks for CO_2 and if they were not decomposed, releasing CO_2 back into the atmosphere, a depletion of the gas with all its consequences for the bio-geochemical cycle would occur. At present, however, many authors believe that all the above-mentioned mechanisms of humic acids formations work together, hand in hand, and that in one and the same place, all three mechanisms may be active, replacing each other at given interfaces according to the physical chemical conditions. Microorganisms mostly provide active biological matter that work as a catalyst.

Fulvic Acid

This is the fraction of humic substances that remains in solution on acidification of the alkaline extract (humic acid precipitates at a pH-value of less than 2).

Figure 2.38 illustrates a partial structure of fulvic acid according to the model proposed by Buffle (1988).

Fulvic acids are less intensively studied than humic acids. Similar to the latter, they are darkly coloured acidic substances lacking the specific chemical and physical characteristics of simple organic compounds. Buffle (1988) suggested a model for fulvic acids (see Fig. 2.38) containing both aromatic and aliphatic structures, extensively substituted with oxygen and characterised by a predominant occupation with –COOH

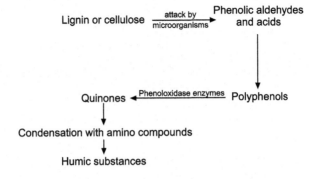

Fig. 2.37 The polyphenol theory (after Stevenson 1994)

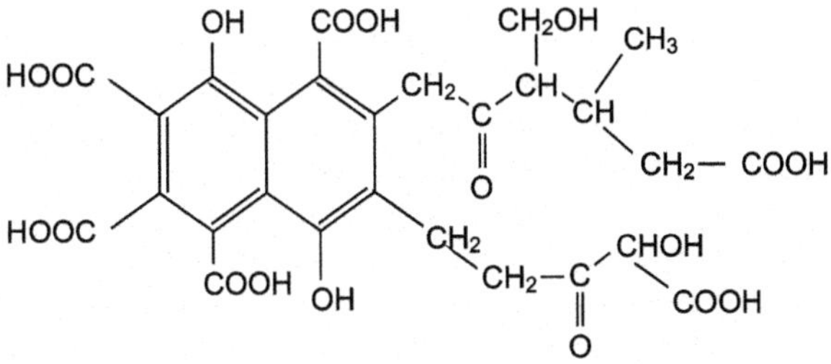

Fig. 2.38 A model for the structure of fulvic acid (according to J. Buffle 1988)

and –OH functional groups. These give them their acidic character and specific activity in adsorbing, complexing and chelating metals.

Fulvic acids modes of formation are similar to humic acids. According to Waksman, (1932 – *lignin theory*), they are formed by fragmentation of humic acids to smaller molecules. The polyphenol theory, however, suggests that they are directly formed from quinones parallel to humic acids. The ratio of humic acids to fulvic acids depends upon the type of soil, ranging from 0.3 in *tundra soils* to 2.0 or 2.5 in ordinary *chernozem soils* (Kononova 1966).

Humin
Humin is the stable fraction of humic materials, insoluble at all pH-values. It usually forms the predominant organic material in most soils and sediments (Kohl and Rice 1996). Its organic carbon represents more than 50% of the total organic carbon in soil. A considerable number of organic pollutants (e.g. pesticides, herbicides, polychlorinated biphenyls – PCBs) bind rapidly, and in many cases, irreversibly to it (*ibid*).

However, the insolubility of humin seems to have made it very difficult to understand its environmental as well as its analytical chemistry, and despite the great advances on the fields of analytical techniques, such as IR and GC, no generally accepted models of structure for all humic substances have been achieved.

Nevertheless, some authors have been trying to establish tentative models for humin (Rice and MacCarthy 1990; Lichtfouse 1999). The model of Lichtfouse (1999) is based on the assumption that humin is mainly formed of three components:

1. Resistant straight chain biopolymers, indicated by the presence of C_{11}–C_{24} compounds, which are typical breakdown products of these biopolymers, C_{27}–C_{33} linear alkanes and alkene doublets, having typical wax structure, encapsulated in the humin matrix.
2. Compounds bound to the humin bulk by chemical bonding (sterol and phytol), probably derived from chlorophyll and other plant material.
3. Despite the fact that no indications were shown for links between two or more molecules, the author suggested a tentative model, shown in Fig. 2.39 to approximately explain the structure of humin.

2.3
The Liquid Phase – Soil Water

Soil water is principally derived from two sources: precipitation and groundwater. Each contributes to the amount of moisture in the soil, within the regime of a sensitive dynamic equilibrium, depending mainly on the climate and the water balance between the atmosphere and the plant-soil system. The amount of water lost to the atmosphere comprises the sum of the water transferred by evaporation and of that transferred by plant transpiration, forming together the *evapotranspiration*. This depends directly on the climatic conditions as well as the properties of the plant-soil system. The evapotranspiration that would take place under optimum precipitation conditions and soil moisture capacity is known as the potential evapotranspiration, or, in short, *POTET*; it can be determined by empirical methods. Under ideal conditions, *POTET* will be equal to the actual amount of water transferred to the atmosphere by evapotranspiration in the region under consideration, which is known as *ACTET*. In other cases when

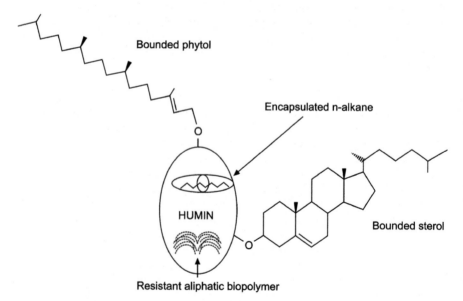

Fig. 2.39 The tentative model of humin suggested by Lichtfouse (1999)

POTET is not met by ACTET, the considered area will suffer a moisture shortage. The difference between POTET and ACTET in this case is known as the deficit or simply DEFIC. Still there may be another scenario when POTET is fully satisfied, yet the soil still receives water input, a situation will arise where water oversupply will lead to the formation of surface puddles and/or a recharge of the groundwater reservoir. This water surplus is shortly known as SURPL. To establish an account or water budget for a given region, all the foregoing quantitative factors are united in the following simple equation, known as *Thornthwaite's water-balance equation* (after C. W. Thornthwaite 1899–1963):

$$PRECIP = (POTET - DEFIC) + SURPL \pm \Delta STRG \times ACTET \qquad (2.1)$$

where PRECIP stands for precipitation and $\Delta STRG$ for the soil moisture, storage or the amount of water that is stored in the soil and is accessible to plant roots. Indeed, not all water stored by the soil can be classified in this category, for there are two main types of soil moisture: the type available to plant roots, which is held in the soil by surface tension and cohesive forces and is known as *capillary water*; and the type inaccessible to plant roots, which is known as *hygroscopic water*. This is made up of thin films of water molecules held tightly to the surfaces of soil grains by hydrogen bonds. Soils containing predominantly hygroscopic water and none or very little amounts of capillary water are not in a position to support plant growth; they are said to be at *wilting point*. Whereas, soils having pores full of capillary water are generally said to be at *field capacity*. In the case of water oversupply to the soil, the amount of water percolating downward to join the groundwater reservoir will be termed *gravitational water*. A soil, in which all gravitational water is drained out, contains the maximum level of capillary water and is consequently at field capacity. Based on the principles of water balance, the U.S. Soil Conservation Service recognises the following five soil moisture regimes:

1. *Aquic.* This is the case, when the soil is constantly wet, like in bogs, marshes and swamps. An aquic environment is a reducing one, virtually without dissolved oxygen. It is characterised by perennial stagnant water.
2. *Aridic (Torric).* This regime occurs mainly in arid or semiarid climates, where soils are dry more than half of the time. Soil temperatures at a depth of 50 cm lie above 5 °C. This regime is characterised by thin soils allowing high rates of evaporation and/or surfaces sealed by dryness, so that infiltration and recharge are inhibited.
3. *Udic.* The water balance in this regime is characterised by a moisture surplus at least during one season of the year. If the surplus exists throughout the year, the regime is called *perudic*. The soil is always in a position to support plants due to its rich reservoir of capillary water.
4. *Ustic.* This regime occurs in semiarid and tropical wet climates. It is between the aridic and udic regimes. The water balance reveals a prolonged deficit period following the growing season.
5. *Xeric.* This is the regime characterising the Mediterranean climate: dry and warm summer, rainy and cool winter. In every six out of ten years, there are (under this regime) at least 45 consecutive dry days during the four months following the summer solstice.

2.3.1
Composition of Soil Waters

Soil solutions contain a wide range of dissolved and/or suspended organic and mineral solid and gaseous substances (Fig. 2.40). Organic substances include all minor amounts of soluble and suspended organic compounds. These, together with minor amounts of dissolved silica and some pollutants, such as heavy metals (lead, zinc, cadmium, etc.), belong to the minor constituents of the soil water. The major constituents of soil water include dissolved salts in mobile ionic form, and gaseous compounds such as CO_2. This may be derived from the atmosphere by being dissolved in precipitation, or from the soil air, as a product of the respiration of soil organisms. It also may be the product of internal chemical reactions (e.g. as a product in the protein decomposition series). Mobile ions resulting from the dissolution of mineral substance include basic cations such as Ca^{2+}, Na^+, K^+, NH_4^+, and anions such as NO_3^-, PO_4^{3-}, Cl^-. These are provided by external sources as well as by internal chemical processes taking place within the soil (weathering, diagenesis, decomposition and synthesis of organic matter). Chloride ions, and to a lesser extent sulphate ions (SO_4^{2-}), may be provided by atmospheric sources including air borne marine salts and acid deposition. Under acidic conditions, cations of iron and aluminum may also constitute a considerable part of the active cations in soil water. The concentrations of ions in the soil solution, however, depend mainly on the soil-pH, its oxidation status and the affinity to processes such as adsorption, precipitation and desorption. Lower pH-values result in diminished metal adsorption capacities, which will in turn lead to higher metal concentrations in the solution. An important factor in determining the soil pH is the concentration of CO_2 in the soil solution. It provides the solution with H^+ through the reaction:

$$H_2O + CO_2 \rightleftarrows H_2CO_3 \rightleftarrows H^+ + HCO_3 \qquad (2.2)$$

Soil water in equilibrium with CO_2 in soil air has pH-values often below five. This acidity arises mainly from the following main sources:

- Organic acids produced during degradation of organic matter;
- Natrification processes, where NH_4^+ is converted into NO_3^-;
- Release of H^+ by plants in exchange for nutritive base cations;.
- Sulphide oxidation;
- Pollution through industrial and urban emissions.

2.4
The Gaseous Phase – Soil Air, Origin, Composition and Properties

Soil air, or soil atmosphere, is the characteristic name given to the mixtures of gases moving in the aerated zone above the water table (see Fig. 10.13 in Sect. 10.4) and filling the soil pores, where these are not already occupied by interstitial water. Mass flow of these gases in the aerated zone will be wholly controlled by atmospheric factors such as temperature, pressure, and moisture conditions. As far as major constituents are concerned, soil air has a composition slightly different from that of atmospheric

2.4 · The Gaseous Phase – Soil Air, Origin, Composition and Properties

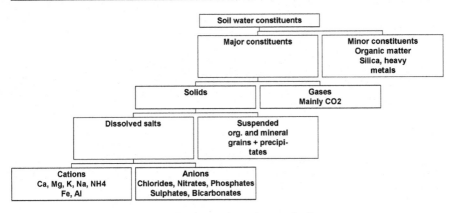

Fig. 2.40 Schematic diagram, showing the principal constituents of soil water

Table 2.5 Major constituents of soil air and atmospheric air in volume %

System	Composition (%)		
	Nitrogen	Oxygen	Carbon dioxide
Atmospheric air	78	21	0.03
Soil air	78	15–20	0.25–5.0

air (see Table 2.5). Although soil air contains 1–6% less oxygen by volume than atmospheric air, we find that it contains about 10 to 150 times more CO_2.

These differences in the concentration of CO_2 and O_2, between soil air and the atmosphere, result in partial pressure gradients between the two systems along which CO_2 moves from the soil to the atmosphere, while the oxygen flow takes place in the opposite direction. Gas exchange between soil air and the atmosphere occurs also along temperature gradients and in sites where rainwater introduces atmospheric gases into the soil. Beside the major constituents, minor or trace amounts of other gases may occur in the soil air, originating from deep-seated sources or as products of organic or mineral reactions in the soil environment. Examples of these are trace amounts of CO (carbon monoxide) and CS_2 (carbon disulphide), occurring in soils overlying geothermal sites in Utah (Hinkel et al. 1978). By contrast, light hydrocarbons such as methane (CH_4), H_2, H_2S, CO, and water vapour, are mainly produced within the soil body as products of degradation of organic material (Schlegel 1974). Traces of the extremely toxic gas *dimethyl mercury*, generated by bacterial activities, may under extremely reducing conditions occur in the soil.

Chapter 3

Soil Properties

3.1 Physical Properties

3.1.1 Colour

Colour is the most obvious trait in a soil profile. It may in many cases be indicative of soil composition. Red and yellow colour hues are generally indicative of enrichment in ferric iron, while grey hues may result of higher concentrations of aluminium oxides and silicates. Black colours are generally caused by an abundance of organic material. However, under warm temperate conditions, soils containing less than 3% of humus may have deep black colours. Soil colours are generally described according to a standard colour chart (Munsel colour chart), having 175 colour hues.

3.1.2 Texture

Texture is the term referring to the size and organisation of the soil particles. Individual particles known as *soil separates*, may be described, according to their grain size, as soil components (less than 2 mm in diameter), or simply as cobbles, pebbles or gravel, if they are larger than this. Soils may be classified according to their textures in different classes depending on the ratio of sand: clay: silt (Fig. 3.1). A hypothetical soil made of equal parts of these 3 components is termed *loam*. A loam can be described more specifically as sandy, clay, or silty loam according to the dominant component.

3.1.3 Structure

Soil structure refers to the aggregation or arrangement of primary soil particles. Structure is important because it can partially modify or overcome aspects of soil texture. The term *ped* (as it was mentioned before) describes an individual unit of soil aggregates; it is a natural lump or cluster, with clay and humus holding the particles together. Peds separate from each other along zones of weakness, creating voids that are important for moisture storage and drainage. Spherical peds have more pore space and greater permeability. They are therefore more productive for plant growth than coarse, blocky, prismatic, or platy peds, despite comparable fertility (see Fig. 3.2).

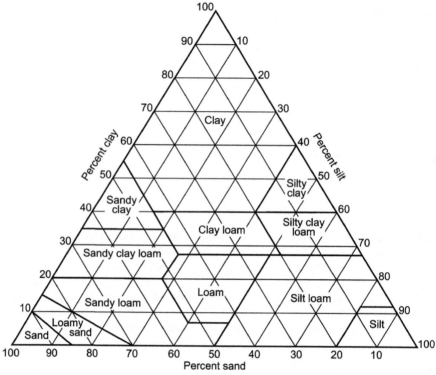

Fig. 3.1 Textural classification of soils (U.S. Department of Agriculture)

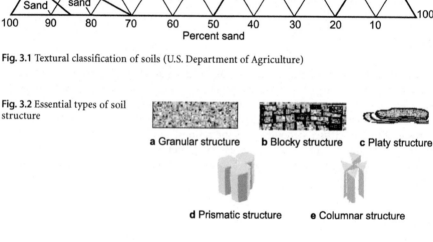

Fig. 3.2 Essential types of soil structure

a Granular structure   b Blocky structure   c Platy structure

d Prismatic structure   e Columnar structure

Terms used to describe soil structure include fine, medium, or coarse, with structural grades of adhesion within aggregates ranging from weak, to moderate, to strong.

3.1.4
Consistence

Cohesive properties of a soil, such as resistance to mechanical stress and manipulation under varying moisture conditions, are grouped under the heading consistence.

3.1 · Physical Properties

Wet soils are variably sticky. Plasticity is roughly measured by rolling a piece of soil between the fingers and thumb to see whether it rolls into a thin strand.

Moist soil implies that it is filled to about half of field capacity, and its consistence grades from loose (non-coherent), to friable (easily pulverised), to firm (not crushable between thumb and forefinger). Finally, a dry soil is typically brittle and rigid, with consistence ranging from loose, to soft, to hard, to extremely hard.

The segmentation that occurs in various soil horizons is a function of consistence and is usually described as continuous or discontinuous. Soils are variously noted as weakly or strongly cemented, or indurate (hardened). Calcium carbonate, silica, and oxides, or salts of iron and aluminium, can all serve as cementing agents.

3.1.5 Porosity

Grain size and pore volume are the most important factors controlling the percolation of water and ventilation within the soil. Porosity, as a measure of the percentage of volume of pore space with respect to the total volume, can be indirectly calculated using the following equation:

$$\text{Porosity} = 1 - \frac{\text{(Bulk) density}}{\text{(Particle) density}} \times 100 \tag{3.1}$$

Porosity can also be directly determined by the volume of water contained in a saturated undisturbed soil core of a given volume. The weight of the saturation water is determined through the difference between the wet and dry volume of the core. Porosity in this case will be calculated as follows:

$$\text{Porosity (\%)} = 1 - \frac{Ws - Wd}{V} \times 100 \tag{3.2}$$

Where Ws stands for the weight of the water-saturated core, Wd for the weight of the dry core, and V represents the volume of the same. Intrinsic to porosity is the pore size distribution; a very important physical parameter, it stands in direct relation to water retention and related properties such as drainage and aeration of a given soil, and hence to its agricultural productivity. Pore size distribution can be determined from the so-called moisture retention curves (see Fig. 3.3).

Moisture retention curves are established by determining the volumetric water content at various points over a range of tensions or suctions applied to an undisturbed soil core (Hall et al. 1977). This depends on the fact that the volume of water removed from an undisturbed soil core, at a given tension, is directly proportional to the pore size.

According to Fig. 3.3, the total porosity of the core under investigation is 50% and of this, 27% of pores are >20 μm in diameter, 15% are 2.0–20 μm and 10% are <0.2 μm.

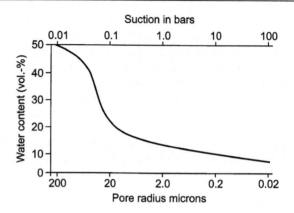

Fig. 3.3 A characteristic moisture retention curve (drawn after Ellis and Mellor 1995)

3.2
Chemical Properties

Based on their chemical composition and the degree of agricultural exploitation, soils may display certain chemical properties, which are not only diagnostic for their origin and environmental parameters, but also very sensitive indicators of their land use value. The most prominent of these are the degree of soil acidity (pH), together with ion exchange and cation adsorption capacities, which are very important factors for plant nutrition potentials of the soil. In fact, these properties are closely related to one another as we may see by looking at them.

3.2.1
Soil Acidity (pH)

Soil-pH is usually measured in a standard suspension of 1:2.5 (weight to volume) of soil in distilled water, or in dilute solution of calcium chloride (0.01 M). Besides being closely related to the ion exchange properties of soil (see below), soil acidity is related to other properties, such as organic content and clay mineralogy of the soil. It also has a direct relationship to the availability of metals as it affects their solubility and their capacity to form chelates in the soil. In this sense, Al^{3+} ions play an important role in controlling the concentration of hydrogen ions in soil waters, and hence the level of soil acidity (pH). This effect is brought about by hydrolysis when H^+ ions would be generated through the following reaction:

$$Al^{3+} + H \rightleftarrows Al(OH)_2^+ + H^+ \qquad (3.3)$$
$$\text{hydroxy-aluminium}$$

Positively charged hydroxy-aluminium species may undergo further hydrolysis to produce additional H^+ ions, leading to higher soil acidity.

$$Al(OH)_2^+ + 2H_2O \rightleftarrows Al(OH)_3 + 2H^+ \qquad (3.4)$$
$$\text{gibbsite}$$

Thus, aluminium hydrolysis in readily acidic soils will promote further acidity. Such a phenomenon is characteristic for humid, clay rich soils.

Soils of acidic character are generally rich in calcium and/or magnesium, because calcium and magnesium may precipitate as carbonate, increasing the buffering capacity of the solution.

$$Ca^{2+} + CO_2(g) + H_2O \rightleftarrows CaCO_3 + 2H^+ \qquad (3.5)$$

As the above reaction demonstrates, sodium does not show the same property as calcium and magnesium and that is why alkaline soils will have a higher sodium presence than that of the latter two. This dependence of soil pH-value on the availability and nature of the base cations makes it transient, i.e. changing in short term as well as in long spans of time, according to the availability and saturation of the base cations Ca^{2+}, Mg^{2+}, Na^+, and K^+. The input of these bases by atmospheric agents (precipitation), geochemical conditions (weathering), or agronomic activities (fertilisers) would eventually lead to fluctuations in the pH-value.

A general pattern, however, is that *water dominated soils* (soils of humid regions) have low pH-values, because their content of organic and carbonic acids is often subject to replenishing and recharge by rain fall. Under these conditions, the acids, in the way mentioned above for alumino-silicates, attack minerals, producing more acidity.

Under arid conditions, however, minerals which are salts of weak acids and strong bases would dominate the system, producing higher levels of alkalinity and causing the soil pH to raise to values between 9 and 10 or even more. Desert soils, having such elevated values of pH, are referred to as *rock dominated soils*. One should however be careful in dealing with soil pH-values above 10, since these may indicate contamination with strong bases such as Na, OH, or $Ca(OH)_2$. Though in some cases, like in some regions of north-western Egypt, this extreme elevation might be due to the dissolution of such minerals as nahcolite ($NaHCO_3$) or natron ($Na_2CO_3 \cdot 10H_2O$).

3.2.2
Ion Exchange

Exchangeable cations are those held between the layers of clay minerals (see Figs. 2.11 and 2.12), as well as those held on the surfaces of clays and organic particles (clay-humus complex or exchangeable complex) by virtue of the high surface energy resulting from the immense surface area of these finely divided substances. In the case of clay minerals, cations occupying interlayer positions possess permanent positive charges that satisfy structural conditions of the crystal lattice.

They may be replaced by other cations of similar size but of lower valency giving rise to isomorphous substitution, which is not directly dependent on the soil pH. In the case of the clay humus complex in general, surface charges are not governed solely by structural considerations. They result largely from surface reactions such as reversible dissociation of surface groups or functional groups of organic compounds, and since these reactions are pH dependent, the charges would vary according to the prevailing pH. This can be illustrated by the reaction of aluminium "oxyacids" under different conditions of Acidity. Two of these compounds are known to occur in soils – $Al(OH)_3$ (gibbsite – some times called simply aluminium hydroxide) and its anhydride $HAlO_2$.

Fig. 3.4 Influence of pH on surface charge in aluminium oxide (White 1997)

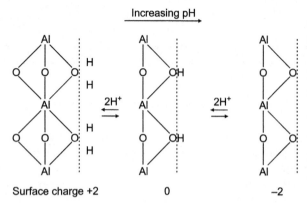

As illustrated by Fig. 3.4, the anhydride, which is amphoeteric, produces different species according to whether it reacts under acidic or alkaline conditions. Under acidic conditions, the edge hydroxyl groups may take up H^+ ions to produce the positively charged hydroxy aluminium with a charge of +2, while removal of H^+ from these groups under alkaline conditions produces a negatively charged species with a charge of –2. Such negatively charged colloidal particles are known as *micelles*. They act like giant anions that attract cations to their surfaces resulting in the formation of an *electrical double layer* (made of the negatively charged layer on the micelle and the positively charged layer of adsorbed cations) around each of them. Cations closer to the negatively charged layer will be more strongly attracted than those at a greater distance; and that is why cations with a smaller radius of hydration will be preferentially adsorbed and less readily replaced.

Another factor controlling adsorption and replacement is the valency. Cations with a high valency have a high energy of adsorption and are therefore adsorbed in preference to cations with a lower valency. The following sequence is generally accepted for preferential adsorption of base cations: $Ca^{2+} > Mg^{2+} > K^+ > Na^+$. These cations will be less readily replaced and that is why they are represented in soil composition by a similar sequence of concentration:

$Ca^{2+} > Mg^{2+} > K^+ > Na^+$
 80%   15%   1.8%   1%

In addition to this, anion adsorption and replacement may occur. It is, under certain conditions, even more common.

3.2.3
Cation Exchange Capacity (CEC)

The cation exchange capacity of soils is generally measured by the cation yield of the soil through extraction with an ammonium acetate solution. It is also a measure of the surface negative charge, as well as the potential for cation adsorption. CEC is generally expressed in meq per 100 g of soil. Its value depends largely on the organic and clay content of the soil as well as the acidity and clay mineralogy of the same.

Anther quantitative attribute of cation exchange in soils is the property known as the *percentage base saturation*, which is simply a measure of the proportion of exchangeable bases on the soil exchange complex. It can be calculated from the following equation:

$$\text{Base saturation (\%)} = \frac{Ca^{2+} + Mg^{2+} + K^+ + Na^+}{CEC} \times 100$$

The difference between CEC and total exchangeable base content provides a measure of exchangeable hydrogen content. (For further information consult Rowell 1994).

$$\text{Exchangeable } H^+ = CEC - (Ca^{2+} + Mg^{2+} + K^+ + Na^+) \qquad (3.7)$$

Metal binding by small organic molecules, which are largely composed of polymeric weak acids of the type of humic and fulvic acids, takes place through the formation of outer sphere complexes by binding the metal to the carboxylic or phenolic groups (see Fig. 3.5).

The CEC of organic soil is largely due to organic matter, and its contribution to the CEC of mineral soil may be >200 meq/100 g of organic matter (Bloom 1981).

3.2.4
The Interaction of Organic Soil Matter with Mineral Components

Chemical changes of sediments at environments near the Earth's surface (at normal temperatures and pressure levels) are generally known in sedimentology under the collective term *diagenesis*. The most important early diagenetic changes of soil organic matter include oxidation, reduction and carbon-carbon coupling. In considering these changes and their varieties, one is immediately struck by the ubiquitous involvement of metals and their derivatives.

Reactions involving metals with organic matter may range from simple salt formation with organic acids, to catalysis of complex formation by metals lending them-

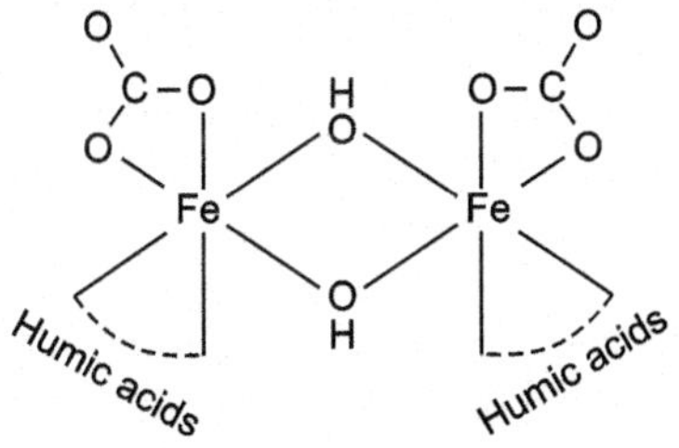

Fig. 3.5 Bonding of metals on humic substances by chelation (from Mirsal 1995)

selves as templates, or even indirectly by metals facilitating the work of enzymes. In the following, some of the well known reactions of metals with soil organic matter will be used to illustrate this.

Formation of Metal Salts

This may occur in the event that metals released from minerals, combine with simple organic acids or humic substances. As an example of this we may consider the action of simple fatty acids on carbonates, characteristic for acidic soils, e.g. the reaction of alkali metals with monobasic aliphatic acids or humic acids, e.g. formation of humates or fulvates.

Formation of Chelate Complexes

Transition metal ions (bivalent or trivalent) may react with organic chelating agents in soils, most probably those containing functional groups such as amine $-NH_2$; azo group $-N=N-$; carboxylate $-COO^-$ or ether $-O-$, whereby the chelating capacity decreases from amines to ethers.

As an example of such compounds, we may take the one in Fig. 3.5, showing a polynuclear chelate complex of iron with humic acid.

Physical and Chemical Adsorption on Clay Minerals

Adsorption caused by Van der Waal's forces. According to Stevenson (1994), adsorption of organic materials on the surfaces of clay flakes by the weak Van der Waal's forces can play a considerable role in attaching organic matter to the surfaces of clay minerals. This is valid for polar as well as for non-polar molecules. Figure 3.6 taken from Jerzy Weber (2005), after Stevenson (1994), shows a model of how Van der Waal's forces may be distributed on clay surfaces.

Adsorption on interlamellar spaces of clay minerals. Schnitzer and Kodama (1977) have shown that fulvic acid may be adsorbed on surfaces at the interlamellar spaces of montmorillonite. Theng (1979) and Theng et al. (1986) confirmed the same opinion. Many recent works have now shown that this mechanism is one of the most important for the retention of proteins and charged organic compounds by expandable layer silicates (see under clay minerals).

Bonding by cation bridging. Clay surfaces are negatively charged – a fact that will cause repulsion of negatively charged organic anions. Polyvalent cations like Ca^{2+}, Fe^{2+}, and Al^{3+} are capable of forming a bridge between such anions and the surfaces of clay, so that they may be attracted to each other, forming clay organic complexes as shown in Fig. 3.7 (Fotyma and Mercik 1992).

Hydrogen bonding. The attachment of organic molecules on clay surfaces, may occur by hydrogen bonding between polar groups of the organic molecules and adsorbed water molecules, or silicate oxygen. A model of such bonding is shown in Fig. 3.8 according to Fotyma and Mercik (1992).

Fig. 3.6 Van der Waal's forces on clays (Stevenson 1994)

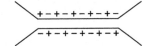

Fig. 3.7 Formation of organic clay complexes by cation bridging according to Fotyma and Mercik (1992)

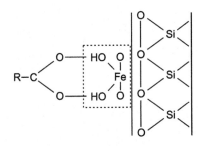

Fig. 3.8 Formation of organic clay complexes by hydrogen bonding (Fotyma and Mercik 1992)

Fig. 3.9 Formation of clay organic complexes with the help of hydrous oxides (Fotyma and Mercik 1992)

Adsorption by coatings of hydrous oxides of iron and aluminium on clay surfaces. Coatings of hydrous oxides of iron and aluminium on the surfaces of clay minerals are capable of adsorbing humic substances, where these may be attached by simple coulombic attraction. Figure 3.9 shows a model of how such complexes may be formed.

3.2.5
Oxidation-Reduction Status

The redox potential (Eh), or oxidation-reduction status, is an indicator of the degree of soil aeration or the amount of oxygen present in the soil atmosphere. It can be directly measured by inserting an appropriate electrode in the soil. In aerobic soils, electrons produced during respiration, combine with oxygen. Under anaerobic conditions, however, oxygen is unavailable, and other chemical compounds must act as electron receptors. Ferric iron compounds often take on this role, undergoing reduction to ferrous iron compounds.

Chapter 4
Soil Types and Classification

The idea of soil classification has always been considered as one of the prerequisites of serious formulation of a theoretical framework for a soil science, which would allow treatment of soil as an independent natural body having a global recognisable pattern. It was the Russian soil scientist V. V. Dukuchaev who published the first system of classification approaching this aim. In his monograph, published in 1883, Dukuchaev introduced a classification of soils into rough groups based largely on genesis as controlled by the factors monitoring soil formation. Later, in 1935, Dr. Curtis F. Marbut introduced the first formal American system of soil classification. He mainly derived his approach from the work of Dukuchaev.

In 1951, the U.S. Soil Conservation Service started a program for constructing a system of classification that takes into consideration, beside genetic factors, descriptive factors observed in the field. This system went under several modifications in 1960 and 1964, culminating in what is called the "seventh approximation". The current version is the *Soil Taxonomy System – Second Edition*, published in 1999. At present, the Soil Taxonomy System, together with the classification system proposed by the United Nation's Food and Agricultural Organisation (FAO), form the most widely used soil classification systems. Despite the different approaches adopted by their authors, both systems have several aspects in common.

4.1
The Soil Taxonomy System – Criteria of Classification

The Soil Taxonomy System broadly classifies soils into twelve orders based on three main groups of criteria: a pedological morphological group based on the morphologic features characterising a soil profile in the field; the environmental conditions controlling the soil formation processes under which the soil profile was developed; and a chemical criterion based on the dominating chemical condition along the profile, or in characteristic parts of it. Each of these items is represented by individual properties, as it will be shown in the following section. Figure 4.1 summarises in a schematic way the relation between all these elements of soil description.

4.1.1
Morphological Criteria of Classification in the Soil Taxonomy System (Diagnostic Horizons)

A diagnostic horizon, within the solum in a soil profile, is one characterised by an individual physical property such as colour, texture, consistence, porosity or moisture.

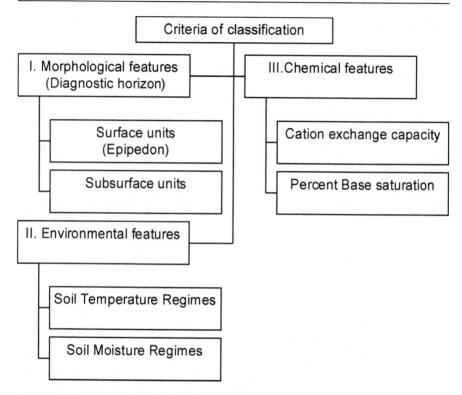

Fig. 4.1 Schematic diagram summarising the criteria of soil classification in the Soil Taxonomy System

It may also be reflecting a dominant process that led to the development of the soil profile.

The use of this criterion, depends on the systematic description of a unit soil series (polypedon), made of an indefinite number of pedons, representing a certain locality, or area, with a distinctive characteristic soil that differentiates it from its surroundings. A profile of such a series, that may or may not be a combination of both A- and B-horizon (Solum), consists of normally two diagnostic horizons: the epipedon and the subsurface diagnostic horizon.

The *epipedon* is the portion of soil formed at the surface and may extend through the A-horizon or even include parts of the B-horizon. It is normally dark in colour and contains high amounts of organic matter. The morphological and chemical characteristics of epipedons, allow their classification into different types that can be used for identification. Table 4.1 shows the most important of these.

The *subsurface diagnostic horizon* is that portion of the solum formed below the surface at varying depths; it may also include parts of the A and/or B-horizons. As in the case of epipedons, subsurface units may be classified into various types (see Table 4.2). These types may be further used in classifying soils, according to their presence or absence in profiles characterising the soil of the area, or locality under consideration.

4.1 · The Soil Taxonomy System – Criteria of Classification

Table 4.1 Surface units. *Source:* Soil Taxonomy, Agricultural Handbook No. 436, and U.S. Department of Agriculture (1975)

Epipedon	Name derived from	Characteristics
Histic A-horizon	*Histos* "tissue"	Peaty or mucky thin layers, organic matter >20%, saturated at least 30 consecutive days a year
Mollic A-horizon	*Mollis* "soft"	A granular dark horizon >18 cm thick, with organic matter >1%, strong structure, Ca is main exchangeable cation
Umbric A-horizon	*Umbra* "shade", hence being dark	A horizon of dark colouration not related to humus. It is identical to mollic, yet having low base cation content
Ochric A-horizon	*Ochros* "pale"	A relatively pale or thin surface horizon, low humus content, hard and massive when dry
Plaggen A-horizon	*Plaggen* "grass, lawn"	Human made >50 cm thick, produced by long, continuing manuring and irrigation
Anthropic A-horizon	*Anthropos* "having to do with humans"	Modified by long and continuous human use, it is similar to mollic

Table 4.2 Subsurface units (B-horizon). *Source:* Soil Taxonomy, Agricultural Handbook No. 436, and U.S. Department of Agriculture (1975)

Name	Name derived from	Characteristics
Argillic B-horizon	*Argilla* "white clay"	Illuviated clay accumulation, clay coatings
Agric B-horizon	*Ager* "field"	Illuvial layer formed below cultivation
Natric B-horizon	*Natrium* "sodium"	Argillic that is high in sodium
Spodic B-horizon	*Spodos* "wood ashes"	Accumulation of organic matter, Fe and Al
Placic B-horizon	*Plax* "flatstone"	Thin, dark-reddish, iron-cemented pan
Cambic B-horizon	*Cambiare* " to exchange"	Physically or chemically altered
Oxic B-horizon.	*Oxide* "oxide"	Highly weathered kaolinite, Fe-Al-oxides, no clay coatings
Kandic	*Kandite* – a collective name for the kaolinite group of clay minerals	Accumulation of iron and aluminium oxides together with clays belonging to the kaolinite group
Ortstein	*arut* for "ore" in old Saxon and *stein* "stone"	Introduced by P. E. Mueller, 1887 to describe weakly indurated B-horizons of red brown or black colours, having high iron content

Description of the Diagnostic Horizons – An Overview

Epipedons, as said before, can on the basis of their lithologic or structural characters be classified into different types. In Tables 4.1–4.3, each of these types will be shortly described, relative to its position in soil profile.

Table 4.3 Diagnostic horizons not essentially bound to a given depth. *Source:* Soil Taxonomy, Agricultural Handbook No. 436, and U.S. Department of Agriculture (1975)

Name	Name derived from	Characteristics
Duripan	*Durus* "hard"	Silica-cemented hard pan; not softened by water
Fragipan	*Fragilis* "brittle"	Weakly cemented brittle pan; loam texture
Albic	*Albus* "white"	Light pale subsurface horizon; clay and iron oxides are leached
Calcic	Calcium or lime	Accumulations of calcite $CaCO_3$ or dolomite $CaMg(CO_3)_2$
Gypsic	Gypsum	Accumulations of Gypsum $CaSO_4 \cdot 2H_2O$
Salic	Salt	Accumulations of more than 2% of soluble salts
Plinthite	*Plinthos* "brick"	Iron-rich horizon that changes to iron stone through wet/dry cycles
Glacic	Glacial "frozen"	Frozen organic material
Densic	Dense "compact"	Unaltered, non cemented material, rupture resistant
Lithic/Paralithic	*Lithos* "stone"	Indurated, coherent unaltered material; weakly cemented unaltered material

Based on the presence, or absence of one or more of these diagnostic horizons, soils are classified in the Soil Taxonomy System into twelve broad categories known as the *Soil Orders of Taxonomy*. Each of these orders is characterised by the occurrence of certain soil horizon(s). A detailed description of these orders, including their representative diagnostic horizons, is given in Sect. 4.1.5.

The twelve orders of soil Taxonomy are further classified, on the bases of a combination of environmental and chemical criteria, into 47 categories, known as suborders. These are again classified according to genesis, environmental and chemical factors into great groups.

4.1.2
Description of the Environmental Criteria of Classification

4.1.2.1
Soil Temperature Regimes (According to USDA, Soil Taxonomy)

The term *soil temperature regime* describes the mean annual soil temperature (*MAST*) at a depth of usually 50 cm. The *MAST* is generally ~1 °C above mean air temperature. One can summarise the categories of soil temperature regimes in the following manner:

- *Pergelic.* High latitudes, where soils may be permanently frozen below a certain depth, thawing only in their uppermost parts during the summer months (Permafrost).
- *Cryic* (0–8 °C, below 15 °C in summer). No permafrost.
- *Frigid* (0–8 °C, above 15 °C in summer). Frigid soils are warmer in summer than soils of the cryic regime. However, *MAST* is lower than 8 °C. The difference between mean summer (June, July, and August) and mean winter (December, January, and February) soil temperatures is greater than 6 °C.
- *Mesic* (8–15 °C). Difference between mean summer and mean winter soil temperatures is greater than 6 °C.
- *Thermic* (15–22 °C). *MAST* is 15 °C or higher but lower than 22 °C. The difference between mean summer and mean winter soil temperatures is greater than 6 °C.
- *Hyperthermic* (>22 °C). *MAST* is 22 °C or higher. The difference between mean summer and mean winter soil temperatures is greater than 6 °C.

In tropical soils, where *MAST* experiences minimal seasonal fluctuations, the prefix "iso-" is added to the qualifying term. Accordingly, a tropical temperature regime with a *MAST* of 20 °C would be described as *isothermic*.

4.1.2.2
Soil Moisture Regimes (According to USDA, Soil Taxonomy)

The term *soil moisture regime* describes the state of presence or absence of water, or water held at a tension of less than 1 500 kPa in the soil or in specific horizons of it, during some periods of the year. Accordingly, soils are described as *dry* if water is held at a tension of 1 500 kPa or higher, and *wet* if the tension is less than 1 500 kPa but still higher than zero.

Soil moisture regimes in pedons are estimated by measurement of wetting certain representative portions of the profile, known as *soil moisture control sections*. These are dependent on the climatic conditions, as well as on the grain size and structure of the soil in the profile. Generally, soil moisture control sections are approximately distributed as shown in Table 4.4).

Description of the Different Classes of Soil Moisture Regime
Aquic Regime (L. aqua – water)
This regime is characteristic for soils saturated with water. They are, therefore, virtually free of dissolved oxygen – a character that renders their environment reducing. In the event that the water saturation occurs in the presence of oxygen, due to any local conditions, the regime will not be considered as aquic.
Aridic and Torric Moisture Regimes (L. aridus – dry, and L. torridus – hot and dry)
Both terms are alternatively used for the same soil moisture regimes. Soils of this regime normally occur in areas with arid climates. A few are in areas of semi-arid climates and, in this case, physical properties such as crusty surfaces, precluding the infiltration of water, keep them dry. Soils of semi-arid regions, occurring on steep slopes where run-off is high, may also attain such an aridity character that they would be placed in this class. Salt accumulations may also occur due to a lack of leaching and low transport of dissolved salts.

Table 4.4 Extension of soil moisture control sections in different soils. *Source:* USDA, Soil Taxonomy

Particle size class	Extension of control section
Fine-loamy, coarse-silty, fine-silty or clayey	From 10 to 30 cm below the soil surface
Coarse loamy	From 20 to 60 cm below the soil surface
Sandy	From 30 to 90 cm below the soil surface
Soil containing rock fragments	Deeper than 90 cm

Udic Moisture Regime (L. udus – humid)
The soil moisture control section is not dry, in any part, for as long as 90 cumulative days in normal years (years having no large moisture fluctuations). The udic moisture regime is common to the soils of humid climates that have well distributed rainfall, i.e. the amount of moisture plus rainfall is approximately equal to, or exceeds, the amount of evapotranspiration.

In climates where precipitation exceeds evapotranspiration in all months of normal years, the water moves through the soil in all months when it is not frozen. Such an extremely wet moisture regime is called *perudic* (L. *per* – throughout in time, and L. *udus* – humid).

Ustic Moisture Regime (L. ustus – burnt; implying dryness)
The ustic moisture regime, intermediate between the aridic regime and the udic regime, includes the semiarid and topical well-dry climates. Under this regime, moisture is available, but limited, with a prolonged deficit period following the period of soil moisture utilisation during the early period of cultivation.

Xeric Moisture Regime (Gr. xeros – dry)
The xeric moisture regime is typical in areas of Mediterranean climates, where winters are moist and cool, and summers are warm and dry. In areas of a xeric moisture regime, the soil moisture control section, in normal years, is dry in all parts for 45 or more consecutive days in the 4 months following the summer solstice and moist in all parts for 45 or more consecutive days in the 4 months following the winter solstice.

4.1.3
Description of the Chemical Criteria of Classification

Soils are also classified by combining all of these criteria, mentioned in this section, with chemical properties that may be characteristic for the soil quality or type at the site of investigation. These include properties like cation exchange capacity and percent base saturation – two properties which are largely dependent on lithologic, morphologic and environmental factors. So we find that:

CEC-values tend to be highest in soils with high clay and organic content, whereby values for organic matter may be virtually twice of those for clay.

Like cation exchange capacity, anion exchange capacity (AEC), is a measure of the total amount of exchangeable anions that a soil can adsorb. It is expressed as milliequivalents per 100 g of soil. This value has already been found to be useful in differentiating tropical soils from those of the temperate regions. Tropical soils often possess significant AEC, unlike soils of the temperate environment (Rowell 1994).

4.1.4
Categories of the Taxonomy System Based on the Criteria of Classification

Soils are classified on the bases of the above-mentioned criteria, from the most general to the most specific, into the following categories:

- Order (orders are largely based on morphological features)
- Suborder (suborders are based on environmental and chemical features)
- Great group (combination of all)
- Subgroup
- Family
- Series

Orders (see below) are divided into suborders based on moisture, temperature or other features. Suborders are divided into great groups, great groups are divided into subgroups, subgroups are divided into families, and families are divided into series.

In the 1975 edition of the *Soil Taxonomy System*, only 10 soil orders were mentioned. The order now called *andisols* was proposed in 1978 and adopted several years later. The twelfth order, *gelisols*, was added in 1997.

In this book, only orders and suborders will be discussed in some detail. The interested reader is referred to the *USA Soil Survey Manual* (Soil Survey Division Staff 1993) for more details on the other taxonomic categories.

4.1.5
The Soil Orders of Taxonomy

Since the Russian soil scientist V. V. Dukuchaev published in 1883 his soil classification system of soil classes, a number of other classifications have been published worldwide. As it was mentioned before, most well known classifications are the soil taxonomy system published by the United States Department of Agriculture's Soil Survey Staff and the system adopted by the United Nations Food Organisation (FAO). The soil taxonomy system of the United States Department of Agriculture has been subjected to many changes since its first version, the last of which was the second edition of *Soil Taxonomy* issued in 1999.

The second edition of *Soil Taxonomy* classifies soils mainly in twelve higher categories known as *orders* (see below). This is largely based on soil properties and the main factors leading to their formation. The placement of a soil within one of the orders depends on the presence, or absence, of one (or more) of the diagnostic horizons (see Table 4.3) and the dominating morphology resulting from all of these conditions (see Fig. 4.2). The twelve soil orders described in the system are distributed worldwide in a recognizable pattern, as the map published by the U.S. Department of Agriculture shows (see Fig. 4.3).

Fig. 4.2 The dominant 12 orders of soil taxonomy (explanation in text). *Source:* U.S. Department of Agriculture, National Resources Conservation Service (NRCS) – Photo Gallery; **a** alfisols; **b** andisols; **c** aridisols; **d** entisols; **e** gelisols; **f** histosols

Description of the Major Soil Orders

Alfisols

These are soils (Figs. 4.2a and 4.4) developed under temperate forests of the humid and subhumid zones of the Earth (see map, Fig. 4.3). Some of them may also be found

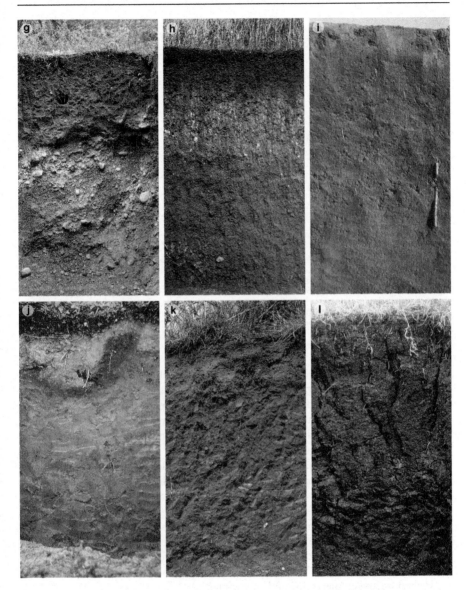

Fig. 4.2 *Continued*; **g** inceptisols; **h** mollisols; **i** oxisols; **j** spodosols; **k** ultisols; **l** vertisols

in the wet/dry tropical climate of Africa. Generally, however, they occur in regions having a temperature range from below 0 °C to above 22 °C.

Alfisols are well developed soils, having >35% base saturation and are characterized by the presence of a diagnostic argillic, or natric B-horizon (see Table 4.2). It is estimated that about 9.7% of the global ice-free land area is occupied by alfisols, supporting nearly 17% of the world population. They are further classified into five suborders:

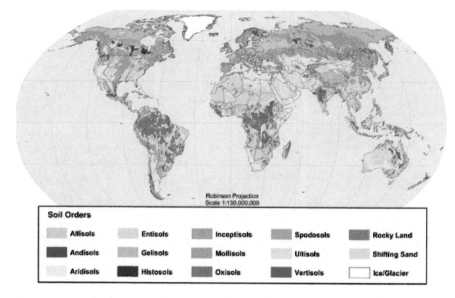

Fig. 4.3 Global distribution pattern of the 12 soil orders of Soil Taxonomy System. *Source:* U.S. Department of Agriculture, National Resources Conservation Service (NRCS)

1. *Aqualfs* (L. aqua – water). These are alfisols which are almost continually wet, due to the fact that the groundwater table lies at or near the surface (aquic regime of soil moisture). Reducing conditions dominate in most years.
2. *Cryalfs* (Gr. kryos – coldness; very cold soils). These are alfisols of the cold regions occurring above 49° north latitude and in some high mountains south of this latitude. They are more or less free drained.
3. *Udalfs* (L. udus – humid). These are alfisols having adequate moisture throughout the year (udic soil moisture regime). Temperature regimes may vary from <8 °C to warmer temperatures.
4. *Ustalfs* (L. ustus – burnt, implying dryness). These are alfisols occurring under conditions between arid regimes and udic regimes (ustic soil moisture regime). Under such conditions, moisture is available but is limited. These environmental conditions are known from semi arid and tropical wet-dry climates.
5. *Xeralfs* (L. xeros – dry). These are alfisols occurring under Mediterranean climates: dry and warm summer, rainy and cool winter.

Formation of Alfisols
The clearest evidence of alfisols is the presence of the argillic horizon, which may be formed as iulluvial accumulations, indicated by clay coatings (see Fig. 4.4), or markedly higher clay content than the overlying horizons. Clays accumulated in such soil horizons may also be of a diagenetic origin, resulting from mineral transformations of primary minerals such as feldspars, micas and other ferromagnesians, as it was explained in Chap. 1 (see Fig. 1.5).

Fig. 4.4 Alfisols

In the tropics, in areas having savannah vegetation and a markedly seasonal distribution of rainfall in the range 60–120 cm yr^{-1}, the formation of alfisols is favoured by high elevations, combined with limited rainfall. Under such conditions, alfisols are formed out of a parent material of felsic to intermediate igneous rocks, or their metamorphic equivalents. In low-lying areas, however, accumulation of clay leads to decreased percolation resulting in water logging. This may present problems leading to compaction by heavy trafficking.

Formation of chelate complexes of metals with organic material leads often to concentration of iron and aluminium in the B-horizon.

Andisols (Jap. ando – black soil)
These are fine textured soils developed from parent material of volcanic origin.

They mostly occur around individual volcanoes created from andesite-rich magma. Andisols have a high ion exchange capacity and often contain considerable amounts of organic matter. Fresh weatherable minerals from the parent material are in most cases abundant.

Also abundant are poorly crystalline colloidal materials such as allophane $[(Al_2O_3)(SiO_3)_{1.3-2} \cdot 2.5-3(H_2O)]$, imoglite $[Al_2SiO_3(OH)_4]$ and ferrihydrite $[Fe_2O_3 \cdot 0.5H_2O]$.

Andisols (Figs. 4.2b and 4.5), are recognised as an independent soil order since 1989; earlier they were considered as a suborder of Inceptisols. They are characterised by a diagnostic cambic horizon (see Table 4.2), occur on steep slopes in mountainous regions in east Africa, New Zealand, Southeast Asia and Central America (see Fig. 4.3). It is estimated that Andisols (the least extensive soil order) occupy nearly 0.7% of the ice-free land area. They are further subdivided into seven suborders:

1. *Aquands* (aqua: see above). These are andisols occurring in aquic moisture regimes (see above). Aquic conditions here are a result of restricted drainage. They occur under various climatic conditions.
2. *Cryands* (kryos; see above). These are andisols occurring within cryic soil temperature regimes. They occur extensively in Alaska as well as in the low lands of Japan (Shoji and Ping 1992). Like cryalfs, they generally occur in the western parts of North America and the northern part of Asia, above 49° north latitude and in high mountains south of this latitude.
3. *Torrands* (L. torridus – hot and dry). These are andisols occurring within an aridic-torric (L. aridis: dry) soil moisture regime. Torrands are known from the western part of North America and some from Hawaii (Soil Survey Staff 1999).
4. *Xerands* (xeros: see above). Andisols occurring under a xeric soil moisture regime.
5. *Vitrands* (L. vitrum – glass). These are more or less well-drained coarse textured andisols having an udic or an ustic (moderate between aridic and udic) moisture regime. They occur at soil temperature regimes warmer than cryic and are characterized by a low content of water held too tightly for plants to use. Parent materials of Vitrands are pumice, cinder or scoria; hence the name glassy.
6. *Ustands* (ustus: see above). These are andisols that have an ustic soil moisture regime. They mostly occur in East Africa, Western part of the United States and the Pacific Islands.
7. *Udands* (udus: see above). These are andisols that have an udic soil moisture regime. They are largely found around the Pacific Rim, in the Western part of the United States, Japan, New Zealand, the Philippines and Indonesia.

Formation of Andisols
As said before, andisols are formed through the weathering of volcanic deposits, mainly pyroclastic material, such as ash, pumice, cinders and lava. These are largely composed of vitric (glassy), poorly crystallised or amorphous substances.

Fig. 4.5 Andisols

Accordingly, they are easily weathered because of their instability in moist landscapes under varying pH conditions. This is demonstrated by the occurrence of colloidal allophane, imoglite and ferrihydrite, which are early products of weathering in volcanic terrains. At advanced stages, metals released by these weathering processes such as aluminium and iron may form complexes with organic substances, causing an accumulation of dark thick surface horizons that help changing the soil colour to darker shades.

Unlike Alfisols, andisols have a low base saturation, due to the high percolation resulting from a high macro-porosity that characterises these soils. The same property leads to high leaching rates of base cations.

Aridisols (L. aridus – dry)
These are soils developed in arid regions including cold polar, cool temperate and warm deserts; they also may occur in semi arid regions. Soil water is restricted (dry in all parts, >50% of the time in most years) and cultivation is impossible without irrigation. They mostly possess a high content of sodium, leading to toxicity of plants and/or aggregate dispersion and consequent breakdown of soil structure.

Aridisols have normally a high content of calcium carbonates and are characterised by subsurface horizons in which clays, calcium carbonates, silica, salts and/or gypsum have accumulated. Predominantly, they possess grey to reddish colours. Under the prevailing shortage of moisture, development of soil profiles is weak, due to the scanty weathering rates. Accordingly, aridisols usually possess indistinct shallow profiles (see Fig. 4.6) with an invariably low content of organic matter. It is estimated that aridisols occupy about 12% of the Earth's ice-free land area.

Soils classified as aridisols are further subdivided into seven suborders:

1. *Cryids.* These are aridisols of cold areas. They have a cryic temperature regime (MAST higher than 0 °C, but lower than 8 °C). They occur at high elevations in mountain valleys.
2. *Salids.* These are aridisols, possessing normally a salic (salty) horizon within 100 cm of the soil surface.
3. *Durids.* These are duripans, having their upper boundary within 100 cm of the soil surface.
4. *Gypsids.* Aridisols containing a gypsic horizon at a depth within 100 cm from the upper surface of soil.
5. *Argids.* Aridisols, having argillic or natric horizon within the upper 100 cm of the soil. They lack calcic or petrocalcic (cemented carbonate) horizons.
6. *Calcids.* Aridisols, having a calcic or petrocalcic horizon within the upper 100 cm of the soil column.
7. *Cambids.* These are aridisols showing none of the characteristics of the other six suborders.

Formation of Aridisols
Weathering of crystalline rocks and marine sedimentary rocks, such as limestones, favour the formation of aridisols on plain terraces and on steep slopes. In major desert regions of the Earth, they are found to develop extensively on fluvial and eolian materials.

Weathering processes leading to the formation of aridisols are predominantly physical. Chemical weathering plays, in the processes leading to the formation of aridic soils, a subordinate role, due to the lack of water. In some cases argillic horizons may develop due to in situ weathering or illuviation of clays.

Fig. 4.6 Aridisols

Entisols
These are shallow, poor soils, occurring under restricted conditions of soil formation, such as in cold and exposed environments. They are of recent origin, developing on unconsolidated parent material and generally possess, except for an A-horizon, no genetic horizons (see Fig. 4.7). They mostly have a low water holding capacity and poor content of organic matter. Entisols (the most extensive soil order) occupy nearly 16.2%

Fig. 4.7 Entisols

of the Earth's ice-free land area. Most characteristic for them is the fact that they form a transition between the other soil orders of Soil Taxonomy and non-soil material. Soils classified as entisols are further subdivided into five suborders:

1. *Aquents.* Aquents, or the wet Entisols, are young soils forming in recent sediments having an aquic soil regime. They mostly occur in low lying areas that have not developed a mollic or histic epipedon. Aquents occur mostly in sandy deposits. They support vegetation that tolerates permanent or periodic wetness and are used mostly

as pasture, cropland, forest, or wildlife habitat. Staining by typical colours of reducing environments indicate lack of dissolved oxygen.
2. *Arents.* Arents are recent soils, having an udic soil moisture regime. They do not have diagnostic horizons due to being continuously and deeply mixed by ploughing, spading, or other methods of moving by humans.
3. *Psamments.* These are Entisols having a loamy sand or sand texture. They develop in outwash, sand dunes, sandstone bedrock, or floodplains.
4. *Fluvent.* Fluvents are the more or less freely drained Entisols that form in recent water-deposited sediments on flood plains, fans, and deltas along rivers and small streams. Stratification of their material is normal.
5. *Orthents.* These are clayey recent soils of shallow profiles, showing a regular decrease of organic matter with depth. They occur under different climatic conditions.

Formation of Entisols
Entisols, that are capable of forming under various climatic conditions, are soils that tend to form on very young surfaces such as alluvium, colluvium or mudflows. One of their suborders, Orthents, occurs mainly on steep surfaces of extremely hard rock. Another suborder, Psamments, is typical for the shifting sands of the Sahara Desert and Saudi Arabia. Entisols may also be associated with salt flats.

Deforestation and other human activities may enhance the formation of Entisols. This is the case in some parts of South Europe and Turkey, where increasing erosion leads to transport of original soil leaving shallow Entisols behind.

Gelisols (L. algare – to freeze)
Gelisols are soils occurring under very cold climates, either having permafrost within 100 cm of the soil surface, and/or gelic materials (mineral or organic soil materials that have evidence of cryoturbation (frost churning)) within 100 cm of the soil surface and permafrost within 200 cm. They are geographically limited to the high-latitude Polar Regions and localized areas at high mountain elevations. Gelisols, with regard to the severe climatic conditions under which they occur, support only ~0.4% of the world's population. This is the lowest percentage of any of the soil orders. It is, however, estimated that gelisols occupy about 8.6% of the Earth's ice-free land area. They normally have a high content of organic matter, which is a result of the slow rates of the decay processes, due to the effect of the low temperature regimes in this environment. Gelisols have normally no B-horizons (Fig. 4.8). They may be black or dark in colour due to the accumulation of organic matter in their upper parts. They are mostly prevalent in Siberia, Alaska and Canada, Tibet, and the ice-free regions of Greenland and northern Scandinavia. Gelisols are further divided into 3 suborders:

1. *Histels.* These are organic soils, similar to *histosols*, having permafrost within two metres below ground surface. Organic matter content may reach values of 80% or more in the upper parts, within 50 cm of the ground surface, or down to a glacic, Densic, lithic or paralithic contact.
2. *Turbels.* These are soils occuring under conditions where cryoturbation (frost churning) have an extensive influence, extending over more than one-third of the depth of the active layer, i.e. the layer at which thawing occurs every one or two years. They may also include irregular, broken, or distorted horizon boundaries and involutions

Fig. 4.8 Gelisols

together with areas of patterned ground. Commonly, Turbels contain tongues of mineral and organic horizons. They occur primarily in the zone of continuous permafrost.

3. *Orthels.* These are soils that show little or no cryoturbation, extending over less than one third of the active zone. They generally occur within the zone of discontinuous permafrost.

4.1 · The Soil Taxonomy System – Criteria of Classification

Formation of Gelisols
Gelisols are, by definition, soils occurring at high latitudes between the −7 °C isotherm (permanent permafrost) and the −2 °C isotherm (intermittent or discontinuous permafrost). Under such low temperatures, processes of pedogenesis, having a predominantly chemical or biological nature, are very slow, whereas physical processes of pedogenesis steered by thermal or mechanical forces are in control of almost the whole system. Under such conditions the soil forming processes are collectively known as *cryopedogenesis*.

Cryopedogenesis starts at temperatures below 0 °C when the moisture contained in rocks freezes, causing a volume increase of about 9% and consequently an increase of the internal pressure, leading normally to a rupturing of the rock material (thermal cracking). This may also lead to characteristic landscape structures known as *frost polygons*, when soil waters or moisture seeping from the upper surfaces into the frost cracks freeze, forming ice wedges that follow these regular patterns; such regular patterns in Gelisols landscapes are known as *patterned grounds*.

Another feature characterizing Gelisols is the Cryoturbation or frost churning. This occurs through mixing due to freezing and thawing, whereby soil horizons may be disrupted and dislocated and some times orientate stones or coarse material in the profile into regular linear patterns. As a matter of fact, through mixing, cryoturbation is also responsible for the translocation of organic matter from the upper horizons to the lower ones, causing an enrichment of organic matter in the lower parts of the profile. Cryoturbation in the soil profiles of Gelisols is manifested by irregular and broken horizons and textural bands.

Histosols (Gr. histos – tissue)
These are organic soils that form in wet terrains (Fig. 4.9), normally peat or bog, containing over 20% of organic matter, or muck with clay, having a thickness of more than 40 cm. Their bulk density is quite low (less than 0.3 g cm^{-3}) due to the high organic content. Histosols have generally no diagnostic horizons and their profile features (Fig. 4.10) reflect the nature of the peat from which they are developed and the amount of mineral matter originally present. Histosols are ecologically important because of the large quantities of organic carbon they contain. Globally, they occupy about 1.2% of the ice-free land area.

Histosols are further subdivided into 4 suborders:

1. *Folists*. These are well drained histosols. They are not saturated with water more than for a few days in the year. Their greatest distribution is in humid regions and in mountain areas almost under all climatic conditions.
2. *Fibrists*. Histosols made up of only slightly decomposed organic materials, often called peat.
3. *Saprists*. These are histosols made up of completely decomposed organic matter. Plant remains show no fibrous textures. They are formed in areas where there are fluctuations in groundwater levels. These fluctuations allow intermittent mixing and solution of atmospheric oxygen in the soil solution and hence an increase of organic matter decomposition rates.

Fig. 4.9 General view of a histosol landscape (*Source: www.epa.gov/owow/wetlands*)

4. *Hemists*. Histosols that are primarily made up of moderately decomposed organic materials. The fibrous textures of plants are still identifiable.

Formation of Histosols

Histosols, also known as peat and muck soils, are formed by incompletely decomposed plant remains, with or without admixtures of sand, silt or clay. They are normally confined to poorly drained basins and decompressions, as well as swamp and marshlands with shallow groundwater tables (Fig. 4.10). Histosols may also occur in highlands having high precipitation rates.

Under such conditions, plant remains are decomposed and transformed into humic substances. These are responsible for the formation of the mould surface layers, characterizing the environment in wetlands. Peat lands are used for various forms of extensive forestry and/or grazing.

Inceptisols (L. inceptum – beginning)

These are soils of intermediate development exhibiting minimal horizon differentiation. They are more developed than entisols, but still lack the features that are characteristic of other soil orders. However, some cambic horizons together with a few diagnostic features may sometimes be present. Inceptisols are generally viewed as embryonic soils (Fig. 4.11).

Inceptisols are widely distributed. They may occur in a wide range of ecological settings, predominantly on fairly steep slopes, young geomorphic surfaces, and on weathering resistant parent materials. A sizable percentage of Inceptisols is found in mountainous areas and is used for forestry, recreation, and watershed.

Inceptisols occupy an estimated 17% of the global ice-free land area, the largest percentage of any of the soil orders. They support about 20% of the world's population, also the largest percentage of any of the soil orders. They are particularly com-

Fig. 4.10 Histosols

mon in floodplain areas of the river Ganges in India and many other parts of South and South East Asia, where they are often used for high yield cultivation of rice. They are also common in Brazil (Amazon region), some parts of the USA and the Sahel zone of Africa.

Inceptisols are further subdivided into 7 suborders:

1. *Aquepts.* These are poorly developed soils, having an aquic moisture regime. They are characterized by cambic diagnostic horizons. Surface units are normally dark,

Fig. 4.11 Inceptisols

while subsurface units display stains characteristic of reducing environments. Natural drainage in aquepts is poor or very poor, and, if the soils have not been artificially drained, groundwater is at, or near, the soil surface at some time during normal years, but typically not in all seasons.
2. *Anthrepts.* These are more or less freely drained Inceptisols that have either an anthropic or plaggen epipedon.
3. *Gelepts.* These are Inceptisols of very cold climates. Mean annual soil temperature is less than 0 °C.

4. *Cryepts*. Cryepts are Inceptisols of cold environments at high mountains or high latitudes. Typically South Alaska and some high mountain regions.
5. *Usteps*. Ustepts are freely drained Inceptisols that have an ustic moisture regime. They are most common under semi-arid and subhumid climates.
6. *Xerepts*. These are mainly freely drained Inceptisols that have a xeric moisture regime (Mediterranean climate: very dry summers and moist winters).
7. *Udepts*. These are mainly freely drained Inceptisols that have an udic or perudic moisture regime.

Formation of Inceptisols
Except for aridic conditions, Inceptisols may be formed under a wide variety of climatic and soil moisture regime conditions; the latter may range from poorly drained soils to well-drained soils on steep slopes. Climates of low temperatures, or those characterised by low precipitation values, favour the development of inceptisols. However, the development of aquepts requires soil moisture conditions higher than the other suborders of inceptisols.

Inceptisols formation is particularly common on geologically young sediments and where highly calcareous parent material or materials resistant to weathering accumulate. They are extensive in areas of glacial deposits, on recent deposits in valleys or deltas. Among other materials, primary and secondary minerals are present.

Mollisols (L. mollis – soft)
These are dark organic soils, having a high base saturation (higher than 50%), a high CEC and a high water holding capacity. They are characterized by argillic or cambic diagnostic horizons that result from eluviation and illuviation and an epipedon formed of a dark, humus rich mollic material. Another characteristic feature of these soils is the blocky prismatic structure of their B-horizon (Fig. 4.12).

Mollisols occur in a variety of climatic zones ranging from cryic, frigid and mesic, to thermic soil regimes. It is estimated that they occupy about 6.89% of the global ice-free land area.

Mollisols are further subdivided into seven suborders:

1. *Albolls*. These are mollisols having an albic diagnostic horizon and an intermittent aquic regime (i.e. occurring for some parts of the year). Argillic or natric horizons may follow on albic horizons in the profile. The main process producing albolls, beside illuviation and eluviation, is the reduction of iron and manganese oxides, due to the wet conditions prevailing for some time in the year.
2. *Aquolls*. These are mollisols formed under aquic conditions. They demonstrate the typical features of wet soils (redoximorphic properties, accumulation of organic matter, histic epipedon, accumulation of $CaCO_3$ near the soil surface).
3. *Rendolls*. These are mollisols formed from calcareous parent material in humid regions. The mollic epipedon is less than 50 cm.
4. *Xerolls*. Mollisols formed under a xeric soil moisture regime. Characterised by accumulations of carbonates in the lower solum.
5. *Cryolls*. Mollisols formed under cryic soil temperature regimes. This is the most extensive mollisols suborder worldwide.

Fig. 4.12 Mollisols

6. *Ustolls.* Mollisols formed under ustic soil moisture regimes. They are normally freely drained mollisols, occurring in semiarid to subhumid climates.
7. *Udolls.* These are mollisols occurring under udic soil moisture regimes, normally in continental climates of the temperate and tropical zones.

Formation of Mollisols
Mollisols occur on deposits of various ages. These may range from the eroded remains of older sedimentary rocks, to deposits associated with glacial sediments of quater-

nary age. In forming this order of soil, the predominant processes, beside eluviation and illuviation, include: melanization, humification and pedoturbation. Melanization, which is the dominant process in the formation of mollisols, is defined as the process by which the darkening of the soil is initiated by direct incorporation of organic matter into the mineral soil. Pedoturbation, however, may include normal bioturbation by soil dwellers mixing the soil through crawling, burrowing or boring, as it is known from other sediments, or cryoturbation, as explained earlier.

Oxisols (Fr. oxide – oxide; Gr. oxide – acid or sharp)
Oxisols (Fig. 4.13) are soils mostly occurring in isotropical soil temperature regimes. They develop in climatic zones with small seasonal variation in soil temperature and no seasonal soil freezing. Oxisols occur in a wide range of soil moisture regimes; these may range from aridic to perudic. It is, however, assumed that oxisols develop under climatic conditions where precipitation exceeds evapotranspiration for some periods of the year. This condition favours the removal of soluble material and residual concentration of kaolinite and sesquioxides, which are essential to form an oxic diagnostic horizon – an important feature of oxisols. They mainly occur on plain or level topographic areas and are unlikely to be present on steep slopes. However, they may occur on floodplains or steep rapidly eroding areas. Oxisols formed on high-lying old erosion surfaces (plateaus), terraces or floodplains, are associated with ultisols (see below), alfisols, or mollisols on the side-slopes.

Oxisols are typically characterised by high contents of oxides and hydroxides of iron and aluminium, and occur most commonly in humid tropical environments, particularly in Africa and South America (rain forest regions of Amazonia and central Africa). They occupy about 7.5% of the ice-free land area.

Oxisols are further subdivided, according to their soil moisture regime, into 5 suborders:

1. *Aquox.* These are oxisols that have aquic conditions for some time in most years and show redoximorphic features or a histic epipedon.
2. *Torrox.* Oxisols with an ustic or xeric soil moisture regime.
3. *Ustox.* Oxisols having low to moderate organic content, normally well drained and dry for at least 90 cumulative days each year.
4. *Perox.* Oxisols with a perudic soil moisture regime.
5. *Udox.* These are oxisols with an udic soil moisture regime.

Formation of Oxisols
Oxisols occur normally on erosion surfaces and old fluvial terraces made of sediments that have been reworked during several cycles of erosion. Sediments favouring the formation of oxisols are typically rich in iron and aluminium oxides, together with quartz and 1:1 clays (kaolinite group). Such a composition contributes later to the formation of oxic diagnostic horizons. These, together with kandic horizons are essential features of oxisols.

Kandic horizons show the same CEC as oxic horizons, but kandic horizons have a clay content increase at its upper boundary of more than 1.2 times within a vertical distance of less than 15 cm, i.e. abrupt or clear textural boundary. Alternating oxidation (reduction caused by fluctuation of the water table in oxisols) may cause formation of plinthite, consisting of red-and-grey mottled material, which in older litera-

Fig. 4.13 Oxisols

ture was designated as laterite or lateritic iron oxide crust. If subjected to repeated wetting and drying, as in exposure through erosion of the overlying material, it becomes indurated to ironstone. This may be subsequently eroded and deposited as ironstone gravel layers in alluvial fans.

Spodosols (Gr. spodos; L. spodus – wood ash)
These are course textured, highly leached acidic soils having a surface of poorly decomposed plant remains (Fig. 4.14). They are widely spread in cool temperate regions

and are rarely used for productive agricultural processes. Generally, however, Spodosols may occur under all soil temperature regimes, but they are restricted to humid, xeric or, in rare instances, aquic moisture regimes. They preferentially form on quartz rich sands overlying a fluctuating groundwater table.

Spodosols form on slopes, ranging from nearly level to very steeply sloping, and on surfaces in which the water table ranges from very deep to fluctuating near the surface. They normally do not form in water-saturated terrains.

Spodosols occupy nearly 2.6% of the ice-free global area and are further classified into 4 suborders:

1. *Aquods.* Aquods are the spodosols occurring under aquic soil regimes characterized either by a shallow fluctuating water table or an extremely humid climate. They normally have a histic epipedon and redoximorphic features in the upper 50 cm. Most of them have a nearly white albic horizon, if the temperature regime is mesic. Yet in very wet conditions, a black surface horizon resting on a dark reddish brown spodic horizon that is virtually free of iron, would lend them a characteristic appearance. In some cases, spodosols may have a placic horizon or a duripan, or are cemented by an amorphous mixture of sesquioxides and organic matter. The Aquods that do not have a placic horizon normally have a transitional horizon between the albic horizon and the spodic horizon.
2. *Cryods.* These are Spodosols occurring at high latitudes and/or high elevations that have a cryic soil temperature regime. Many Cryods have formed in volcanic ash or glacial drift, and some in residuum or colluvium on mountain slopes. They are commonly characterised by an O horizon, lying on a very thin or intermittent albic horizon, with the latter overlying a well-developed spodic horizon. Some cryods may have a placic horizon, ortstein, or any other cemented soil layers within 100 cm of the soil surface.
3. *Humods.* These are relatively freely drained Spodosols that have a large accumulation of organic carbon in the spodic horizon (not less than 6%). They have either a thin intermittent, or a distinct continuous albic horizon. This overlies a spodic horizon, which in its upper part is nearly black and has a reddish hue. The hue normally becomes more yellow with depth. Humods are derived predominantly from Pleistocene or Holocene sediments.
4. *Orthods.* These are also relatively freely drained Spodosols that have a horizon of accumulation containing aluminium and/or iron, together with organic carbon. Most orthods are formed in coarse, acid Pleistocene or Holocene deposits under coniferous forest vegetation. They normally have an albic, and a spodic horizon, and may have a fragipan.

Formation of Spodosols

By definition, spodosols are course textured, highly leached acidic soils having a surface of poorly decomposed plant remains. Such material can readily occur in humid regions, where carbonates are leached and soil is generally of low base saturation. Under such conditions, bacterial decomposition of organic matter would be retarded, allowing the accumulation of poorly decomposed plant material, while enhanced weathering of mineral matter releases iron and aluminum, together with other elements contributing to the formation of a spodic horizon. Soluble organic matter dissolves, joining the soil solution.

Fig. 4.14 Spodosols

Ultisols (L. ultimus – last)
These soils are similar to alfisols and share some properties with them (e.g. a high content of clay in the B-horizon), yet they are of low base saturation and normally show strong weathering in the argillic horizons. Their red colouration (Fig. 4.15) is more intensive than that of alfisols and their soil temperature regime is generally higher. They occur generally under xeric to aquic soil moisture regimes, yet may also form in frigid soil temperature regimes. Xeric, perudic, udic, ustic, and aquic soil moisture regimes are present in various ultisols.

Fig. 4.15 Ultisols

Soils classified as ultisols are principally characterised by low base saturation (less than 35%) at a depth of 180 cm below the surface of the soil; an argillic or kandic soil horizon; low nutrient status; and relative high acidity, with pH-values at surface horizons rarely higher than 5.5. Their CEC is generally low with slightly upper values in the upper horizons.

Ultisols occupy about 8.5% of the global ice-free land area and support nearly 18% of the world's population. They occur most commonly in southeastern parts of the USA, eastern and southern parts of Brazil, and in parts of Southeast Asia.

On soil moisture regime basis, ultisols are further subdivided into five suborders:

1. *Aquults.* These are ultisols having aquic soil moisture conditions, characterised by redoximorphic features.
2. *Humults.* These are organic rich ultisols, satisfying the required criteria, yet having no particular wetting conditions.
3. *Udults.* These are ultisols, of low organic content, formed in humid regions, where dry periods are short. They may, occasionally, have high water table levels, but no distinct redoximorphic features.
4. *Ustults.* Ultisols having an ustic soil regime. Moisture is seasonably available in adequate amounts for at least one crop per year.
5. *Xerults.* Ultisols having xeric soil moisture conditions.

Formation of Ultisols
There is no limitation on the relief upon which ultisols may be formed. They may be formed on steep slopes as well as on plains. An important prerequisite, however, is that the climatic situation should be such that precipitation exceeds potential evapotranspiration, with an adequate degree of freedom for water percolation, that allows leaching and initiation of low basic saturation. Base depletion is one of the basic attributes of ultisols. Accordingly, ultisols can mainly be formed on parent material possessing low content of basic cations. This material may be siliceous crystalline rock such as granite, or sedimentary rocks poor in bases, as it is the case in sediments of highly weathered coastal plains.

Vertisols (L. verto – to turn)
Vertisols are soils rich in expanding clays (>30%, 2:1-type), that make them shrink and swell with changes in moisture content. During dry periods, deep wide cracks (see Fig. 4.16) are formed due to shrinkage. This is followed by swelling and expansion on wetting. Successive shrinking and expanding of the soil may prevent the formation of well-developed horizons. Vertisols have normally a medium to low organic content (0.5–3.0%) and may be further characterized by an ochric epipedon.

Vertisols occur in almost all major climatic zones, yet their largest occurrence is in Australia with 80 million ha, mainly under aridic moisture regime conditions. They are also particularly common in India and Sudan. Globally, they occupy nearly 2.4% of the ice-free land area.

Vertisols are further subdivided into six suborders:

1. *Aquerts.* These are vertisols formed under aquic soil moisture conditions. High water saturation conditions may be prevailing due to the limited water circulation in the clay rich fine-grained medium of the soil profile. Black or dark colouration may appear due to the reduction of manganese compounds originally found in the mineral material.
2. *Cryerts.* These are vertisols formed under cryic soil temperature regime conditions.
3. *Xererts.* These are vertisols formed under thermic, mesic or frigid temperature regime conditions. Their shrinkage cracks are open for 60 consecutive days in summer. They are normally closed for, at least, the same number of consecutive days in winter.

Fig. 4.16 Vertisols

4. *Torrerts.* These are vertisols having shrinkage cracks that are closed for less than 60 consecutive days when the soil temperature at a depth of 50 cm is higher than 8 °C. This suborder forms the most extensive suborder of vertisols in Australia.
5. *Usterts.* These are vertisols having cracks that are opened for at least 90 cumulative days a year. They form the most extensive suborder of vertisols on a worldwide scale.
6. *Uderts.* Vertisols having open cracks during a period of less than 60 consecutive days during summer. Annually, they are open for less than 90 cumulative days.

Formation of Vertisols
Vertisols form on a wide range of terrains. These may be alluvial, colluvial or lacustrine in origin. The parent material may be a basic volcanic rock, a calcareous sedimentary rock or a shale deposit. Generally, they develop readily on recent material of allochthonous or autochthonous origin. The expandable clays (smectites), responsible for their essential characters, may be derived from the original rock or are of diagenetic origin.

4.2
The FAO-UNESCO Soil Classification System

The FAO Soil Classification System was first published in the form of the UNESCO Soil Map of the World (1974). Various modifications, based on the first classification and other systems, such as the Russian and American ones, followed since then. In 1980 the IRB (International Reference Base for Soil) was launched. This was modified in 1982, 1988 and 1990 by introducing the concept of twenty major groups of soils based on their representation of the world soil cover.

Since 1998, the WRB (World Reference Base for Soil Resources) was established as a FAO World Soil Classification System. It is a much-improved system of classification based on soil properties, their physiographic settings and relation to geology, without any special reference to climatic conditions. Practically, it has proved to be a simple system, useful in mapping soils on a continental scale, rather than on a local one.

The WRB-System comprises 30 Reference Soil Groups, distinguished by the presence (or absence) of specific diagnostic horizons, properties and/or materials. The reference groups may be further subdivided into Soil Units, the number of which, however, has been deliberately left open for further modifications.

4.2.1
Description of the Reference Soil Groups of the World Reference Base for Soil Resources (WRB)

1. *Histosols.* Soils having a histic or folic diagnostic horizon, lacking an andic or vitric soil horizon starting within 30 cm of the soil surface.
2. *Cryosols.* Soils having one or more cryic horizons within 100 cm from the soil surface.
3. *Anthrosols.* Soils, the formation of which has been conditioned by man.
4. *Leptosols* (Gr. *leptos* – thin). Very shallow soils (<25 cm) over hard rock or highly calcareous material, but also deeper soils that are extremely gravelly and/or stony.
5. *Vertisols.* Soils having periodic cracks (see vertisols in the USDA taxonomy system), vertic horizon, >35% clay.
6. *Fluvisols* (L. *fluvius* – river). Soils formed mainly of alluvial material.
7. *Solonchaks* (Rus. *sol* – Salt). Soils having a salic horizon starting within 50 cm of the soil surface.
8. *Gleysols* (Rus. *gley* – mucky soil mass). Soils having excess water (gleyic properties) within 50 cm of the surface.
9. *Andosols* (Jap. *an* – dark; *do* – soil). Soils having an andic (dark rich in volcanic glass) or vitric horizon.

10. *Podzols* (Rus. *pod* – under; *zola* – ash). Soils having a spodic horizon starting within 200 cm from the soil surface (strongly bleached).
11. *Plinthosols* (Gr. *plinthos* – brick). Mainly mottled clayey material that hardens on exposure.
12. *Ferralsols* (L. *ferum* – iron). Soils having a high sesquioxides content.
13. *Solonetz* (Rus. *sol* – salt; *etz* – strong). Soils having a natric horizon within 100 cm from the soil surface.
14. *Planosols* (L. *planus* – flat, level). Soils having an eluvial horizon, the lower boundary of which is marked, within 100 cm from the soil surface, by an abrupt textural change associated with stagnic (stagnation) properties above that boundary.
15. *Chernozems* (Rus. *chern* – black; *zemelia* – Earth, land). Soils having a mollic horizon, secondary carbonates, no uncoated silt and sand grains.
16. *Kastanozems* (L. *castana* – chestnut; Rus. *zemelia*). Organic rich brown soils, mollic horizon.
17. *Phaeozems* (Gr. *phaeos* – dusky; Rus. *zemelia*). Soils having a mollic horizon, base saturation 50% or more, calcium carbonate-free soil matrix at least to a depth of 100 cm below the soil surface.
18. *Gypsisols* (L. *gypsum*). Gypsic or petrogypsic horizon within 100 cm from the soil surface.
19. *Durisols* (L. *durus* – hard). Soils having a duric (silica cemented) horizon.
20. *Calcisols* (L. *calx* – lime). Soils having a calcic or petrocalcic (hardened) horizon within 100 cm of the soil surface. Calcium carbonate rich.
21. *Albeluvisols* (L. *albus* – white; *eluere* – wash out) – old name *podzoluvisols*. These are acid soils having a bleached horizon penetrating into a clay rich subsurface horizon (*albeluvic tonguing*).
22. *Alisols* (L. *aluminium*). Soils characterised by high aluminium content, having an argic (clay rich) horizon, a CEC of 24 mol kg^{-1} clay or more.
23. *Nitisols* (L. *nitidus* – shiny). Soils having a nitic (shiny) horizon starting within 100 cm from the soil surface.
24. *Acrisols* (L. *acer, acetum* – acid). Soils having low base saturation (less than 50%) in the major part between 25 and 100 cm below the surface. Argic horizon having a CEC less than alisols.
25. *Luvisols (L. eluere* – wash). Soils having high accumulations of clay in an argic horizon.
26. *Lixisols* (L. *lixivia* – washing). Weathered clay rich soils. Argic horizon.
27. *Umbrisols* (L. *umbra* – shade, dark). Soils characterised by an umbric horizon (see Table 4.1).
28. *Cambisols* (L. *cambiare* – to change). Soils characterised by change in colour, structure and consistence. A cambic or mollic horizon.
29. *Arenosols* (Gr. *arena* – sand). Soils characterised by a texture which is loamy sand or coarser, either to a depth of at least 100 cm from the surface, or to a plinthic or salic horizon between 50 and 100 cm from the surface.
30. *Regosols* (Gr. *regos* – blanket). Soils characterised by a loose mantle of material, showing none of the characteristics of other soils in this list.

Based on the foregoing descriptions, an approximate scheme of classification according to the WRB, suitable to be used in the field, would be constructed as shown in Fig. 4.17.

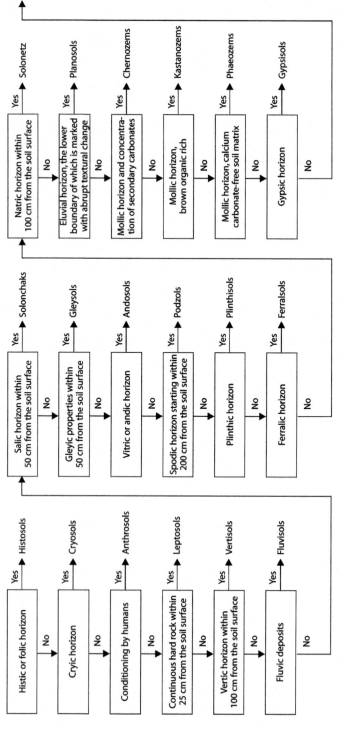

Fig. 4.17 Scheme of soil classification according to the WRB, suitable to be used in the field

4.2 · The FAO-UNESCO Soil Classification System

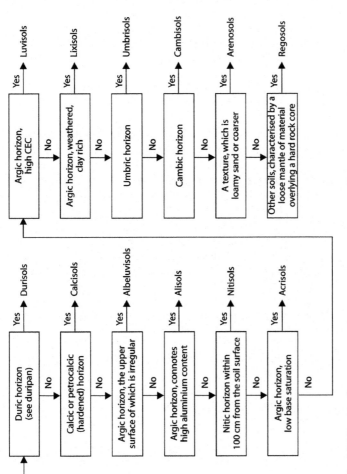

Fig. 4.17 *Continued*

4.2.2
Relation Between the WRB System and the USDA Taxonomy System

Soil classification in the USDA system has put primary emphasis on soil properties that can be observed in the field (or later measured in field laboratories), whereby quantitative aspects are preferred. It is thus one of those scientific systems based on existing quantifiable characteristics. The UNESCO-FAO classification system strongly resembles the USDA Taxonomy system in the sense that it is also a hierarchical system based on existing properties, despite being descriptive in some cases. It does not put stress on climatic parameters, yet it heavily borrows from the USDA system in naming its categories on the levels of representative groups and diagnostic horizons. Essential correlating aspects are shown in Table 4.5.

4.3
Other Systems of Classification

A classification system, for any scientific data, is always dependent on the main purpose set by the author or authors designing the concept upon which the system is based. This is mainly the reason why we have various soil classification systems existing parallel to each other. Sometimes, different nations would design their own systems depending on old vernacular names for local soils, or adapted to one of the world systems, as it is for example the case in developing countries working together with experts of the UNESCO-FAO organisation.

Beside national differences, it is always observed that different professional groups (e.g. engineering geologists, civil engineers, etc.) may also have their own classification systems, designed specially to fulfil their needs and used parallel to other systems, even

Table 4.5 Approximate relationship between the Soil Survey Staff classification (USDA) and the FAO-UNESCO major soil groupings

Soil Survey Staff soil orders	FAO-UNESCO main soil groupings
Alfisols	Luvisols
Andisols	Andosols
Aridisols	Glacisols, gypsisols, solonchaks, solonetz
Entisols	Arenosols, fluvisols, leptosols, regosols
Histosols	Histosols
Inceptisols	Cambisols
Mollisols	Chernozems, greyzems, katstanozems, phaeozems
Oxisols	Alisols, ferralsols, nitosols, plinthosols
Spodosols	Podzols
Ultisols	Acrisols, lixisols
Vertisols	Vertisols

4.3 · Other Systems of Classification

within the same country or nation. As an example, we may give the different special systems used within the United States of America. The following list represents a selection of them:

- The system of the American Association of State Highways and Transportation (AASHTO), which is used by state and country high way departments for construction engineering purposes.
- The system of the Federal Aviation Department (FAA), which is used by airfield designers.
- The United States Army system (Unified Soil Classification System), which is the system preferred by engineering geologists.

Examples of National Systems

In the following some of the major national systems will be shortly mentioned:

- *Australian system – Australian Soil Classification System* (Isbell 1996). This is a hierarchical system based on diagnostic properties, horizons or materials.
- *British system – Soil Classification for England and Wales* (Avery 1980). This is also a hierarchical system having 4 levels of classification – Major Groups (10), Groups (41), Subgroups (109) and an open number of Series.
- *French systems* (Duchaufour 1988; Finkl 1982). The most well known of the French systems is the French Soil Reference System – "*Référentiel pédologique français*". The French systems are mainly hierarchical, based on soil formation processes and morphology. The French system was widely used in the colonial time in many countries. It is now mostly used in France and its overseas territories.
- *German system* (Mückenhausen 1965; Finkl 1982). The German system of classification was originally developed by Walter L. Kubiena (1953) and later modified by Mückenhausen (1965). It is a hierarchical system based on soil genesis, pedogenic processes and morphologic features. The German system, which was revised in 1994, is composed of four levels of classification: 4 Divisions, 14 Soil Classes, 40 Soil Types and an open number of Sub Soil Types.
- *Russian system – Russian Soil System* (Rozov and Ivanova 1967). Hierarchical sytem, comprising 4 levels: Genetic Types, Subtypes, Species, Subspecies.

Chapter 5

Soil Degradation

5.1
Soil Degradation and Soil Quality

Soil degradation is defined as the decline in soil quality caused through its misuse by human activity (Barrow 1991). The term *soil quality* itself may be further defined as, *"The capacity or capability of a soil to produce safe and nutritious crops in a sustained manner over a long term, and to enhance human and animal health, without impairing the natural resource base or adversely affecting the environment"* (Parr et al. 1992).

Degradation or decline of soil quality may occur due to physical or chemical processes triggered off by natural phenomena, or induced by humans through misuse of land resources. Processes such as soil erosion, nutrient run-off, water logging, desertification or compaction, may give examples of physical degradation processes, while acidification, organic matter loss, salinization, nutrient depletion by leaching, or toxicants accumulation, are all processes that can be classified as being agents and indicators of chemical degradation of soil.

Indicators of soil quality are numerous. They may be biological indicators, assessed by interpreting the figures indicating density of microorganisms' populations, or through measuring some of the basic biological activities, such as respiration or intensity of biogeochemical reactions. They may also be physical indicators measured by investigating some of the fundamental physical characteristics of soil, such as bulk density, water infiltration or field water holding capacity. Chemical indicators of soil are normally assessed through pH-values or the concentration of certain ions such as nitrates. Indicators of soil quality may also be just visual indicators, obtained from observation or photographic interpretation, as for example exposure of subsoil, change in soil colour, ephemeral gullies or blowing soil. Visual indicators can deliver very important indications for the change of soil quality.

5.1.1
Biological Indicators of Soil Quality – Soil Respiration Rates

The main indicator of biological activities in soil is the soil respiration rate, which can be assessed through measuring the carbon dioxide evolution resulting from the decomposition of organic matter. In other words, it is closely related to the efficiency of biological processes taking place within the soil. Biological processes in soils depend on several factors, among which moisture, temperature, oxygen, partial pressure, and availability of organic matter are of central importance. We find that soil pores filled with moisture up to 60% will provide the most optimum conditions for organic activities, because pores, which are more saturated with water (>60%), cannot provide sufficient oxygen concentration for the flourishing of aerobic decomposers, which carry out most of the work in organic matter decomposition. A similar pattern of behaviour is also displayed by temperature, as it is observed that biological activities double for every increase of 18 °F in temperature; yet this trend ends at a certain maximum (specific for different groups) after which they decline once more until they reach a minimum.

Adding organic matter of low C/N-ratio to the soil (e.g. manure, leguminous cover crops) may increase soil respiration rates since these substances are easily decomposed. The addition of pesticides or similar agricultural chemicals may impair or directly kill soil organisms, eventually leading to lower soil respiration rates and diminished soil quality. Table 5.1 shows a ranking of soils according to their soil respiration rates (Woods End Research 1997; Evanylo and McGuinn 2000).

5.1.2
Physical Indicators of Soil Quality

The availability of oxygen and the mobility of water, into or through the soil, are all attributes very closely related to physical properties, as for example texture, structure, density and porosity. Some of these properties (e.g. texture) are immutable, i.e. they cannot be modified by technical efforts. Others, such as density or water holding capacity, may be improved using appropriate soil management techniques. In the following, a short account of these properties and their use as indicators for soil quality will be considered.

Soil Bulk Density

Soil bulk density is defined as the mass of soil per unit volume in its natural field state, including air space and mineral matter, plus organic substance. High values of bulk density may restrict the movement of surface waters through the soil, leading to a loss of nutrients by leaching. It may also increase erosion rates. Continuous tilling may

Table 5.1 Class ratings of soils according to their respiration rates (Woods End Research 1997)

Soil respiration (CO_2-lb ac^{-1} day^{-1})	Class	Soil condition
0	No soil activity	Soil has no biological activity – virtually sterile
<9.5	Very low soil activity	Very depleted of organic matter; little activity
9.5 – 16	Moderately low activity	Some what depleted; low biological activity
16 – 32	Medium soil activity	Approaching or declining from an ideal activity
32 – 64	Ideal soil activity	Ideal state of activity
>64	High soil activity	Very high level of microbial activity

increase the bulk density of soils, yet continuous cropping, adding of organic conditioners and trafficking on wet soil may largely reduce it. Bulk density measurements are very important for assessing soil quality, since root growth and penetration of soil, together with the ease of soil aeration, are largely controlled by this factor.
Table 5.2 (Arshad et al. 1996) shows some values of bulk densities for several textural classes, together with an assessment of their observed effect on root growth.

Water Infiltration

Infiltration is defined as the process by which water enters the soil. Its rate depends on soil type, soil structure and soil water content (Lowery et al. 1996). Infiltration is important for reducing run-off and consequent erosion. Increased soil compaction and loss of surface structure (reduced aggregation) are the main factors in reducing water infiltration rates in soils. Such rates are normally dependent upon the occurrence of large pores occupying the upper surfaces of the soil; therefore they depend on soil texture in the first place. Table 5.3 (Hillel 1982) shows the steady infiltration rates in inch per hour for some soil textural groups.

Table 5.2 General relationship between soil bulk density and soil quality based on soil texture (Arshad et al. 1996)

Soil texture	Ideal bulk densities ($g\ cm^{-3}$)	Bulk densities that may affect root growth ($g\ cm^{-3}$)	Bulk densities that restrict root growth (g/cm^3)
Sands, loamy sands	<1.60	1.69	>1.80
Sandy loams, loams	<1.40	1.63	>1.80
Sandy clay loams	<1.40	1.60	>1.75
Silts, silt loams	<1.30	1.60	>1.75
Silt loams, silty clay loams	<1.40	1.55	>1.65
Sandy clays, silty clays, some clay loams (35-45% clay)	<1.10	1.49	>1.58

Table 5.3 Steady infiltration rates for different soil texture groups in very deeply wetted soils (Hillel 1982)

Soil type	Steady infiltration rate (in h^{-1})
Sands	>0.8
Sandy and silty soils	0.4 – 0.8
Loams	0.2 – 0.4
Clayey soils	0.04 – 0.2
Sodic clayey soils	<0.04

Field Water Holding Capacity

Field water holding capacity is the amount of water a soil can hold after having been saturated and then allowed to drain for a period of one to two days. Field water holding capacity depends also on the soil texture. It increases by adding organic amendments and by conservation tilling; conventional tilling results in reducing the water holding capacity.

Loamy soils provide the most plant-available water, whereas clayey soils hold their water content so tightly that it might not be available for plants. Adding organic matter to the soil may increase its water holding capacity, due to the fact that organic matter physically holds more water than mineral substances.

5.1.3
Chemical Indicators of Soil Quality

Soil chemical properties are determined by given parameters arising from the presence of certain amounts and certain types of soil colloids (clays and organic matter). Most important among these are parameters related to acidity or alkalinity (pH), parameters related to mineral solubility and nutrition availability, and parameters determining such important properties as mineral solubility and CEC. In the following, some of these parameters and their relation to soil quality will be briefly discussed.

- *Soil pH.* Soil pH is originally inherited from components forming the parent material. It tends, however, to decrease with time. Addition of limestone and other basic materials is normally used to maintain soil pH at a desirable range. A severe decrease or excessive increase in pH-values, may cause nutrient deficiencies or elemental toxicities, resulting in adverse effects on crop yield. This makes pH-values to be among the most important indicators of soil quality.
- *Soil nitrate.* Nitrates (NO_3^-) and ammonium are the only nitrogen forms that can be used by plants. This is why chemical nitrogen fertilisers are added to soil, they are capable of adding nitrate or ammonium directly to the treated site.

    Nitrate concentration at the top 12 inches of the soil has been used successfully to predict the sufficiency of nitrogen for the crop (Evanylo and McGuinn 2000).

In 1996, the United Nations Food and Agriculture Organization (UN FAO), in its Global Assessment of Human Induced soil degradation (GLASOD), declared around 15% of the global land area to be in a condition of degradation, 13% suffering light degradation and 2% being severely degraded (see Fig. 5.1). On a national level, soil degradation in some countries has reached threatening dimensions. Turkey, a country where the soil supports a population of about 72 million people, serves as an example. Turkey represents a serious case as 21.6% of the total land area is severely damaged, and the state of degradation in 14.86% of the whole country's area is described as very severe (see Fig. 5.2). This is unfortunately not the only case, many other countries are suffering of very serious problems including desertification, deforestation and loss of agricultural areas due to intensive urbanisation (e.g. Egypt). According to the Land and Plant Nutrition Service of the UN-FAO, the soil condition in Egypt can be summarised in the

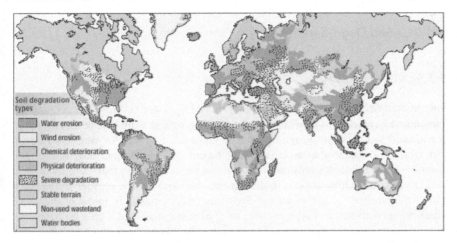

Fig. 5.1 An ISRIC/UNEP/FAO – map published by the World Food Summit, Rome, 13–17 Nov. 1996, showing soil degradation types around the world

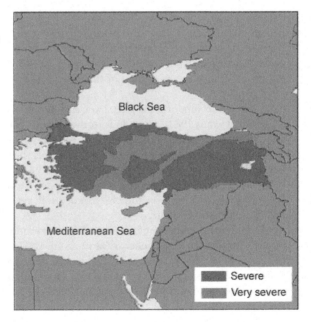

Fig. 5.2 Degrees of soil degradation in Turkey. according to UNESCO Food and Agriculture Organisation (FAO)

following: "*The productivity of only 5 percent of the present cultivated land is classified as excellent, 40 percent is classified as good and the remaining 55 percent as either medium or poor*" (FAO-AGL 2000). This situation arises predominantly from hydrological and chemical factors, such as water logging, sodification and soil fertility decline, due to the use of low quality water for irrigation and environmental pollution. Biological exhaustion of soil fertility, as a result of intensive cropping, worsens the situation.

5.2
Physical Soil Degradation

5.2.1
Soil Erosion

Soil erosion occurs when the rate of soil removal by water and/or wind exceeds the rate of soil formation. Soil formation is generally a very slow process with rates ranging around 1 cm/100–400 years; this makes about 0.1–1.3 t ha^{-1}. In areas with intensive land use or deforestation, erosion may be enhanced by human activities, yet we should bear in mind that erosion in general is a natural process that has always been taking place on Earth's surface. To differentiate between natural erosion and erosion induced by human activities, we call the former *background erosion*. It is almost in equilibrium with soil formation (<1.0 t ha^{-1}) in plain areas and a little bit higher in mountain regions. The harm caused by human induced erosion is that it seriously disturbs this balance.

Detachment of soil material may occur because of running water or winds, whereby, in both cases, the severity of erosion will depend upon the capacity of the eroding agent to transport it. Following factors generally determine the intensity of erosion.

- *Soil erodiblity.* This is a measure of the soil resistance to detachment and transport. It depends particularly on soil texture, organic content, structure, and permeability. Generally, soils with low contents of clay and organic matter are more readily eroded than soils with a higher content of the same. The rule shown in Fig. 5.3 holds for most soil types.

    This relation supports the idea that clay content may provide a measure for soil erodibility (Evans 1980), since clods formed by clay minerals together with organic matter stabilize the soil and make it more resistant to erosion. Soils with a high content of base minerals are generally more stable as these contribute to the chemical bonding of the aggregates (Morgan 1995).

    Resistance of soils to wind erosion will depend more on the moisture content of the soil, wet soil being less erodible than dry soil; yet all other factors remain as in the case of water erosion.
- *Erosivity.* This is a measure of the potential of the eroding agent to erode and is commonly expressed in kinetic energy (Morgan 1995). With regards to rainfall, this will be related to the intensity of the rainfall as well as to the size of raindrops. The most widely used index is the one known as EI_{30}. This is a compound index of kinetic energy and the maximum 30 minutes rainfall intensity. Wind erosivity indices are based

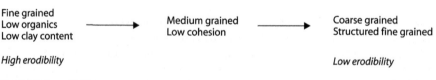

Fig. 5.3 Soil erodibility

largely on the velocity and duration of the wind. Erosion induced by human activities depends upon various factors such as land use, overgrazing in pasturelands and deforestation. In the developing world, however, socio-economic factors add to the reasons of enhanced erosion.

- *Vegetation cover.* Plant cover on soils may function as a buffer between the soil surface and the rest of the environment, including all natural forces such as wind, rain and running water. It dissipates the energy of raindrops, or run-off, allowing water to enter the soil. This was experimentally demonstrated by Hudson and Jackson (1959) and more recently by Zanchi (1983). In both cases, soil loss by erosion on a plot covered with fine wire gauze was compared with that recorded on an identical neighbouring bare plot. In the former experiment (Hudson and Jackson), which was carried out on a clay loam soil, the mean annual soil loss over a 10-year period was 126.6 t ha^{-1} for the bare plot and 0.9 t ha^{-1} for the covered one. Similar results were obtained by Zanchi, who carried out his experiment on a clay soil in Tuscany.
- *Topography.* Increases in slope steepness and slope length are factors that make soil more susceptible to erosion due to the respective increases in velocity and run-off volume. The relation between erosion and slope can be mathematically expressed in the following equation:

$$E \propto \tan^m \theta L^n$$

where E is soil loss per unit area, θ is the slope angle and L is the slope length. According to Zingg (1940), this relation may be written as:

$$E \propto \tan^{1.4} \theta L^{0.6}$$

More recent works suggest that the exponents in the last equation are generally more or less valid, despite the fact that they are apt to change due to several factors.

Flat flood plains, where deep and fertile soils occur, may also be eroded by concentrated flood flows (Carry and Harris 2001).

5.2.1.1
Measuring Soil Erosion

Data on soil erosion can be collected under simulated conditions in research laboratories or under realistic conditions using field investigation methods. The main advantages of these methods are that they are cheap, simple and allow the collection of enough data to assure a degree of reliance on the results.

Field methods may be designed to collect data on soil loss from relatively small representative plots (erosion plots), or to carry out serial investigations and measurements on wide areas, as for example a drainage basin. The so-called field reconnaissance methods, which are normally used to get first approximation of the amount of erosion in a given situation, are very popular among soil conservation workers. These are cheap methods and can be carried out by semi-skilled staff and need little maintenance. Reconnaissance methods may be classified into two main types: measuring change of surface levels and volumetric measurements.

Measuring of Surface Levels

For sites of intensive erosion such as on steep slopes and areas of extensive run-off, the direct measurement of level changes may provide an appropriate method of predicting soil loss amounts. Changes in soil level may be determined by:

- Point measurements to determine the level change in a single dimension.
- Profile and cross section measurements in two dimensions.
- Volumetric measurements in three dimensions.

Point Measurements

Several methods have been designed for measuring soil level in single points, and, since these methods are mostly inexpensive, a great number of measuring stations are usually used to collect a large volume of data. Of these methods the following are most popular:

- *Erosion pins.* In this method a pin is driven into the soil. The top of the pin gives a datum from which change in the soil level can be determined. The pins (also known as pegs, spikes, stakes or rods) may be made of wood, or any other material, which will not rot or decay (see Fig. 5.4).

  As seen in Fig. 5.4, the pin has a moving metal washer sliding over the whole pin and providing better hold for measurement up to the pin top. A typical length of pins is 300 mm, which allows a firm and sufficient penetration into the soil. Readers interested in more details on this method, together with an illustration of its use, may refer to Takei et al. (1981).
- *Paint collars.* In locations of high erosion rates such as a stream bed or gully floor, change of soil level may be made visible by painting a collar just above soil level around rocks, fence posts or anything stable. Erosion reveals an unpainted band below the paint line.
- *Bottle tops.* This is a very simple way of making changes in soil level visible and thus available to measurement. Pressing bottle tops into the soil surface provides a protection for the initial soil level at the beginning of the measurement process. The depth of subsequent erosion is shown by the height of the pedestals where the soil is protected by the bottle tops.

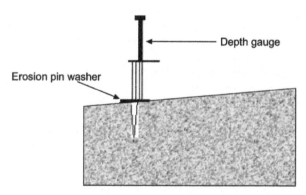

Fig. 5.4 Erosion pin to measure surface level (Hudson 1993)

Profile and Cross Section Measurements in Two Dimensions

Measuring small topographic changes or roughness properties along a straight line can be achieved using simple profile meters such as the one shown in Fig. 5.5, or by using advanced methods such as analysing air photographs or laser scanning graphs that may be specially useful in case of measuring surface roughness caused by wind erosion. The simple tool shown in Fig. 5.5 (Hudson 1993) is made of a horizontal bar, along which erosion pins are inserted to measure the changes of the soil level in several points along a straight line. These may later be used to construct an approximate profile of the surface. More precise measurements may be achieved using a laser scanner to determine small undulations of the surface (Huang and Bradford 1990), or by interpreting satellite surveys.

Volumetric Measurements

Various methods are used to determine volumetric changes related to the volume of soil loss by erosion. Measurements in the field that may render such calculations feasible include: measuring the changes of cross-sectional area of rills and road cuts, or estimating the rate of depth and cross-sectional increase of gullies. Measuring or estimating the volume deposited as an outwash fan, or in a catchpit or reservoir, may also provide an easy approach to calculate the volume of soil lost by erosion.

5.2.1.2
Modelling Soil Erosion

Many models have been proposed to quantify the extent of soil erosion. One should, however, be aware of the fact that models may differ according to geographical, climatic, and geologic conditions. An example may be given by the *universal soil loss equation* used by many authors (see Wischmeier 1976). It provides a general tool for modelling and predicting erosion phenomena. In this equation, soil erosion rate is considered a function of five factors:

$$E = RKLSCP \tag{5.1}$$

where E stands for mean annual soil loss,. R stands for rainfall erosivity index, K stands for soil erodiblity index, L stands for slope length, S stands for slope steepness, C stands for cropping factor (the ratio of soil loss under a given crop to that from bare soil), and P stands for conservation practice factor.

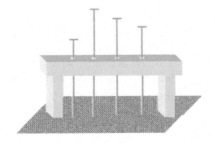

Fig. 5.5 Simple profile meter (Hudson 1993) (This figure is only for demonstration, the number of pins is usually higher)

Besides ignoring the human factors and socio-economic conditions, this equation, as said before, cannot be used without adaptation to special hydrologic and morphological conditions. Morgan (1995) discusses other models; some of these will be shortly mentioned:

- *The Soil Loss Estimator for Southern Africa (SLEMSA)*. This model was originally developed in Zimbabwe to evaluate the erosion resulting from different farming systems, in order to develop adequate conservation measures. Mathematically, the following equation (Elwell 1978) expresses the essential factors of the model:

    $Z = KXC$

    where Z stands for the mean annual soil loss (t ha^{-1}); K is mean annual loss (t ha^{-1}) from a standard field plot, 30 m long, 10 m wide, at a slope of 2.5°; for a soil of known erodibility of value F under a weed-free bare fallow; X is a dimensionless factor combining slope length and steepness; and C is a dimensionless crop management factor.
- *Wind Erosion Prediction Equation*. The same line of reasoning, as in the universal soil erosion equation, has been used to develop an equation for modelling soil erosion. In this equation (Woodruff and Siddoway 1965; Skidmore and Williams 1991), WE, the mean annual wind erosion (t ha^{-1}) is considered a function of five parameters as follows:

    $WE = f(I, C, K, L, V)$

    where I is soil erodibility, C is a climatic factor expressing wind energy, K is surface roughness, L is length of open wind blow, and V is vegetation cover.

5.2.1.3
Control of Soil Erosion and Conservation Strategies

As said before, the harm caused by human induced erosion, if taken together with natural background erosion, is actually based upon the fact that the equilibrium needed to conserve soils is seriously disturbed due to the slow rates of soil formation. This gives the impression that all conservation strategies should be oriented to the aim of keeping, or re-establishing, this equilibrium. However, a re-establishment of equilibrium is a non-realistic aim and no strategies, no matter how good they are, will achieve it, as all estimations of soil formation rates indicate very low values (Table 5.4).

The maximum permissible rate of soil loss, at which soil quality can be conserved for periods between 20 and 25 years, is known as the *soil loss tolerance* and it is within this soil loss tolerance that adequate strategies for soil conservation are designed. Hudson (1981) recommends values of soil loss as low as 2 t ha^{-1} for sensitive areas, where soils are highly erodible, while Morgan (1995) considers a mean annual soil loss of 11 t ha^{-1} to be generally acceptable.

5.2 · Physical Soil Degradation

Table 5.4 Estimations of soil formation rates by different authors for different regions

Author(s)	Rate estimated (mm yr^{-1})	Region
Buol et al. (1989)	0.01 – 7.7	World average
Zacher (1982)	0.1	World average
Kirkby (1980)	0.1	USA, northeast
Kirkby (1980)	0.02	USA, Great Plains
Kirkby (1980)	0.1	USA, arid southwest
Dunne et al. (1978)	0.01 – 0.02	Kenya, humid areas
Dunne et al. (1978)	<0.01	Kenya, semi arid regions
Johannesson (1960)	0.05 – 0.15	Prehistoric soils of Iceland

Main Soil Conservation Strategies
Management and remediation of soil erosion can be achieved according to one of the following soil conservation strategies:

- *Agronomic practices.* These aim to minimise the period of exposure to erosion when the soil is left bare, by encouraging the cultivation of dense vegetation cover and plant root network.
- *Soil management techniques.* These aim to increase the resistance of the soil to erosion, by following techniques that improve and maintain the soil structure. Such methods mainly apply processes such as mulching, reduced or zero tillage and addition of synthetic soil conditioners, e.g. PVA (polyvinyl alcohol), PAM (polyacryl amide) and PEG (polyethylene glycol).
- *Mechanical techniques.* The main strategy here is to reduce the energy of the eroding agent through modifying the surface topography. This is attained by geotechnical methods such as bunding, terracing or constructing diversionary spillways to direct water away from areas that are highly susceptible to erosion.

Table 5.5 shows an overview of the effectiveness of the various soil conservation strategies mentioned above.

5.2.2
Soil Compaction

Compaction is the mass reduction of soil and is generally expressed in dry bulk density, porosity, and resistance to penetration. Compacted soils are normally of a higher bulk density than comparable non-compacted ones (>1.5 g cm^{-3} compared to 1–1.5 g cm^{-3} in non-compacted soils). The resulting reduction of pore space lends the soil a compact dense character (see Fig. 5.6). This leads to a reduction of both water

Table 5.5 Effectiveness of some soil conservation strategies (Morgan 1995)

Practice	Control over					
	Rain splash		Runoff		Wind	
	D	T	D	T	D	T
Agronomic measures						
Covering soil surface	++	++	++	++	++	++
Increasing surface roughness	–	–	++	++	++	++
Increasing surface depression storage	+	+	++	++	–	–
Increasing infiltration	–	–	+	++	–	–
Soil management						
Fertilizers, manures	+	+	+	++	+	++
Subsoiling, drainage	–	–	+	++	–	–
Mechanical measures						
Contouring, ridging	–	+	+	++	+	++
Terraces	–	+	+	++	–	–
Shelterbelts	–	–	–	–	++	++
Waterways	–	–	–	++	–	–

– = no control; + = moderate control; ++ = strong control; D = detachment phase; T = transport phase.

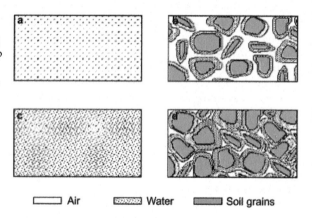

Fig. 5.6 a Non-compacted soil; **b** close up from (a) – observe the proportional volume of soil air; **c** compacted soil; **d** close up from (c) – observe the increase of the proportional volume of water-filled pores at the cost of soil air. The legend directly under the figure is applicable only for (b) and (d)

infiltration and drainage from the compacted layer. Aeration problems may also arise due to the reduction of the proportional volume of air in the interstitial space (Fig. 5.6d).

Physical changes arising from compaction may have far reaching consequences for biological and chemical conditions prevailing in the soil. The reduction of the rate of decomposition of organic matter and consequent reduction of nutrients may provide an example. This occurs when the proportion of water-filled pore space is increased due to the total reduction of interstitial pore size (Fig. 5.6d), which in turn causes a temperature drop reducing the activity of soil organisms.

Compaction, being a result of high vertical pressures, is controlled by many factors pertaining to soil properties, such as texture clay content and clay mineralogy. Clays, due to their sheet structures, are highly susceptible to volume reduction. This leads, in clay rich soils (on application of vertical high pressures), to the formation of highly compacted horizons at shallow depths (20–30 cm). These horizons, known as *cultivation pans*, impede drainage and hamper the formation of plant root networks. They also delay germination and may even completely stop it.

Causes of Soil Compaction

Soil compaction may occur as a result of various human induced factors as well as natural causes, however, the steady increase of machinery used in farming, as well as the increase in weight of field equipment, makes human induced compaction a matter of growing concern in many countries today. Some of the factors causing soil compaction are briefly discussed in the following:

- *Tillage and other farming operations.* Compacted layers (also known as tillage pans) may occur as a result of continuous ploughing at the same depth. Tillage pans (2.5–5 cm thick) may be easily alleviated by varying the depth of ploughing over time.
- *Wheel traffic.* As a matter of fact, this is the major human induced cause of soil compaction. Farm machines are continuously becoming bigger and heavier to meet the increase in farm size. This increases the danger of compaction for farm soils; especially in spring planting, which is often done before the soil is dry enough to support the heavy weight of farm machines (up to 20 tons).
- *Minimal crop rotation.* Crop rotation, if practiced frequently, helps break subsoil compaction due to variation in rooting systems of the different crops, and makes it essential to carry out frequent tilling operations, which may reduce the probability of compaction.
- *Raindrop impact.* This is the major natural cause of soil compaction. It aids the formation of crusts on the soil surface (~1.5 cm thick). Such crusts may prevent germination and cause the same harms mentioned above.

Testing for Soil Compaction

Soil compaction may not be easy to locate, due to the fact that it mostly occurs at the subsurface. Compaction testers such as manual rods (Fig. 5.7a) or dial probes (Fig. 5.7b) may provide the best help to deal with this task. Compaction testers pushed into the soil with steady, even pressure, to a depth of 90–125 cm, will indicate compaction, if they are met with a resistance at any depth within this range.

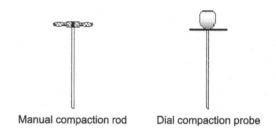

Fig. 5.7 Compaction testing tools; **a** manual compaction rod; **b** dial compaction probe

Manual rods provide evidence of compaction, yet they provide no concrete measurement of strength or degree of compaction. For this, dial probes and digital recording compaction meters may be used. Many companies are now offering digital compaction meters (penetrometers) with integrated data loggers that can work together with GPS-tools, at relatively low prices.

Alleviating Soil Compaction

One of the best methods for remediation of soil compaction is the technique known as *subsoiling* or *chiselling*. This is done by breaking the compacted subsoil, without inverting it, with a special knifelike instrument (chisel), which is pulled through the soil usually at a depth of 30–60 cm and a spacing of 60–150 cm. Drilling holes under the drip line of trees and back filling with mulch or compost adds organic matter to the soil and preserves soil moisture. This reduces the adverse effects of compaction, since soils rich in organic matter are less susceptible to compaction than mineral ones. Most important in the avoidance of compaction, however, is limiting traffic in areas in the field with a compaction problem and reducing the pressure exerted by agricultural machines through inflating their tyres to the least permissible psi, while using tyre types that lessen the negative effects of wheel traffic.

5.2.3
Soil Crusting and Sealing

Crusting is the name given to the phenomenon of thin crust formation on soil surfaces, whereby wet crusts are collectively known as *seals*; this is the reason why crusts and seals are sometimes treated as being one and the same object. Crusts are sometimes known as *cappings* and are generally formed on exposed soil in brittle and semibrittle (seasonably dry) environments.

Chen et al. (1980) classify crusts into two main types according to their mode of formation. These are:

- Structural crusts, which are developed in place (in situ crusts).
- Crusts, which have been transported from their original place of formation.

To these we may also add salty crusts, which are thin layers of salt formed at soil surfaces. Short descriptions of these types may be given as follows:

Structural Crusts

These are crusts formed in place by raindrop impact. They evolve through gradual coalescing of aggregates under the impact of raindrops and may form broken clods that lie above the surface (*slaking crusts*: Valentin 1991), or may be formed in several (generally 3) well-sorted layers. The uppermost layer is normally composed of loose, coarse sand, the middle one is formed of fine densely packed grains, and the lower one is almost totally made of fine particles with reduced porosity (*sieving crusts*: Casenave and Valentin 1989 and 1992).

Depositional (Sedimentary) Crusts

These are crusts that are caused by the transport and deposition of particles, due to water flows. Micro-aggregates are washed away by such flows, to be deposited at low lying topographic parts of the soil, resulting in the spreading of thin sheets that may later coalesce to form seals or crusts. Sources of fine aggregates are various, they may be derived from flood and furrow irrigation water, from raindrop impact splash of loose soil particles, or from run-off and sheet erosion.

Some authors classify crusts according to their nature into physical/chemical and biological crusts. In this respect we may consider salty crusts as chemical ones, whereas biological crusts may be widely varying, both in composition and mode of formation.

Biological Crusts

Biological crusts (also known as cryptogamic, microbiotic, cryptobiotic or microphytic crusts) are, as the name says, crusts formed through biological activities of plants and algae. Different assemblages of blue green algae, diatoms, golden brown algae, lichens, mosses and few xerophytes (salt loving plants) that can flourish in presence of limited water supply may form such crusts. Their effect on soils is not entirely agreed upon. Some authors consider them useful for soil quality development due to their role in helping to fix nitrogen (blue green algae) and offering protection against erosion. Other authors, however, consider their role in soil protection against erosion as being minimal, due to their fragility. Generally, though, biological crusts may hamper germination and affect soil cultivation.

Mechanisms of Crust Formation

Crust formation results from four main processes that work hand in hand and complement each other (Agassi et al. 1981; Morin et al. 1981). These are:

- physical dispersion of soil grains followed by compaction due to raindrop impact;
- chemical dispersion of clay particles;
- formation of continuous dense layers of clay particles by interface suction forces (Morin et al. 1981);
- increases in soil acidity making soil more susceptible to crust formation.

Whatever the mechanisms and processes involved in crust formation are, the results are very harmful to soil, for crust formation markedly reduces the micro-porosity of the soil surface layer. The results can be devastating for the soil, since water infiltration varies as the fourth power of the diameter of the pores.

Avoidance and Alleviation of Crust Formation

Soil crusting is mostly induced on soils of higher acidity and those having a high content of fine clay material. On such soils, raindrop impact destroys the structure of the surface layer. Accordingly, alleviation and avoidance of crusting may include the reducing of sodicity and acidification, as well as finding ways to induce flocculation of fine materials. This may partially be achieved by adding gypsum to the soil. Gypsum increases the flocculation of fine mineral and organic matter, reduces the amount of shrinkage and swelling of clays, by moderating change of water content, and increases the activity of soil organisms through adding calcium to the soil. Using soil conditioners such as polyacrylamide (PAM) mixed with gypsum was found to be very effective in decreasing crust formation, run-off and erosion (Tang et al. 2006).

The addition of gypsum is more effective in volcanic soils rather than in alluvial ones. The best results are achieved in soils with a high clay content (30% or more) and exchangeable sodium of more than 5%. Sometimes accumulation of litter on the surface would also lead to sealing and crusting, since broadleaf litter (especially if the leaves lie horizontally) reduces infiltration. Accordingly, removal of leaf litter is considered as one of the measures to avoid crust formation.

5.3
Chemical Soil Degradation

5.3.1
Acidification

Like all other problems of soil, acidification has an anthropogenic dimension as well as a natural one, arising from background factors. Natural processes that lead to acidification range from long-term base leaching and microbial respiration to nitrification and plant growth.

Base leaching attains a maximum in regions where precipitation exceeds evapotranspiration. It is also enhanced by higher concentrations of CO_2 in rainwater and by decomposing materials such as humic and fulvic acids in the soil. Wide parts in eastern North America, as well as northwestern parts of Europe, suffer of base leaching due to the prevalence of wet climatic conditions. Plant growth contributes to soil acidity because plant-nutrition depends upon the exchange of nutrient base-cations for H^+ and thus increases the soil acidity. Nitrifying bacteria also helps in lowering the pH of the soil by oxidation of ammonium, according to the equation:

$$NH_4^+ + 1.5O_2 \rightleftarrows NO_3^- + 4H_2O \tag{5.2}$$

Apart from the natural processes leading to soil acidity, some factors inherent in the soil itself may make a given soil susceptive (if not predestined) to soil acidity. Most important here is the effect of parent material: soils derived from granites and sandstones are normally acidic, while those derived from limestone may be alkaline.

Anthropogenic processes leading to acidification are principally related to land use practices such as needle leaf afforstation and excessive use of inorganic nitrogen fertilisers. Considering the effect of human activities on the development of atmospheric conditions on the Earth, one may attribute almost all atmospheric deposits to this factor. Therefore, acidic deposits are mainly due to the solution of industrial and transport emissions (SO_2, NO_x) in rainwater, forming sulphuric acid (H_2SO_4) and nitric acid (HNO_3) that eventually contribute to the increasing acidity of soils. In some parts of North America and Europe, pH-values less than 4 were recorded in acid deposits.

The Impact of Acidity on Soil Quality

Soil acidity has an impact on many essential properties that determine soil quality. Most important in this respect is its effect on the availability of nutrients. Nutrients such as phosphorus, potassium, magnesium and calcium decrease through acidity, whereas the availability of metallic micronutrients increases. This applies for metallic ions such as zinc, manganese, copper, and iron.

Mobilisation of aluminium, zinc and manganese may result in toxicity to plant roots. The following problems may arise:

- Aluminium and/or manganese toxicity
- Deficiency of phosphorus, due to its tying up by iron and aluminium
- Calcium and magnesium deficiency

In addition to problems related to nutrient availability, the prevalence of lower pH-values may affect the environmental conditions required for bacterial growth, thus hampering the evolution of organic matter in the soil and lowering its biological activity.

Avoidance and Alleviation of Soil Acidity

The ideal values of pH for soils depend upon the climatic conditions and the mineralogical and chemical composition of the soil, as well as the type of agricultural management and crops. In temperate regions, a pH-value of about 6.5 is preferred for arable crops and 6.0 for grassland, whereas a pH-value of 5.5 is considered to be ideal in tropical areas. Soils not fulfilling the required pH conditions for a given purpose are normally treated to meet the conditions desired. The remediation process known as *liming* is considered as the most effective remediation technique in this respect. It reduces the harmful effects of low pH, such as aluminium and manganese toxicity, while adding calcium and magnesium to the soil.

Liming is carried out by adding finely ground limestone to the soil in amounts capable of raising soil pH to the required level. These amounts, however, depend on the pH of the soil and its buffering capacity. The buffering capacity, which will be discussed in more detail in Chap. 9, is a measure of the amount of calcium carbonate that should be added to the soil to raise its pH to a given level. It is related to the CEC and is directly proportional to the amount of exchangeable acidity (hydrogen and aluminium) held by the soil colloids. This means that it is also directly proportional to the amounts of organic matter and clays in the soil.

Buffering capacities of soils can be measured in the laboratory using buffered solutions (see p. 201).

Liming Materials

Liming materials most commonly used include limestone, chalk, marl and basic slag. They may contain minor amounts of quick lime (CaO), slaked lime (Ca(OH)$_2$), and magnesium carbonate (MgCO$_3$).

5.3.2
Salinization and Sodification

In regions where evapotranspiration is higher than precipitation, soil water flow will be driven by capillary action in an upward direction and eventually, due to evaporation, saline precipitates will be formed in the interstitial pores of the soil, leading to salt accumulation between the soil grains. This phenomenon is known collectively as *salinization*. In cases where the parent material of the soil is rich in sodium, salinization will also lead to the increase of sodium in the soil water, leading to *sodification*, as this is termed by soil scientists.

Saline soils are generally formed in low lying flood plains, where the water table is high. Sodic soils, in contrast, occur commonly on slopes immediately above valley floors and flood plains. Both types are widely present in semiarid regions.

The classification of soils as saline, sodic, or both, mainly depends on physical properties such as Ec$_e$ (the electrical conductivity of a saturated extract of the soil) and ESP (the exchangeable sodium percentage, which is related to soil acidity).

Generally, however, all soils having high contents of soluble salts are called *saline soils*. They may be identified through morphological features such as exhibiting whitish surface crusts when dry. The term *soluble salts*, used here in the definition of saline soils, is standardised with reference to the solubility of gypsum (Seelig 2000). Accordingly, only salts more soluble than gypsum are considered soluble; examples are sodium sulphate (Na$_2$SO$_4$) and sodium chloride (NaCl). Insoluble salts (less soluble than gypsum) are, according to this definition, considered not to cause salinity.

Practical Classification of Saline and Sodic Soils

Depending on the laboratory classification of the U.S. Salinity Laboratory Staff (1954), one can classify saline and sodic soils according to their physical properties, as shown in Table 5.6. The table provides convenient criteria for classifying soils as saline, sodic, or sodic saline.

As the table shows, sodic soils are alkaline and have generally a lower electrical conductivity relative to saline ones. In contrast, saline soils have a wide range of pH-values ranging from acidic to slightly alkaline and are characterised by higher electrical conductivity. Soils having pH-values in ranges similar to saline soils, yet have high ESP-values, are classified as sodic-saline.

Table 5.6 Relation of soil types to salinity and sodicity

pH	ECE	ESP	Soil type
≤8.5	≥4	<15%	Saline
≤8.5	≥4	≥15%	Sodic-saline
8.5–10	<4	≥15%	Sodic

The extremely high pH-values of sodic soils (>8.5) are due to reactions of sodium clay, bicarbonate and carbonate with water to produce OH^--ions, according to the following equations:

$$Na\text{-clay} + H_2O \rightleftharpoons H\text{-clay} + Na^+ + OH^-$$

$$HCO_3^- + H_2O \rightleftharpoons H_2CO_3 + OH^-$$

$$CO_3^{2-} + H_2O \rightleftharpoons HCO_3^- + OH^-$$

Soils of such pH-values may be unsuitable for plant growth because the availability of some plant nutrients is normally reduced under these conditions. Phosphates are mainly available to plants at a pH between 6 and 7. Micronutrients, such as iron, manganese, zinc, copper and cobalt, are all far less available at pH greater than 7 (Seelig 2000).

Also, soils having a high content of soluble salts (saline soils) increase the problems of nutrient uptake, since high degrees of ion concentration mostly block nutrient uptake and hamper certain physiological processes. Besides, salinity enhances electrochemical reactions, through which metal corrosion may be enhanced. Corrosion of iron and other metals may cause different problems for farmers and other people working on soils.

Indicators of Salinity and Sodicity

In addition to analytical methods of diagnosis, some reconnaissance methods may be very useful in identifying saline and sodic soils. These are generally classified into two main groups: methods of diagnosis through topographic features and methods depending on plants as indicators.

1. *Identification through topographic features.* Prevailing topographic features provide very good indicators for salinity and sodicity. Saline soils are often seen as bare grounds covered by whitish crusts, having a scabby appearance. Sodic soils may be identified through the occurrence of dense clay pans at or near their surfaces. These pans occur as a result of the dispersion of Na-clays.
2. *Identification through plant indicators.* Tolerance of plants to salinity or sodicity varies widely according to species. This fact provides a criterion to compare the degree of salinity or sodicity of soils according to the type of plant communities growing upon them. *Halophyte* (salt loving) plant communities, for example, are very reliable indicators of saline soils. They flourish on soils having more than 0.5% (by weight) soluble salts.

Amendment of Soils Affected by Salinity and Sodicity

Amendment and alleviation methods of soils affected by salinity, sodicity, or both, depends upon the type of soil and the economic feasibility of the process. Different methods are used according to the problem and the type of degradation that should be alleviated.

- *Control of salinity.* The main factors for salinity occurrence in soils are a relatively high water table and a high degree of evapotranspiration; this is why the controlling of salinity must be carried out using technical methods that can mitigate the effect of these two factors. Subsurface drainage may provide an effective means of lowering the water table, while surface mulches may effectively help reduce evapotranspiration and stop salt accumulation.
- *Amendment of sodic soils.* In addition to drainage management techniques, in the case of saline soils, two main techniques are essential for the amendment of these soils, namely the disruption of clay pans and the use of soil chemical conditioners to replace the adsorbed sodium with calcium. A popular chemical conditioner used for this purpose is gypsum ($CaSO_4 \cdot H_2O$). It helps replace the exchangeable sodium ions in the upper 10–15 cm of soil by calcium ions. The amount of gypsum required depends upon the amount of exchangeable sodium, expressed in centimole per kilogram of soil. Thus, it is estimated that, for a soil containing 1 centimole of exchangeable sodium per kilogram, the amount of gypsum required to replace the exchangeable sodium in one hectare of soil to a depth of 10 centimetres is about 1.3 tonnes.

Sulphur can also be used as an amendment in calcareous soils because the added sulphur is oxidized to sulphuric acid, which is capable of dissolving calcium carbonate to produce gypsum.

Part II
Soil Pollution – An Overview

Chapter 6
Major Types of Soil Pollutants

6.1
Heavy Metals and Their Salts

Natural concentrations of heavy metals in soils depend primarily on the type and chemistry of the parent materials from which the soils are derived. However, anthropogenic inputs may lead to concentrations highly exceeding those from natural sources. Average concentrations of some heavy metals in the Earth's crust, in some sediments and generally in soils, are shown in Table 6.1. From the table, we can conclude that lead, cadmium, tin, and mercury, are the most abundant metallic pollutants introduced into soil by anthropogenic activities. The mean concentration of cadmium in soils is six times its mean concentration in the crust. Concentrations of lead, mercury, and tin in soils attain double their mean values in the Earth's crust.

Lead is one of the oldest metallic pollutants introduced by man into the environment. It was called *plumbum nigrum* by the Romans to differentiate it from *plumbum candidum* – tin. Water drainage pipes made of lead and carrying the insignia of some Roman emperors are still in use, despite the fact that they introduce lead as a dangerous pollutant into the soil when they become porous. The use of lead to make water pipes is no longer allowed due to their dangerous effect on the environment and the harm they cause for humans. Lead can, if ingested with food or water, cause severe damage to the nervous system, the urinary system and the reproductive system. It may cause abortion in females and reduce the fertility of males by severely lowering their capability of sperm production. It may also cause anaemia by obstructing the biosynthesis of haemoglobin.

Cadmium – a heavy metal, which is mainly used in the production of nickel/cadmium batteries or that of pigments and stabilisers for PVC; in metallurgical and electronic industries, it is one of the most frequently registered metallic pollutants. It passes into the environment through emissions in the metallurgical industries or application of phosphate fertilisers in agriculture. Detergents and petroleum products may also contribute to its circulation as an environmental pollutant.

In humans, long-term exposure to cadmium may lead to renal dysfunction and obstructive lung disease. It is also linked to bone disorders such as demineralisation of the bone substance *osteomalacia* and *osteoporosis*.

Daily intake of cadmium for humans is estimated to be somewhere between 0.15 and 1.5 µg. Smokers may risk a considerable increase; for every cigarette, about 2 to 4 µg cadmium is additionally inhaled.

Tin, a soft, pliable, silvery-white metal, which is used for can coating and as solder for joining pipes or electric circuits is, as said before, one of the most abundant metallic pollutants introduced by man into soils. For a long time it was also used for the

Table 6.1 Elemental composition of the Earth's crust and sediments. Note: Only iron and titanium are in percent, all other elements are in µg/g (modified after Salomons and Förstner 1984)

Element	Mean crust	Mean sediment	Average shale	Deep-sea clay	Shallow water sediments	River suspended sediments	Sandstone	Limestone	Soil
Iron	4.1%	4.1%	4.7%	6.5%	6.5%	4.8%	2.9%	1.7%	3.2%
Titanium	0.6%	0.4%	0.5%	0.5%	0.5%	0.6%	0.4%	0.03%	0.5%
Vanadium	160	105	130	120	145	170	20	45	108
Chromium	100(?)	72	90	90	60	100	35	11	84
Nickel	80(?)	52	68	250	35	90	9	7	34
Zinc	75	95	95	165	92	350	30	20	60
Copper	50	33	45	250	56	100	30	5.1	26
Cobalt	20	14	19	74	13	20	0.3	0.1	12
Lead	14	19	20	80	22	150	10	5.7	29
Tin	2.2	4.6	6.0	1.5	2	–	0.5	0.5	5.8
Cadmium	0.11	0.17	0.22	0.42	–	1	0.05	0.03	0.6
Mercury	0.05	0.19	0.18	0.08	–	–	0.29	16	0.1

production of tinfoil, which has been replaced by the aluminium ones. It has also many applications in the industry such as ceramics and electronic industries, yet it is mainly used for plating steel containers used to transport food and beverage.

The main source of tin is the mineral cassiterite (SnO_2), widely occurring in the geographical region known as *the tin belt*. This is a zone extending from China through Thailand, Burma and Malaysia to Indonesia.

Tin enters the environment as a pollutant in different forms, the most dangerous of which is the organic one (methylated tin), used in various industrial processes such as the production of paints, plastics and pesticides.

In humans, long-term exposure to methylated tin compounds, especially trimethyl tin, may cause disorders as severe as depression, liver disease, malfunctioning of immune system and chromosomal damage.

Organic tin compounds may be adsorbed on sludge particles and thus find their way into surface waters, where they may cause great damage by poisoning a wide variety of aquatic organisms.

Mercury also known as *Hydrargyrum* – liquid silver – (observe the relation to chemical symbol Hg) and *argentum vivum* – *quick silver* – is characterised among heavy metals by its low melting point, the existence of a whole series of organic compounds and the formation of amalgams. In the elemental state, it is known since the early times of history. According to reports by Schliemann, metallic mercury was found in vessels discovered by him in an archaeological site in Asia Minor, dating back to the 16th century B.C. This might be due to the ease of its extraction from the mineral cinnabar (HgS) by treating the powdered mineral with vinegar in a copper vessel. Theophrastus mentioned this method of extraction in his book on *Stones,* about 300 B.C.

Mercury passes into soils either naturally through the parent material or deposition of transported weathering products of pre-existing rocks, or through human activities, as it is the case in metallurgical industries or the disposal of mining products and old devices having mercury-related components or parts. Mercury may also be deposited from acidic waters.

The interaction of mercury and other heavy metallic pollutants with the soil environment depends upon the following four factors:

- pH-values and redox potential of the ambient waters
- microbial activities
- concentration of any complexing materials in their direct vicinity
- prevailing temperature

The uptake of mercury by humans may occur through the ingesting of contaminated material or, inhalation of volatile mercury compounds (circulating in the air) resulting from the combustion of fossil fuels. Fish and crustaceans that accumulate mercury from acidic waters, in a metal or organic form, may help in transporting it to human beings. This results in severe disorders such as a decrease in sperm production and eventually male sterility, chromosomal and DNA damage, disruption of the nervous system, and damage to the brain functions.

Beside the three above-mentioned heavy metals, copper and zinc provide serious reasons for concern due to the ease with which they may pass into the environment and their ruinous effects on human and animal life.

Copper is perhaps the oldest metallurgical product known and extracted by humans. Its metallurgy was widely known around four thousand years before the Christian era. Copper weapons and toilet articles belonging to the early Minoan age (3 000–2 000 B.C.) have been excavated in Crete. The name *copper* has been applied to the metal presumably in relation to the classical region of its mining on the island of Cyprus, where it was known as *aes Cyprium* – brass of Cyprus (later modified into *cuprum*). The metallurgical operations involved in its extraction and refining, however, seem to have been developed in Egypt a long time ago (3 000–2 550 B.C.). The coppersmith profession was at that time – as we may conclude from a poem discovered in an ancient tomb – a commonly recognised trade. In the that poem, an old anonymous Egyptian poet describes the coppersmith in the following words:

I saw the smith at his work,
Standing before the vent of his oven
His fingers were like crocodile-skin:
He stank worse than fish roe.

Copper is used in many industries, predominantly for wiring in the electrical industries, because of the ease of manufacturing wires out of it. It is also used in metallurgical industries for making alloys. Examples are the alloys used for low denomination coins and copper-zinc alloys known as brass. Copper-zinc-tin alloys are very hard and were once used for making guns and cannons. Copper passes into the environment as air born particulate pollutants, as organic complexes in water and as a metal or inorganic compound. This is causing a steady increase in the copper content of soils and consequently increasing damage for soil organisms. Its effect on human health is also severe and some studies link it to serious diseases and disorders as liver and kidney damage. Speculations about its role as a carcenogenous agent, however, have not been emphasised.

Zinc (also *zinckum*), the name of which was probably invented by Paracelsus (1493–1541), is an abundant element produced principally in the United States, Germany, Great Britain, Poland and Belgium, where it occurs as zinc spar or calamine ($ZnCO_3$). Zinc is principally used for galvanising steel; it also forms a very important component of many alloys. An example of such alloys is the one used to make American pennies, which contains zinc as a principal component. Other important applications of zinc include casting in the automobile industry, making some construction elements, production of photocopier paper and watercolours, making negative electrodes of some batteries, as well as using it as activator and catalyst in the rubber industry. Zinc is added to soils, either through the weathering of zinc containing minerals and rocks, or through human activities including disposal of industrial waste.

Zinc is linked to some health disorders, such as pancreas disease and disturbed metabolism.

Relations between metal concentration in soils and their parent material were investigated by Bowen (1979) and later discussed by Martin and Coughtrey (1982).

Figure 6.1 shows that iron, titanium, manganese, zinc, gallium, silver, and mercury show a close relationship between parent material and soil concentrations, while thallium, cobalt, copper, nickel, chromium and vanadium are somewhat depleted in soils relative to parent rock, while indium, bismuth, cadmium, tin, and arsenic seem to be

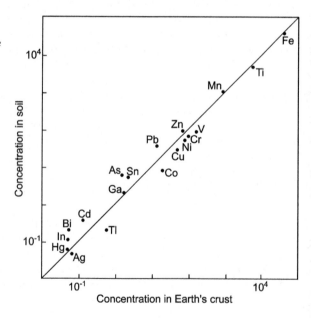

Fig. 6.1 Relationship of recorded concentrations of various metals in soils with average concentrations in the Earth's crust (Martin and Coughtrey 1982, p. 155)

enriched in soil relative to the original material. Most studies dealing with vertical distribution in soil profiles report a tendency for several metals to be concentrated in the upper layers or horizons of many soils of open land and woodland. This was reported by Burton and John (1927), who studied the vertical distribution of lead, cadmium, copper, and nickel in some soils in Wales, as well as by Wilkins (1978), who reported higher concentrations of lead in surface soil layers of 500 sites in the United Kingdom.

The surface enrichment of metals in soils may be the result of fallout of wind-transported pollutants, concentration of metals by plants from lower horizons or chemical complexing of metals by organic compounds.

6.1.1
Heavy Metals and the Soil System

As it was mentioned before, soil is a three-dimensional system composed mainly of a solid-state component beside the other two components – the solid phase and the gaseous one. Heavy metals may have, according to the physical condition and chemical form the metal or its compound may be taking, an important effect on any of these phases. To understand this, a short discussion of the behaviour of heavy metals within the soil system will be given in the next few paragraphs.

The interaction of metals with the solid state does not occur isolated from the liquid phase, for the pH, together with other properties of soil water, will control solution, adsorption on the solid surfaces and feasibility of chemical reactions within the system. In concrete terms the interaction of heavy metals with the solid state will be within the framework of any of the following processes:

- *Adsorption.* Adsorption in the soil system depends principally on the physicochemical parameters, especially pH and the metal adsorbent. So we find that clay mineral will in general adsorb less amounts of metals than oxides and organic material. Dissolved organic matter can increase metal solubility and activate reactive mineral surfaces, thus enhancing adsorption.
- *Diffusion into soil material.* Metal ions are capable of diffusing into the soil in rates that will depend upon their ionic diameters as well as the prevailing pH of the medium. Increases in pH, however, may lead to the formation of hydroxocomplexes that could cause an increase of the ionic diameters, thus decreasing the rate of diffusion and hence immobilising the heavy metals aspiring to diffusion.
- *Complex formation.* As mentioned before, metal ions can be coordinated to soil organic substances, in particular humic and fulvic acids (see soil organic matter).

6.1.2
Transport of Heavy Metals within the Soil System

In fact, diffusion and dispersion of heavy metals may be in the form of an active transport process depending on many factors, the most important of which is related to the movement and characteristics of the soil solution and infiltrating waters. In summary, these factors are rain intensity, evaporation, soil water retention and soil hydraulic properties such as conductivity and diffusion properties.

6.1.3
Bioavailability of Heavy Metals

The term *bioavailability* (originally taken from pharmacology) stands for the measurement of the rate and the extent of the uptake of an active substance that reaches the systemic circulation in an organism – in short, the extent of its uptake and systemic circulation by the organism. It is in this sense that bioavailability of heavy metals will be discussed in the present book.

Uptake (of heavy metals) is a term describing the entrance of these metals, their salts or compounds into an organism, such as by breathing, swallowing, or absorbing it through the skin. Uptake is a complex process, depending mainly on the chemical potential – the force that makes metal ions move passively along concentration gradients in soil solution. Along these gradients various chemical forms of metal ions move from outside to inside an organism. This essentially explains the process of passive uptake. However, uptake may occur in an active way, i.e. when organisms use the metal-containing fraction of soil, in whatever form it may be available, as food.

Following uptake, the penetrating substance is first circulated and later excreted, stored or metabolised. Excretion, storage, and metabolism, decrease the concentration of the chemical inside the organism, thus increasing the potential of the chemical in the outer environment and eventually enhancing movement into the organism. This may lead, if persistent environmental exposure exists, to a state of dynamic equilibrium between the amount of the chemical accumulated inside the organism, and the amount leaving it. So long as this equilibrium remains undisturbed, the amount of chemical inside the organism remains constant. But if the environmental concentration of the chemical outside the organism increases, the equilibrium will be disturbed and the concentration inside the organism starts once more to increase until the system reaches a new dynamic equilibrium. Repeating this situation for a long time may lead to very high concentrations of the chemical inside the organism, leading sometimes to a collapse of the system followed by severe health problems or even death.

The elevated concentrations of a substance, arising through the above-mentioned processes, may be temporarily stored in the body of the organism. One describes the process in this case simply as *storage*, such as the storage of fat in hibernating animals or the storage of starch in seeds. In the event that the concentration inside the organism becomes anomalously higher than that in the air or water around the organism and seems not to be of a temporary nature, one speaks of *bioconcentration* or *bioaccumulation*. Although the process is the same for both natural and manmade chemicals, the term *bioconcentration* usually refers to chemicals foreign to the organism. Accumulations of xenobiotic compounds in fish and other aquatic animals, after uptake through the gills (or sometimes the skin), represent one of the most well known bioaccumulation processes.

Preference in bioaccumulation is for hydrophobic substances, i.e. substances soluble in fats (*lipophilic*); these will have the best chance to be accumulated, while water-soluble ones (*hydrophilic*) remain, in most cases, in the solution. However, heavy metals like mercury and certain other water-soluble chemicals form an exception to this rule, because they bind tightly to specific sites within the body. When binding occurs, even highly water-soluble chemicals can accumulate. An example in this respect may be given by cobalt. It binds very tightly and specifically to sites in the liver where it accumulates despite its water solubility. Similar accumulation processes are known to occur with mercury, copper, cadmium, and lead.

The rate and extent of the uptake of heavy metal ions and the degree of their contribution to the systemic circulation of an organism (bioavailability) depend on various factors, the most important of which is the oxidation state (valency) of the metal ion. This is known as the *speciation* of the ion. Speciation describes the various chemical forms, i.e. valence states of the same element in the solution (e.g. FeII, FeIII or CrIII and CrVI – each of these is known as an ion species of the same element; hence the name). Speciation is very important in environmental toxicology; since speciation can affect both the mobility of metals in the environment, and their toxicity (bioavailability). An example may be given by chromium (Cr) – the hexavalent form (Cr^{6+}) is more toxic than the trivalent form (Cr^{3+}); also trivalent arsenic (As^{3+}) is more mobile in the aquatic environment than its pentavalent form (As^{5+}).

Determining the speciation of a metal in aquatic media is a very difficult process and depends on analytical methods which are imprecise and can only be described as semi quantitative.

Another factor affecting the mobility and bioavailability of heavy metals is the constellation and nature of complex ions. Normally, organic complexes of solute forms have a reduced bioavailability (i.e. not readily taken up and cannot easily join the systemic circulation of an organism), yet reports about opposite behaviour are also known. Examples of this are mercury and arsenic. Organic forms of both metals have different toxicity to the inorganic ones; organic mercury is more toxic than its inorganic counterpart (i.e. more readily up taken and circulated in the living system); yet, in the case of arsenic, the reverse is true. Therefore, the rule is quite ambiguous and cannot be generalized.

As a rule, however, all metals (except molybdenum) are most soluble and bioavailable at low pH-values and therefore toxicity problems are likely to be more serious in acidic media.

6.1.4
Biochemical Effects of Heavy Metals

In the last few sections, different health disorders caused by long exposure to heavy metals were mentioned, giving perhaps the false impression that heavy metals under all conditions may be harmful or even lethal for living organisms. As a matter of fact, heavy metals may, under certain conditions, be of great benefit for animals and plants, since some of them are constituents of enzymes and other important proteins involved in key metabolic pathways (see Chap. 9). In a biochemical sense, heavy metals may be classified into two groups:

1. *The group of biologically essential heavy metals.* This is the group also known as the micronutrients or essential trace metals. They contribute to the structure of enzymes and important proteins steering principal metabolic processes. Deficiency of these elements, below certain critical concentrations, leads under normal conditions to serious diseases and metabolic dysfunctions.
2. *The group of biologically non-essential heavy metals.* This group is not essential like the first one, and has no critical concentrations below which deficiency disorders

occur; instead, they may be toxic to organisms if concentrations go beyond a given limit. Toxicity by these elements occurs due to their disturbance of metabolism by replacing the essential metals in enzymes or reacting with the phosphate group in ADP and ATP (see Sect. 9.4). Examples of this class may be given by arsenic, cadmium, mercury, lead, plutonium, antimony, thallium, and uranium. Figure 6.2 represents in a schematic way the important biochemical properties mentioned in this section.

6.1.5
Major Environmental Accidents Involving Pollution by Heavy Metals

6.1.5.1
The Minamata Disaster, 1950–1956

> *When I thought I was dying*
> *and my hands were numb*
> *and wouldn't work –*
> *and my father was dying too – when*
> *the villagers turned against us –*
> *it was to the sea*
> *I would go to cry.*
> (A fisherman from Minamata, inflicted with the Minamata Syndrome)

At the beginning of the 1950s – a time when the production of acetaldehyde was booming to satisfy the needs of the world market for plastics – a mysterious disease causing bizarre behaviour in animals, especially cats, started to haunt the population of the small town of Minamata in southern Japan (see Fig. 6.3). Loss of motor coordination caused cats to move in an abnormal way that resulted in their falling into the sea and dying. The villagers started calling it the cat suicide syndrome or even the disease of dancing cats (Smith and Smith 1972, 1975; Ishimure 1990). In the meantime, wholesale dying of fish started in the Minamata bay, followed by what seemed to be a transfer of the disease to humans. The physiological effects, including successive loss of motor control, were devastating, and resulted sometimes in partly paralysed and contorted bodies. This serious development was played down by the local authorities and the Chisso Corporation, the owner of the acetaldehyde-producing factory near the town. Nobody seemed to connect this mysterious disease with the spill of industrial waste into the sea. It was not before 1956, after the disease started taking epidemic scale among humans, that it was identified as heavy metal poisoning caused by draining industrial liquid waste containing mercury into the bay. Research results showed that $HgSO_4$ reaching the bay in industrial drained effluents was first precipitated as insoluble HgS, which was later on methylated through the biological action of bacteria in the sediments to form CH_3Hg^+. This form of mercury (monomethyl), being lipophylic and readily bioavailable to sea creatures, was bioaccumulated by fish and other sea organisms. Since the Minamata residents relied almost exclusively on fish and other aquatic organisms as a source of protein, organic mercury was transferred to their bodies, triggering off the disaster.

Mercury is indeed a neurotoxin, having teratogenic effects (causing birth defects arising from damage to embryonic or foetal cells), and this is perhaps why it came to

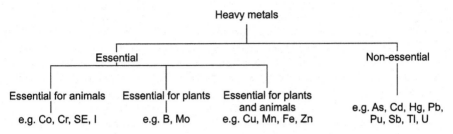

Fig. 6.2 Diagrammatic representation of the biochemical classification of heavy metals

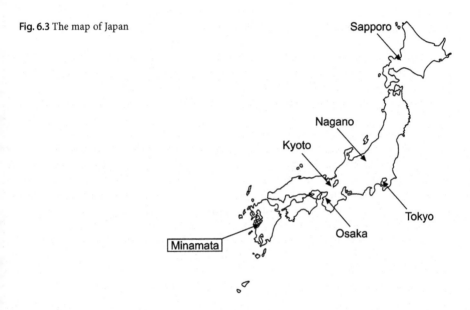

Fig. 6.3 The map of Japan

all the bizarre symptoms and health disorders that appeared at that time in the surroundings of Minamata.

6.1.5.2
The Rhine Pollution at Basel, 1986

The incident in Basel involved a spillage of mercury into the river Rhine in the course of fighting a fire that broke into a warehouse at a location along the river. The warehouse, used for the storage of pesticides, mercury and other highly poisonous agricultural chemicals, was a part of a chemicals factory belonging to the Sandoz works near Basel, Switzerland. As a result, tons of toxic chemicals went into the nearby river, turning it red. People in the city of Basel and the surrounding region on the border between Germany and France were told to stay indoors. Witnesses reported a foul smell of rotten eggs and burning rubber.

It is now known that the chemicals were washed into the river with the water used to fight the fire. About 30 tons of pesticides were discharged into the Rhine, Western

Europe's most important waterway that flows through Switzerland, Germany, France and Holland before flowing into the North Sea.

The harm caused to the bottom living (*benthic*) organisms in the river was disastrous. Most of them were completely eradicated over a distance of 400 km. In addition, despite the fact that after one year most aquatic life had returned to the river, the groundwater of the Rhine alluvium posed a problem for years afterwards.

Both incidents (Minamata and Basel) reveal the extremely poisonous nature of mercury, making it essential in the future to monitor its bioaccumulation along the North Sea in Holland, to be on the alert to prevent any further damage.

6.1.5.3
The Doñana Spill – Southern Spain, 1998

The *Coto Doñana* national park in southern Spain (see Fig. 6.4), an area of 50 000 ha, is one of the most well known bird sanctuaries in the world. It lies at the bottom of the wetland region of the *Marismas del Guadalquivir*, totalling 230 000 ha, most of which is protected under the EU Birds Directive and the regional Spanish administration of natural parks.

On April 25, 1998, a dyke, specially built to retain waste from pyrite exploitation at the *Alnazcollar* mine, collapsed causing a massive spill of 5 000 000 m³ of acid sludge into the *Guadiamar* River. The spill contained high concentrations of arsenic, zinc, lead, copper, thallium, and cadmium (Green and Chase 1998). The incident resulted in great havoc for the environment and caused great loss in tourism – one of the major economic resources of the area. In addition, diversion of the toxic sludge into the Guadalquivir estuary caused great harm to fish and other aquatic organisms. Mechanical removal of the sludge, six months after the incident, was not very successful and could not remove completely the hazardous material spilled into the estuary (Santamaría and Amézaga 1999).

Fig. 6.4 The map of Spain, showing the Geographic position of the Doñana National Park (rectangular frame near Seville)

6.2
Other Inorganic Pollutants

Elements like aluminium, beryllium, and fluorine are considered environmental pollutants if they occur in excessive quantities in soils. Attention was drawn to aluminium, when in July 1988, 20 tons of aluminium sulphate were accidentally introduced into the water supply of the Camelford District of Cornwall. This led to serious toxic effects on fish and other aquatic biota. Excessive concentrations of aluminium and fluorine in soils lead to abnormal concentrations of these two elements in some plants. The tea plant is an example of natural accumulators of aluminium as well as fluorine. Aluminium is linked to some diseases such as Alzheimer (senile dementia) and kidney dysfunction. Since hydrous aluminium oxide is used for water treatment, the World Health Organisation (WHO) set a tolerance level of 200 µg l^{-1} for the concentration of aluminum in drinking water (WHO 1993). Fluorine, despite being added to many public water supplies to improve dental health, can, at concentrations higher than 3–6 mg l^{-1} in drinking water, cause a toxic condition called *skeletal fluorosis*, which deforms the limbs. Crippling occurs at concentrations higher than 10 mg l^{-1}.

Another example is given by arsenic – a metal that has been always extracted and used for different purposes by humans in different parts of the world. Jabir, the Moslem alchemist (8th century) described its extraction from realgar (AS_4S_4) and orpiment (AS_4S_6) – both are still considered as the main ore minerals of arsenic. Pilny (1st century) described the formation of white arsenic (As_2O_3).

Arsenic compounds are used in the glass industry and as a wood preservative. In modern times it is used in the semiconductor gallium arsenades, which can convert electric current into laser light. Arsine gas (AsH_3) is an important dopant gas in the microchip industry. It was sometimes used in the pharmaceutics industry as additive in some medicines.

High exposure to inorganic arsenic can cause various health problems including irritation of the stomach and intestines and decreased production of red and white blood cells. It may also increase the chances of developing skin, lung, liver and lymphatic cancer. It is also linked to infertility and miscarriage in females. A lethal dose of arsenic oxide is generally regarded to be 100 mg. Links of organic arsenic to cancer or DNA damage have not been proven.

Arsenic passes into the environment mainly through emissions from the metallurgical industries; especially from the copper, zinc and lead industries. Once in the environment, it persists unchanged for a long time and can be easily absorbed by plants that in turn pass it to other organisms feeding on them and thus bringing it into the food chain.

Similar to arsenic, chromium is a metal that often accumulates in aquatic life, adding to the danger of eating fish that may have been exposed to high levels of it. *Chromium* – the name deriving from its characteristic property of forming coloured compounds (Greek: χρωμα – colour) – is a metal used in different industries such as preparing metal alloys and making pigments for paints, cement, paper, rubber, and other materials.

In soils, chromium strongly attaches to soil particles and, as a result, it will not move towards the groundwater. It also absorbs on sediments in water and becomes immobile. Only a small part of the chromium that ends up in water will dissolve.

The health hazards associated with exposure to chromium are dependent on its oxidation state. Metallic chromium is of a low toxicity, while its hexavalent form is toxic. According to studies published within the National Toxicology Program (NTP), there is sufficient evidence linking the following hexavalent chromium compounds – calcium chromate, chromium trioxide, lead chromate, strontium chromate, and zinc chromate – to carcinogenic effects in animals.

6.3
Radionuclides

Since 1954 there has been an increasing pollution of soils with radioactive nuclides. These are natural or technically produced elements with unstable nuclei that spontaneously alter their compositions through radioactive decay in a series of successive nuclear reactions, ultimately leading to a stable configuration. Decay of unstable nuclei takes place according to one of the following mechanisms:

- gamma decay: emission of gamma ray
- alpha decay: emission of alpha particle
- beta decay: emission of electron by nuclear neutron
- electron capture: capture of electron by nuclear proton
- positron emission: emission of positron by nuclear proton

The rate at which the nuclei of a sample decay is known as the *activity* of the considered radioactive nuclide. If N is the number of nuclei present in the sample at a given time, its activity R would be given by: $R = -dN/dt$

The SI unit of activity is named after Henri Becquerel, who discovered radioactivity in 1896:

1 Becquerel = 1 Bq = 1 event s^{-1}

In practice, however, the activities encountered are so high that it is more appropriate to use MBq (10^6 Bq) and GBq (10^9 Bq).

The *Curie* is the traditional unit of activity. It is arbitrarily defined as:

1 Curie = 1 Ci = 3.70×10^{10} events s^{-1} = 37 GBq

1 Ci is nearly the activity of 1 g of radium $_{88}Ra$ (few % less).

The majority of radioactive nuclides occurring in nature are members of exactly four radioactive series, with each series consisting of a succession of daughter products all ultimately derived from a single parent nuclide. The reason for this is that α-decay reduces the mass number of a nucleus by 4. Accordingly, nuclides, of mass number

A (mass number) = $4n$

where n is an integer, will be in a position to decay into one another in descending order of numbers. The same holds for nuclides with mass numbers:

Table 6.2 The four radioactive series

Mass number	Series	Parent	Half-life (yr)	Stable end product
$4n$	Thorium	$^{232}_{90}Th$	1.39×10^{10}	$^{208}_{82}Pb$
$4n+1$	Neptunium	$^{237}_{93}Np$	2.25×10^{4}	$^{209}_{83}Pb$
$4n+2$	Uranium	$^{238}_{92}U$	4.51×10^{9}	$^{206}_{82}Pb$
$4n+3$	Actinium	$^{235}_{92}U$	7.07×10^{8}	$^{207}_{82}Pb$

$A = 4n+1, A = 4n+2$ and $A = 4n+3$

This allows them to have exactly four series named after their parent nuclides. Table 6.2 summarises the data of the four series.

Apart from artificial sources of radioactivity in soils, natural sources of radionuclides, such as cosmic radiation or terrestrial radiation derived from radioactive decay of some elements in the Earth's crust, contribute to the radiation emission from landscapes. It is estimated that the natural radioactive emissions from soils have intensively increased during the last 200 years since the beginning of the industrial revolution. This increase is mainly due to the displacement of sediments in the upper crust by mining activities, road building, and other construction activities. Also industrial activities, such as cement production and metal treatment industries, intensify the release of radionuclides into the atmosphere and the upper soil. The more or less even distribution of background values of radioactivity observed in certain regions is due to the absorption of naturally occurring radioactive particles by humans and living organisms through water, air, and nutritive substances; Fig. 6.5 shows these relations in a schematic way. The intensity of background radiation (radiation from natural sources) in a given region depends generally on the geologic conditions as well as on the topographic relations prevailing in the given region. However, soil pollution by artificially produced radioactivity occurs through handling, transporting, testing, and the use of nuclear materials in warfare and industry. The main sources are nuclear power stations, atomic tests, belligerent activities, and major nuclear accidents.

The contribution of atomic nuclear weapon testing to the pollution of the environment with radionuclides is enormous, as it may be clear from Table 6.3, comparing the release from weapon tests to the activity release from the Chernobyl accident.

It is estimated that weapon tests supply most of the radio activity passing into the environment – about 1.2×10^{16} Bq of $^{239,240}Pu$ and 2.8×10^{14} Bq of ^{238}Pu (Hardy et al. 1973). In addition, 9.5×10^{17} Bq of ^{137}Cs were also added (UNSCEAR 2000a,b). A considerable increase of radionuclides in the environment has been caused by the Chernobyl incident 1986. Table 6.3 is a listing of some of these figures.

6.3.1
Speciation and Behaviour of Radionuclides in the Soil System

Like in the general case of other heavy metals, speciation plays an important role in the behaviour, distribution and uptake of radionuclides in the soil environment. An

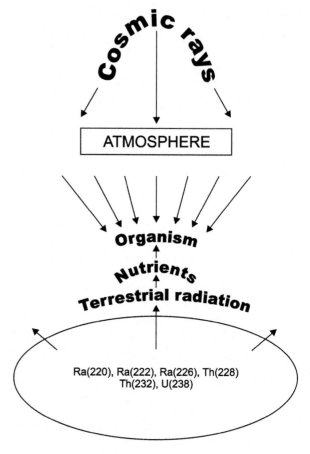

Fig. 6.5 Natural sources of radioactivity

Table 6.3 Nuclides release from weapon tests and the Chernobyl major incident

Nuclide	Release from weapon tests (Bq)	Release from Chernobyl accident (Bq)	Reference
^{238}Pu	2.8×10^{14}	3.5×10^{13}	Hardy et al. (1973), UNSCEAR (2000b)
^{239,240}Pu	1.2×10^{16}	7.2×10^{13}	Hardy et al. (1973), UNSCEAR (2000b)
^{137}Cs	9.5×10^{17}	8.5×10^{16}	UNSCEAR (2000a,b)

example may be given by plutonium; it exists in the soil environment in four oxidation states: III, IV, V and VI, whereby the III and IV generally take the form of simple hydrated cations, while V and VI form linear dioxo species Pu(V)O$_2^+$ and Pu(VI)O$_2^{2+}$ (Choppin 1988). The major part of plutonium in environmental systems is associated with soil, especially the layer forming the uppermost 10 cm of the topsoil.

Generally the distribution of radionuclides in soils depends on several factors including mineralogy, type of soils, as well as the microbiological processes taking place within the profile. So while most plutonium is largely associated with organic matter and colloidal soil oxides (Lee and Lee 2000), we find that ^{137}Cs is preferentially associated with clay minerals in the interlayer exchange sites (Oughton et al. 1992).

6.3.2
Uptake of Radionuclides by Plants

Uptake of radionuclides depends, among other things, on individual physiological characteristic of the plant, the intensity of pollution, bioavailability of the nuclides, as well as the soil type and the pathway followed in transport from soil to the vegetation (surface pathway or deep transport through the root system). Similarly, we find that, according to Watters et al. (1983), the transport of plutonium to plants via surface absorption is 10 times more than that taking place through the root membranes. The effect of soil mineralogy has been substantiated by the work of Vyas and Mistry (1981), who reported a considerable reduction in the uptake of plutonium and americium with the increase in concentration of organic matter and clay minerals. Other factors pertaining to the soil and to the type of microbiological processes taking place within the profile are evident with respect to the uptake of ^{137}Cs, which seems to be considerably influenced by the availability of potassium in the soils and is enhanced by the richness of soil in organic matter. This may be a result of the distribution of organic matter around clay minerals, which form the main accumulation sites for ^{137}Cs.

6.4
Nuclear Debris from Weapon Tests and Belligerent Activities

The distribution of nuclear debris has increased on a worldwide scale since thermonuclear devices first contaminated the Earth's atmosphere, following the nuclear bombardment of Hiroshima and Nagasaki by the United States Air Force at the end of the Second World War.

Pierson (1975) studied the abundance of ^{90}Sr in fallout from the atmosphere from 1955 to 1970. The study revealed (Fig. 6.6) that the distribution curve represents three phases: a phase up to 1959; another from 1961 to 1962; and the third one from 1966–1970. Peaks of deposition rate are observed immediately after the first and second phases, and a levelling during the third phase when the rate of injection into the stratosphere has been in fortuitous balance with the rate of depletion of the reservoir.

A more detailed study was the one carried out by the Federal Agency of Environment (*Umweltbundesamt*) in Berlin, Germany. The study, involving a series of measurements of atmospheric radioactivity between the years 1958 and 1992 (Fig. 6.7), clearly shows a relation between the average daily radioactivity in the atmosphere and surface nuclear tests. The year 1963, for example, was a year of exceptionally high radioactivity in the atmosphere. This is interpreted to be a result of the nuclear tests of the years 1961/1962 in which highly explosive hydrogen bombs were tested on the surface. The retardation of one year, between the increase of radioactivity in Berlin and the tests, can be understood if we consider the residence time of the debris in the

Fig. 6.6 Annual deposition of ^{90}Sr in the period 1955–1970 taken from Fortescue (1980), after Pierson (1975)

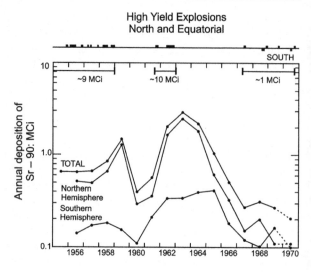

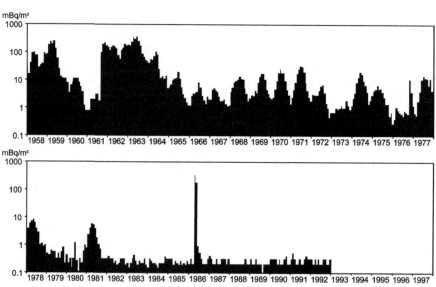

Fig. 6.7 Radioactivity in soils between the years 1958 and 1992 (Umweltbundesamt, Berlin)

upper atmosphere. In the 1970s, the higher values of radioactivity can be attributed to the Chinese and French nuclear tests. In the 1980s, first came a period of decreasing atmospheric radioactivity, where we can observe, for example, that the radioactivity in 1963 was a thousand times higher than that of the years 1982–1985. Later, in the year 1986, an increase of radioactivity, up to the level of the nuclear tests years (1961/1962), was caused by one single incident – namely the major accident of Chernobyl.

6.5
Nuclear Debris from Major Nuclear Accidents

Between the years 1957 and 1993, major nuclear accidents have occurred, contributing to the pollution of soil and atmosphere with nuclear debris. Some of these accidents are listed in Table 6.4 (Alloway and Ayres 1997), together with their level of risk, as given by the scale of the International Atomic Energy Agency. The scale is from 1 to 7 in increasing order of risk.

Inspection of Table 6.4 shows that the accident in Chernobyl was the most serious and perhaps the most investigated, due to its proximity to many European centres with highly sophisticated research facilities. Accordingly, the Chernobyl disaster will be discussed here in some detail.

On April 26, 1986, the core of one of the reactor-blocks in Chernobyl caught fire causing a great increase in temperature and the emission of radioactive nuclides reaching an altitude of up to 1500 m in the atmosphere. As a result, radioactive debris was dispersed, aided by the high wind speed, over an immense area in central and southern Europe, thousands of kilometres away from Chernobyl. Table 6.5 shows the distribution of ^{137}Cs and ^{131}I in some of these countries, together with their distance from the region of the catastrophe.

Table 6.4 Some nuclear accidents as placed on the IAEA scale of risk (after Alloway and Ayres 1997)

Level	Nuclear accident
7	Chernobyl, 1986
6	Ural mountains, waste explosion, 1958
5	Fire at Windscale (Sellafield), 1957 and Three Mile Island, USA, 1979
4	Fatal accidents at Los Alamos, Wood River, and Idaho Falls (1945–1964)
3	Unauthorised release at Vandellos, Spain, 1989; Tomsk-7, Russia, 1993
2	Incidents at UK Magnox stations, 1968, 1983, 1989
1	Management deficiencies in waste reprocessing, Windscale, 1986

Table 6.5 Spread of ^{137}Cs and ^{131}I from Chernobyl. *Source:* Umweltbundesamt, Berlin

Country affected	Distance from Chernobyl (median km)	Activity in soil (kBq m^{-2})	
		^{137}Cs	^{131}I
Austria	1 250	23	120
Norway	2 000	11	77
United Kingdom	2 250	1.4	5.0
France	2 000	1.9	7.0
Ireland	2 750	5.0	7.0

During the first two days, prevailing easterly winds carried the nuclear debris mainly to Hungary and Austria. However, as the wind changed its direction to the south on May 2, parts of Germany were also affected (Fig. 6.8). Measurements of atmospheric radioactivity in Berlin, between April 27 and May 12, 1986, showed that the accident had brought the burden of radioactive nuclides in the atmosphere to maximum values, comparable to those measured in the years 1962/1963 (Fig. 6.9).

Washout by rainfall brought an additional load of radionuclides to the soil. Following the accident, measurements after the first rainfall in some stations in Berlin showed that additional burdens of ^{137}Cs, depending on precipitation intensity, varied between 860 and 5 600 Bq m^{-2}. Additional burdens of radioactivity were between 430 and 2 600 Bq m^{-2}. According to a report published by the World Nuclear Association (London) and the Uranium Information Centre (Melbourne) in 2002, radioactive contamination of the ground was found in practically every country of the northern hemisphere. The European Commission published, on the basis of local measurements, an atlas of contamination in Europe.

Deposition patterns of radioactive particles depended mainly on physical and meteoric parameters, such as the particle sizes and the occurrence of rainfall. This led to deposition of the largest particles within 100 km of the site, whereas fine particles

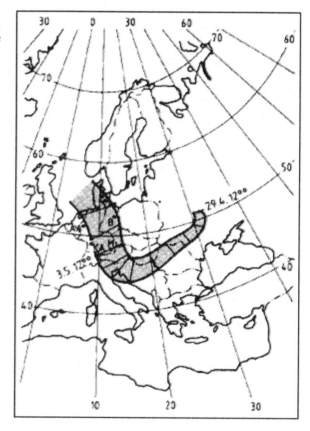

Fig. 6.8 Diffusion of emissions on April 29, 1986, at an altitude of 1 500 m. *Lines* marked in the *shaded area* indicate intervals of 12 hours (Neider 1986)

6.5 · Nuclear Debris from Major Nuclear Accidents

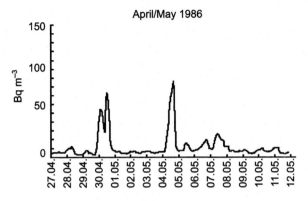

Fig. 6.9 Daily maximum values of natural and artificial radioactivity in the air from April 27 to May 12, 1986. *Source:* Umweltbundesamt, Berlin

Table 6.6 Current estimate of radionuclide releases during the Chernobyl accident. *Source:* NEA (2002)

Core inventory on April 26, 1986			Total release during the accident	
Nuclide	Half-life	Activity (PBq)	Percent of inventory	Activity (PBq)
^{33}Xe	5.3 d	6500	100	6500
^{131}I	8.0 d	3200	50 – 60	~1760
^{134}Cs	2.0 yr	180	20 – 40	~54
^{137}Cs	30.0 yr	280	20 – 40	~85
^{132}Te	78.0 h	2700	25 – 60	~1150
^{89}Sr	52.0 d	2300	4 – 6	~115
^{90}Sr	28.0 yr	200	4 – 6	~10
^{140}Ba	12.8 d	4800	4 – 6	~240
^{95}Zr	1.4 h	5600	3.5	196
^{99}Mo	67.0 h	4800	>3.5	>168
^{103}Ru	39.6 d	4800	>3.5	>168
^{106}Ru	1.0 yr	2100	>3.5	>73
^{141}Ce	33.0 d	5600	3.5	196
^{144}Ce	285.0 d	3300	3.5	~116
^{239}Np	2.4 d	27000	3.5	~95
^{238}Pu	86.0 yr	1	3.5	0.035
^{239}Pu	24400.0 yr	0.85	3.5	0.03
^{240}Pu	6580.0 yr	1.2	3.5	0.042
^{241}Pu	13.2 yr	170	3.5	~6
^{242}Cm	163.0 d	26	3.5	~0.9

PBq = Petabequerel = 10^{15} Bq, yr = year, d = day, h = hour

were carried by wind to far distances to be eventually deposited by rain. The Soviet experts at the time constructed maps showing the distribution of different radionuclides, depending upon the amounts of ^{131}I to deduce the amounts of ^{137}Cs, due to the fact that ^{131}I was easier to measure. These maps are now considered imprecise, even though during the first 10 days ^{131}I was considered as the most important of radionuclides deposited.

The most recent report of the Nuclear Energy Agency (NEA), in 2002, identifies the radionuclides emitted and deposited as shown in Table 6.6.

According to the same report (NEA 2002), the principal physicochemical forms of the deposited radionuclides are dispersed fuel particles, condensation-generated particles, and mixed-type particles. The report points out that, on the forest podzolic soils, migration of ^{137}Cs is pronounced, with increased amounts in the mineral layers.

Chapter 7
Sources of Soil Pollution

The pollution of soil may arise from a wide range of sources. These might be discrete point sources, or diffuse sources, and the pollution process itself may be deliberate, as in fertilisation processes or following an accident, as in the case of radio nuclear accidents or oil spills. Figure 7.1 summarises the main sources of soil pollution.

7.1
Pollutants of Agrochemical Sources

Pollutants from agrochemical sources include fertilisers, manure, and pesticides. We may add to these the accidental spills of hydrocarbons used as fuels for agricultural machines. As it was mentioned before, the main pollution effect, caused by fertilisers and manure, is the introduction of heavy metals and their compounds into the soil. Examples of these are the introduction of arsenic, cadmium, manganese, uranium, vanadium and zinc by some phosphate fertilisers, or soil contamination with zinc, arsenic and copper when poultry or pig manure materials are used. Organic compounds used as pesticides, however, have more far reaching effects for the whole community depending on soil ecology. The use of pesticides in agriculture has been steadily increasing in the last 40 years. Figure 7.2 shows that, except for a short decrease in worldwide sales at the beginning of the 1990s, the market has been growing since 1992 (Taradellas et al. 1997). Pesticide sales reached a market peak in 1998, followed by a period of steep decline that halted in the year 2003 with sales of U.S.$29 390 000. In 2004, a surge in the pesticide market led to record global sales of U.S.$32 665 000. This corresponded to a rise of 4.6% after inflation – the largest single year growth for 10 years.

According to the British Food and Environmental Act, 1985, a pesticide is defined as: *"any substance or preparation prepared or used for any of the following purposes"*:

- Destroying organisms harmful to plants or to wood or other plant products
- Destroying undesired plants
- Destroying harmful creatures

Pesticides applied to plants, or harmful organisms living on soil, may (by successive adsorption and elution) move down the soil column, where they would be bound within the latticework of clay minerals or adsorbed on to soil organics. They may also join the soil water or the gas phase in the interstitial space, if the active ingredients are of suitable volatility.

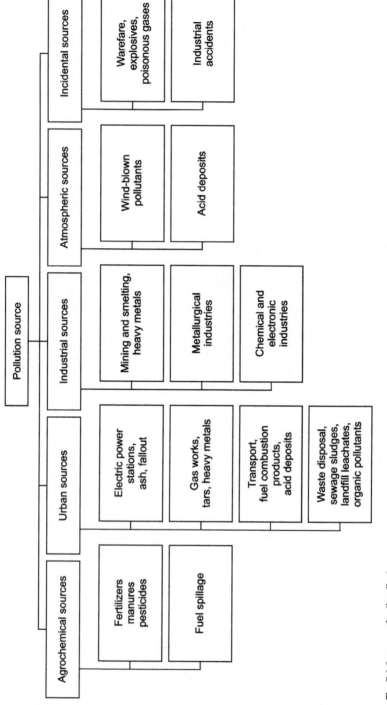

Fig. 7.1 Sources of soil pollution

7.1 · Pollutants of Agrochemical Sources

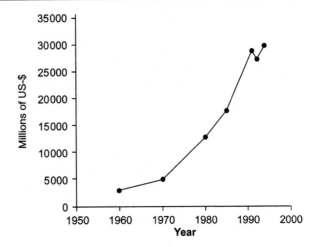

Fig. 7.2 Evolution of pesticide market in millions of U.S.$ between 1960 and 1994

The degree of penetration or sorption of pesticides into the tissues of their living targets, whether animals or plants, provides one of the bases for their classification. According to this, pesticides that remain as superficial deposits, exerting only a local contact action, are known as *contact pesticides*, while those with a local internal movement within the cuticles of leaves, or the epidermis of animals, are known as *quasi-systemic*. Pesticides that directly penetrate through the outer layers and are transported around the organisms of their targets are classified as *systemic pesticides*.

Pesticides are generally classified into the following groups according to their mode of action and the specific organisms they are used to combat:

- *Insecticides*. These are chemical compounds used to kill insects, whether specifically for a particular type or generally for a variety of insects.
- *Herbicides*. Chemicals used to combat or suppress the growth of all or certain types of plants.
- *Fungicides*. Chemicals used to kill or suppress the growth of all kinds or of a certain type of fungus.

7.1.1
Insecticides

The worldwide use of insecticides has been greatly increasing in agriculture and in other fields since the end of the Second World War. Nowadays, there are a great number of commercial formulations for these products, yet they belong principally to four groups of organic compounds, providing a fundamental scheme for their classification. These are *the organophosphorus compounds, the organochlorines, the carbamates*, and *the pyrethroids*.

Organophosphorus Compounds

These are technically nerve poisons, the basic technology of which was developed during the Second World War in Germany and Britain. They are used in many differ-

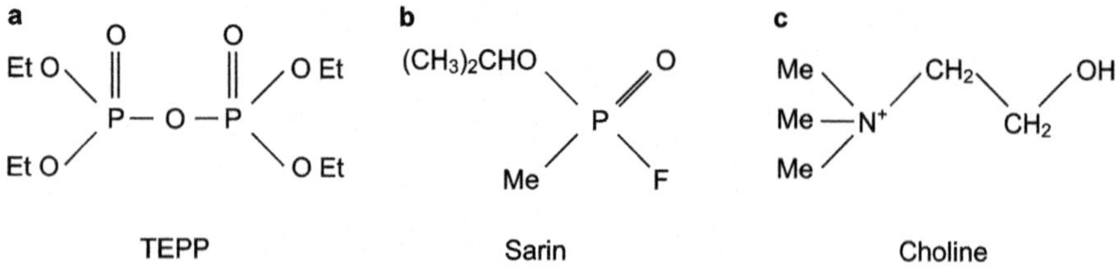

Fig. 7.3 Organophosphorus pesticides

ent ways in agriculture and animal hygiene. Some of them are used as fumigants, others as contact poisons, while others are used as systemic pesticides. Two prominent examples of this group are tetraethyl pyrophosphate – TEPP (Fig. 7.3a) and the warfare agent sarin (Fig. 7.3b), both of which are highly toxic for mammals.

The toxic action of organophosphates arises from their disruption of the nervous system by inhibiting the enzyme cholinesterase, responsible for the establishment of nervous transmission. To this category, we may add a group of organophosphates with an ester function (phosphorothionates) known as *proinsecticides*. These are only toxic to animals, producing high levels of special enzymes, known as *mixed function oxidases* (MFO).

Most organophosphorus pesticides have the general structural formula shown in Fig. 7.3a, where the two alkyl groups R may be methyl or ethyl but they are the same in any given molecule. X, the leaving group, is generally a complex aliphatic cyclic group. Table 7.1, taken from Hassal (1982), shows 6 possible variations of the general formula, with examples of the commercial products related to each of them.

Organochlorines

During the Second World War, a group of organochlorine compounds were found to be very effective in controlling pests responsible for diseases such as malaria and yellow fever. These compounds, being cheap, easy to produce, and (at that time thought to be) safe to man and other warm-blooded animals, were hailed as the best pesticides ever discovered by man. They belong to three chemical families: the DDT (dichlorodiphenyl trichloroethane) family (Fig. 7.4a), the BHC family (Fig. 7.4b) and the cyclodiene family (Fig. 7.4c).

DDT was first described by Othmar Zeidler in 1874, yet its use as insecticide was only established 60 years later by the Geigy chemical industries. The principal representative of the BHC family is often called *Lindane* (after van der Linden, who discovered some of the BHC isomers). It is prepared by adding three molecules of chlorine to benzene activated by UV irradiation and is superior to DDT in controlling soil pests. Aldrin (Fig. 7.5a), dieldrin (Fig. 7.5b) and heptachlor (Fig. 7.5c) are stereochemically related compounds belonging to the cyclodiene family, which were effectively used for controlling locusts.

Despite the fact that organochlorine compounds have been effectively used in the past in agriculture and hygiene, the later discovery (in the late fifties) of their persistence in the environment and their indiscriminate killing of beneficial as well as harm-

7.1 · Pollutants of Agrochemical Sources

Table 7.1 Chemical groups of organophosphorus insecticides after Hassal (1982)

Compound	Structural formula	Example (commercial product)
Organophosphates	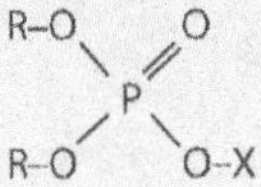 R-O, R-O–P(=O)–O-X	Chlorfenvinphos, dichlorvos, mevinphos, phosphamidon
Thionphosphates (phosphothionates)	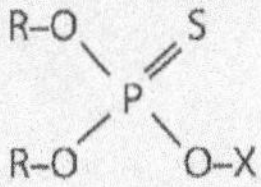 R-O, R-O–P(=S)–O-X	Bromophos, diazinon, fenitrothion, parathion, primiphos (methyl and ethyl)
Thiophosphates (phosphothiolates)	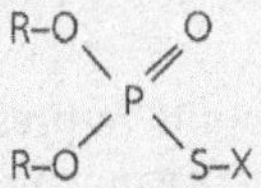 R-O, R-O–P(=O)–S-X	Demiton-S-methyl, oxydemeton-methyl, vamidothion
Dithiophosphates (phosphorothiolothionates)	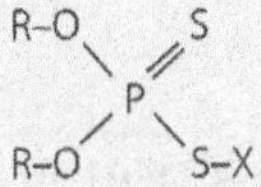 R-O, R-O–P(=S)–S-X	Azinophos-methyl, dimethoate, disulfoton, malathion, menazon, phorate
Phosphonates	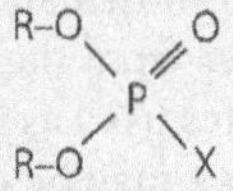 R-O, R-O–P(=O)–X	Trichlorophon, butonate
Pyrophosphoramides	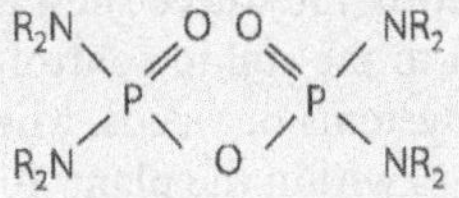 R_2N, R_2N–P(=O)–O–P(=O)–NR_2, NR_2	Schradan

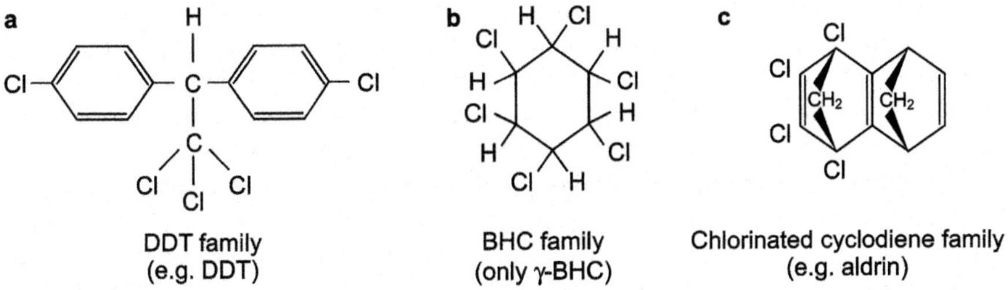

Fig. 7.4 The three families of organochlorine pesticides

a — DDT family (e.g. DDT)
b — BHC family (only γ-BHC)
c — Chlorinated cyclodiene family (e.g. aldrin)

ful insects, has led to an emotional discussion about their use. This ended with a ban on their application in many developed countries. The ban is justified by the fact that the stability resulting from the inactive nature of the C–C, the C–H and the C–Cl bonds forming these compounds, makes them very persistent and hence dangerous for humans and animals. To this, we should also add the observation that due to their partition coefficients that favour the accumulation in biolipids, they tend to accumulate in

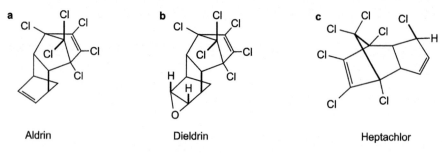

Fig. 7.5 Members of the cyclodiene family

body lipids of organisms exposed to their action. At present, organophosphorus and carbamates insecticides are largely replacing organochlorines.

Carbamates

These are derivatives of carbamic acid NH_2–COOH, of which about 40 commercial compounds, used as insecticides, molluscicides, or nematocides, are on sale. Their toxic effect, like that of the organophosphates, arises from their disruption of the nervous system by inhibiting cholinesterase. Carbamates used as insecticides possess the general structure shown in Fig. 7.6. However, they may be classified according to their mode of action and chemical structure into three sub-groups as shown in Fig. 7.7.

Carbamates are directly applied to the soil to control nematodes and snails, or in order to be absorbed by the root systems of weeds, where they operate as systemic pesticides after being translocated to within the plant. Toxic and health damaging effects of carbamates insecticides have been reported by many authors, e.g. Anger and Setzer (1979).

According to Hassal (1982), mild carbamate poisoning can affect behavioural patterns, reduce mental concentration and slow the ability to learn; protein deficiency accentuates these symptoms. This renders their effect highly dangerous, especially for poor farm workers and children in third world countries, where food shortage and protein deficiencies are always a result of the bad economic conditions.

Beside the above-mentioned synthetic carbamates, some naturally occurring carbamates, e.g. *physotigmine*, were used in studying the toxic effect of carbamate compounds on insects and other organisms; physotigmine is extracted from the Calabar bean.

Natural and Synthetic Pyrethroids

Pyrethroids were originally quite effective natural pesticides, extracted from *Chrysanthemum cineraria folium* – a plant that was for centuries grown specially in Persia to obtain these substances. Nowadays the main producers of natural pyrethrum are Kenya and Tanzania. This is simply because pyrethrum plants give larger yields of pyrethrin

Fig. 7.6 General structure of the carbamate insecticides

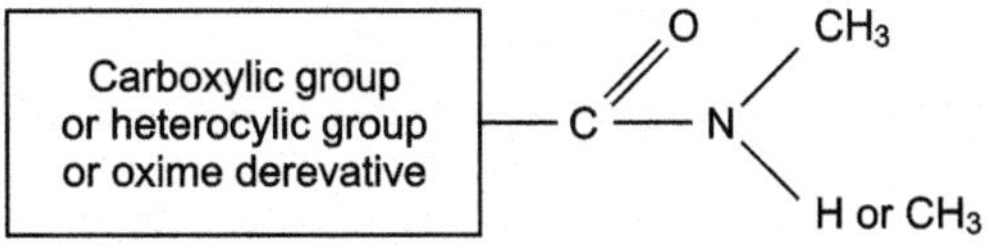

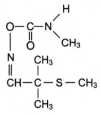

Carbaryl

Subgroup 1:
Aryl N-methyl carbamates
example:
1-naphthyl N-methyl carbamate

Carbofuran

Subgroup 2:
Heterocyclic mono- or dimethyl carbamates;
example:
2,3-dihydro-2,2 dimethyl benzofuran-7-yl N-methyl carbamate

Aldicarb

Subgroup 3:
Oxymes, the OH-group of which has been carbamylated;
example:
2-methyl-2(methylthio) propionaldehyde O-(methyl carbamoyl) oxime

Fig. 7.7 Subgroups of carbamate insecticides

Fig. 7.8 Nicotine structure

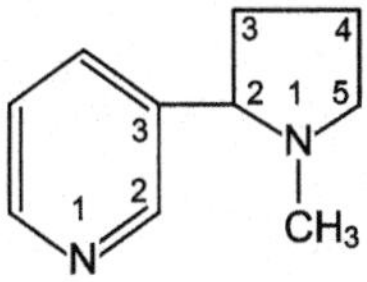

when grown on volcanic ash at high altitudes (1 500–3 500 m) in tropical zones. Natural pyrethroids extracted from the dried pyrethrum flowers comprise of four active ingredients known as *pyrethrins* I and II, and *cinerins* I and II. By comprehending the structure of the natural pyrethroids, it became possible to produce synthetic substances related to the pyrethroids, possessing similar or even higher insecticidal characters than the natural compounds. Some of these are preferred due to their lower toxicity, lower persistence and higher tolerance to light. Synthetic pyrethroids belong to four groups known as the *alethrin, bioresmethrin, permethrin,* and the *fenvalerate* groups.

Some other Natural Insecticides

Beside natural pyrethroids, some other plant-derived compounds were used as insecticides in the Far East and South America. Of these, we may mention nicotine, 1-methyl-2(3'-pyridyl)pyarrolidine (Fig.7.8) and rotenone.

7.1.2
Herbicides

The use of chemical weed-control agents is a disputable problem among environmentalists since selectivity of these agents has never been completely achieved. After 1945, however, a considerable number of commercial organic compounds with some degrees of selectivity have replaced the older traditional herbicides, such as copper sulphate solutions, dilute sulphuric acid and petroleum oil. The main herbicides belong to one of the following groups:

- *Organochlorine compounds.* In this group, one principally encounters derivatives of phenoxyacetic acid (Fig. 7.9a), such as 2,4-dichlorophenoxyacetic acid, known as 2,4-D (Fig. 7.9b); 2,4,5-trichlorophenoxyacetic acid, known as 2,4,5-T (Fig. 7.9c); or 2-methyl-4,6-dichlorophenoxyacetic acid, known as MCPA (Fig. 7.9d).

  Organochlorine derivatives of phenoxyacetic acid mimic natural growth hormones in weeds, leading to over-production of RNA and death of the plants, because their roots will not be able to deliver sufficient nutrition to support their abnormally induced growth. During the war against Vietnam, the US Army sprayed millions of acres of woodlands with an equal mixture of 2,4-D and 2,4,5-T, code-named *Agent Orange*, causing persistent environmental damage.

  Beside derivatives of phenoxyacetic acids, derivatives of aniline major high among organochlorine herbicides. Examples of these are propanil (Fig. 7.10a) and alachlor (Fig.7.10b).

  Both propanil and alachlor are organochlorine derivatives of acetanilide (Fig. 7.9a). The U.S. EPA prohibited the use of alachlor in 1987 due to its carcinogenic character.
- *Organophosphorus herbicides.* Organophosphorus herbicides (known as *glyphosates*) are widely used in agriculture due to their effectiveness against weeds and their non-carcinogenic character. A glyphosate (Fig. 7.11b) is a modified glycine (Fig. 7.11a) – it mimics glycine and hence can be accepted by peptides, where it works as a synthesis inhibitor. It has a half-life in soil of about 60 days and is excreted by mammals unchanged.
- *Derivatives of carbamic acid.* Examples include several derivatives of urea (Fig. 7.12a), such as diuron (Fig. 7.12b); fluometuron (Fig. 7.12c); linuron (Fig. 7.12d); and chlorobromuron (Fig. 7.12e).
- *Triazine derivatives.* Triazines are compounds in which 3 nitrogen atoms are incorporated into the benzene ring (Fig. 7.13a). Derivatives of these, like atrazine (Fig. 7.13b) and simazine (Fig. 7.13c), are used as systematic weed control agents of relatively low toxicity for mammals.

  Water solubility of both atrazine and simazine is enhanced by enzymatic action of soil organisms leading to replacement of the chloro-substituent by a hydroxyl group. The same was also found to occur through dealkylation of these compounds by UV radiation. Accordingly, after discovering that the use of triazine based herbicides polluted water supplies in the Thames Valley, the UK government has banned the use of both compounds. Some EU countries have also done the same.
- *Pyridine derivatives.* In pyridine, one nitrogen atom is incorporated into a benzene ring (Fig. 7.14a). Bipyridyl (Fig. 7.14b), known under the name *Diquat*, is used as systemic herbicide.

7.1 · Pollutants of Agrochemical Sources

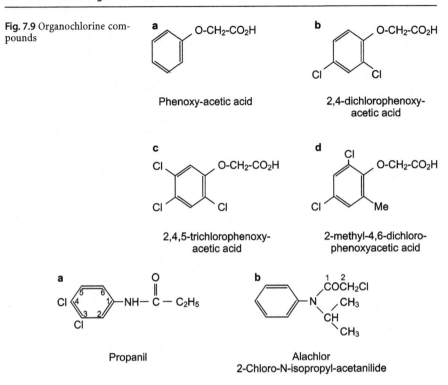

Fig. 7.9 Organochlorine compounds

a Phenoxy-acetic acid
b 2,4-dichlorophenoxy-acetic acid
c 2,4,5-trichlorophenoxy-acetic acid
d 2-methyl-4,6-dichloro-phenoxyacetic acid

Propanil

Alachlor
2-Chloro-N-isopropyl-acetanilide

Fig. 7.10 Aniline derivatives used as organochlorine herbicides

Fig. 7.11 Glycine and its glyphosate derivative

a Glycine
b Glyphosate

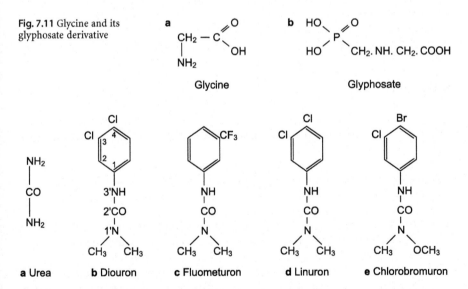

a Urea   b Diouron   c Fluometuron   d Linuron   e Chlorobromuron

Fig. 7.12 Urea herbicides (ureides); a urea; b 3'-(3,4-dichlorophenyl)-1',1'-dimethyl urea; c 3'-(3-trifluoromethylphenyl)-1',1'-dimethyl urea; d 3'-(3,4-dichlorophenyl)-1'-methoxy-1'-methyl urea; e 3'-(4-bromo-3-chlorophenyl)-1'-methoxy-1'-methyl urea;

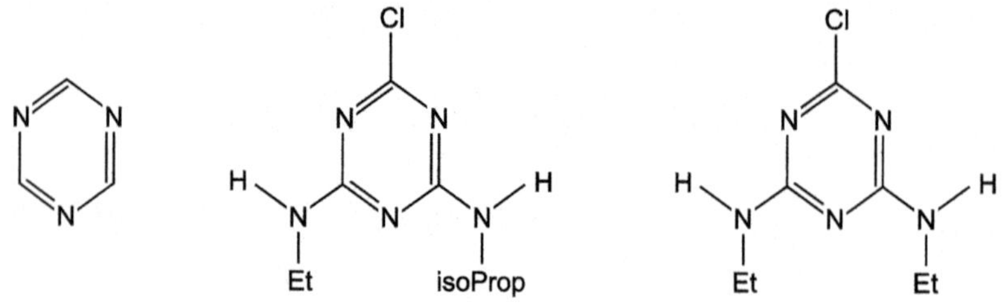

Fig. 7.13 Triazine derivatives

a Triazine   b Atrazine   c Simazine

Fig. 7.14 Pyridine derivatives

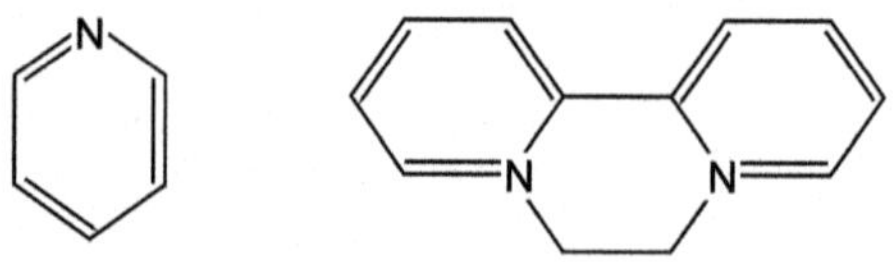

a Pyridine   b Diquat

Fig. 7.15 Dalapon

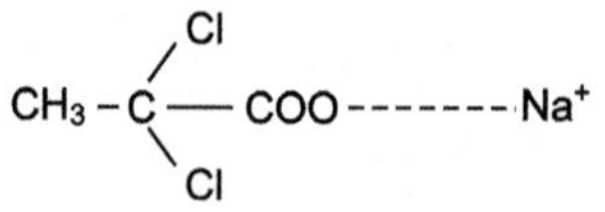

- *Aliphatic compounds.* There are few aliphatic compounds used as herbicides. Of these, the product known under the commercial name *Dalapon* (Fig. 7.15) was found useful in controlling the couch grass. It is not persistent because of it being readily hydrolysed to pyruvic acid (Fig. 9.21).

7.1.3
Fungicides

Fungicides are a group of chemicals, ranging from inorganic to organic compounds of comparable structures as the previous pesticides. Of these, the followings are examples:

- *Inorganic and organic compounds of heavy metals.* Examples are mixtures of copper bearing inorganic compounds (e.g. Bordeaux mixture), or organometallic compounds such as organotins, which may be represented by tributyltinacetate (Fig. 7.16a) or triphenyltinacetate (Fig. 7.16b).
- *Derivatives of phthalic acid.* Example here is given by phthalimide (Fig. 7.17), which is a compound produced by the reaction of phthalic acid with ammonia. This is marketed under several commercial names (e.g. Captan, Captafol).
- *Benzimidazoles.* Benzimidazole (Fig. 7.18), a compound related to histamine, known for its blood pressure reducing character, is used as a systemic fungicide. The pentagonal ring in histamine is known as an *imidazole ring*. Its fusion with a benzene nucleus gives the benzimidazole.

Fig. 7.16 Structure of some organotins

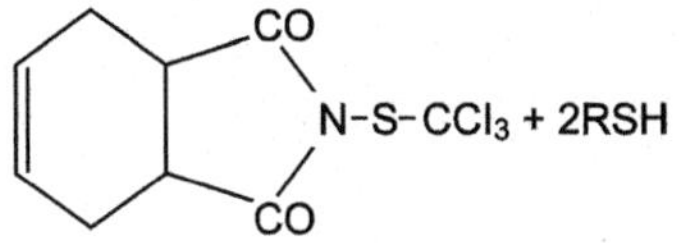

$(But)_3-Sn-O-\overset{\overset{O}{\|}}{C}-CH_3$ **a** Tributyltinacetate

$(Ph)_3Sn-O-\overset{\overset{O}{\|}}{C}-CH_3$ **b** Triphenyltinacetate

Fig. 7.17 Phthalimide

Captan (N-S-CCl$_3$ + 2RSH)

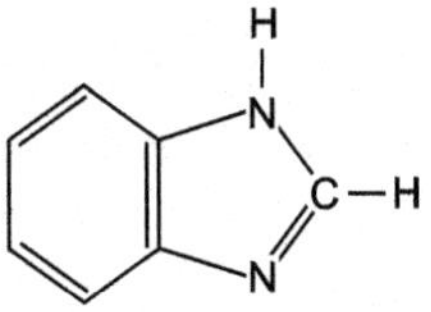

Fig. 7.18 Structure of benzimidazole

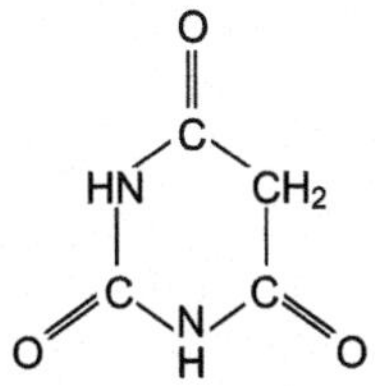

Fig. 7.19 Structure of barbituric acid

- *Derivatives of barbituric acid.* Barbituric acid (Fig. 7.19), on treatment with phosphorus oxychloride followed by reduction with hydroiodic acid, gives a group of compounds known as *the pyrimidines*; these are used as fungicides.

7.1.4 Fuel Spills in Farms

Through accidents or the careless handling of fuel in farms, soils may be polluted. Fuels used in agricultural machines are mostly petroleum products that may contain organic contaminants like benzene, heptane, hexane, isobutane, toluene, phenol, tetraethyl, and tetramethyl lead and zinc (anti-knocking compounds). Soil pollution by petroleum hydrocarbons will be discussed later under a separate heading.

7.2 Soil Pollutants of Urban Sources

Soil pollution by materials of urban sources is a problem as old as urbanisation itself. Archaeological studies show that, through the construction and demolition of domestic concentrations and public centres of human activities (temples, sport arenas, etc.), a great deal of polluting substances were always dumped, or disposed of, on soils, resulting in their physical or chemical degradation. The damage of soil in those ancient days was of a limited scale, yet since the beginnings of the industrial revolution it has taken dimensions that are hardly controllable in modern times. According to Bridges (1991), a considerable quantity of construction materials (concrete, gypsum, asbestos, etc.) may come into contact with the water table and ultimately lead to changes in the chemistry of the soil waters. The main sources of urban soil pollution, however, are power generation emissions, releases from transport means and waste disposal.

7.2.1
Power Generation Emissions

Emissions from power generation plants include Co_x, NO_x, SO_x, UO_x and polycyclic aromatic hydrocarbons (PAHs, see Fig. 7.20) from coal-fired power stations and radionuclides from nuclear power plants. These may be introduced into the soil either directly as fallout (dry deposition) or in a wet form after being dissolved in precipitation.

A number of organic and inorganic soil pollutants, including tars, cyanides, spent iron oxides cadmium, arsenic, lead, copper, sulphates and sulphides, may be released in sites of abandoned gas stations. The most abundant radionuclides found in soils, originating from nuclear power generation, are ^{137}Cs and ^{134}Cs.

In soils with a high CEC and pH-values near 7.0, these radionuclides are normally absorbed onto clays and humic materials.

Electric power generation in coal-fired power plants contributes not only to the addition of inorganic and organic pollutants to the soil through air born fly ash, but also adds to the radioactive nuclide content of the soil. In the USA, many studies have been done on the concentration of uranium in fly ash, showing that uranium in fly ash may reach concentrations of between 1–10 ppm (see Fig. 7.21)[1]. Despite the fact that these concentrations may not represent severe danger to individuals and life in general, chemical conditions under which uranium may be leached from fly ash and be concentrated in soil are still not completely understood.

Studies done in Germany show a high potential of pollution by heavy metals through deposition of fly ash on soils. Table 7.2 shows the ratio of heavy metals in emissions from coal-fired power stations in comparison to the content of the same in total emissions in West Germany.

7.2.2
Soil Pollution through Transport Activities

Transport activities, in and around urban centres, constitute one of the main sources of soil pollution, not just because of the emissions from internal combustion engines and petrol spills, but rather from these activities and their accompanying changes as a whole. To explain this, we should consider the breathtaking increase in highway con-

---

[1] Central Region Energy Team- Fact Sheet FS-163-97, October 1997

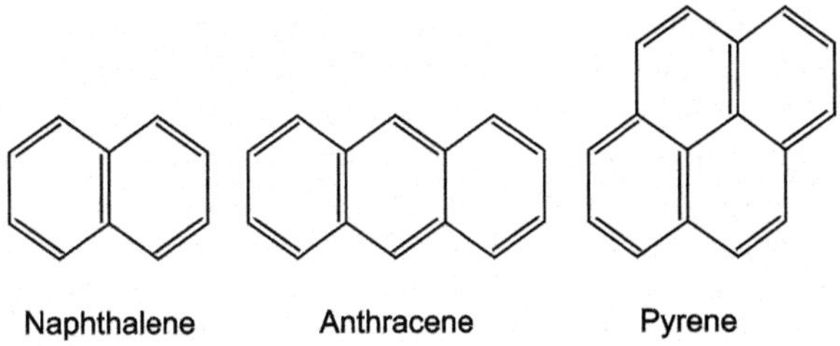

Fig. 7.20 Some polycyclic aromatic hydrocarbons (PAHs)

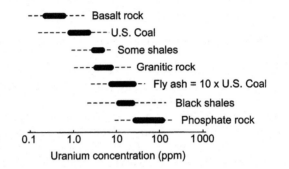

Fig. 7.21 Uranium in fly ash as compared to other Earth materials

Table 7.2 Contribution of heavy metal emissions from coal-fired public power plants to total emissions in the western part of Germany

Pollutant	Contribution (wt%)	
	1982	1990
As	38	27
Cd	7	7
Cr	12	4
Cu	22	8
Hg	11	14
Ni	5	4
Pb	8	1
Se	1	1
Zn	7	6

struction projects all over the world. One also should not ignore the secondary or satellite land use activities attracted to the sites of newly constructed highways, such as gas stations, shopping centres and all other services offered to car owners and commuters.

In fact, the impact of highways on the hydrogeologic environment may cause considerable transformations on the terrain, leading to the physical and/or chemical degradation of soil. According to Richard R. Parizek (1971) these may be summarised in the following:

- Water quality changes due to sediment damage to surface and groundwater supplies.
- Pollution due to highway activities such as accumulations of oils, chemicals, and hazardous substances through accidental spills.
- Pollution resulting from maintenance activities requiring the use of chemicals, such as weed and insect control compounds, as well as salts used to control the formation of ice in winter.
- During highway construction, road cuts may expose pyrite-bearing strata that in turn would produce acid and other chemically polluted waters.
- Enhanced new economic activities attracted to the highway site may result in producing huge amounts of roadside litter and debris.

The principle contribution of transport activities to soil pollution is caused by the emissions from vehicles and aeroplanes, especially supersonic ones. Emissions from all transportation means, driven by internal combustion engines, include oxides of carbon, nitrogen and sulphur as well as some heavy metals. These pollutants may be transported to the soil by deposition of particulate matter or by being washed from the atmosphere. Table 7.3 shows, as an example, the yearly amount of pollutants emitted by vehicles in the region of Berlin, Germany (reference year 1993).

On oxidation by photochemical reactions in the atmosphere, sulphur and nitrogen oxides react with water droplets in the air to produce strong acids such as HNO_3 and H_2SO_4. These acids produce, by reaction with bases (existing in the atmosphere mainly as particulate matter), a mixture of basic and acid radicals that dissolve in the rain, forming what has been known as the phenomenon of acid rain, causing great devastation in soils and plants.

As the concentration of these radicals, together with carbon dioxide in surface and pore water, approaches equilibrium, a great deal of change in the chemical environment of the soil takes place, leading to a drop in pH and to an increase in the acidity of the soil. As a result, an increase in the intensity of weathering, combined with the release of toxic aluminum ions from clay minerals, as well as the leaching of nutrients from the upper soil, may take place. Figure 7.22 shows a summary of the process involved.

Table 7.3 Yearly total emissions by vehicles (motorcycles are not included) in the region of Berlin: Total transportation capacity 12151.8 million vehicle-km yr^{-1} and total fuel consumption of 1138046.1 tons (reference year 1993). *Source:* Umweltbundesamt, Berlin

Emission	Total weight (t)
Hydrocarbons	25 461.7
Benzene	1 279.5
Carbon dioxide	3 424 519.4
Carbon monoxide	144 196.5
Nitrogen oxide	19 024.8
Exhaust particles	1 135.5
Abrasion dust (tyres)	1 290.9
Elemental carbon (exhaust + tyre abrasion dust)	873.5
Sulphur dioxide	1 398.2

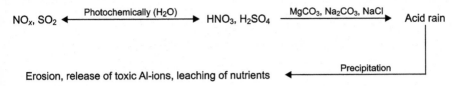

Fig. 7.22 Formation of acid rain

7.2.3
Soil Pollution by Waste and Sewage Sludge

Of all urban sources contributing to soil pollution, waste and sewage sludge disposal occupy a central role in this environmental problem. In highly developed OECD countries, in spite of retreating rates of population growth, the production of waste is still increasing, especially in the industrial sector. In developing and under-developed countries, high rates of population growth and increasing waste and sludge production, combined with lack of municipal services, create a dangerous situation. Even for some of the OECD countries like Poland and Hungary, this is still posing a problem. The percentage of the population served by municipal waste services in these two countries, during the early nineties, was around 55% for Poland and 36% for Hungary compared to 100% in most of the EU countries and the United States. Waste produced by households is known collectively as municipal waste, in order to differentiate it from waste originating from industrial processes. It includes various types of materials that may contribute to changing the environment of soil. Table 7.4 shows the composition (%) of municipal waste in both France and Turkey in the year 1993 as published by the OECD.

Municipal waste disposal by landfills and incineration may in both cases lead to a concentration of heavy metals, such as cadmium, copper, lead, tin and zinc, either directly from landfill leachates that may be polluting soil and under groundwaters, or by ash fallout from incinerating plants. To this we may add the effect of landfill gases that may pass to neighbouring soils, causing a change in their soil air environment.

The disposal of sludge produced by sewage treatment poses a great problem as well, since in almost all developed countries the disposal of this sludge by dumping it at sea is being phased out and the principal method of disposal is now shifting to land use. In fact, the mere use of sludge to amend soils is an advantageous process in itself. It adds essential organic matter as well as useful nutritive elements like phosphorus and nitrogen to the soil. Yet pollutants such as heavy metals, which are normally concentrated in the sludge, may accumulate within the soil and eventually be taken up by food crops such as leafy vegetables, which are known to preferentially take up cadmium – one of the heavy metals that are normally abundant in sewage sludge.

To reduce the hazard of soil pollution through sewage sludge, the EC-Directive 86/278/EEC has set the maximum permissible concentrations of heavy metals and other elements in sewage sludge amended soils. Table 7.5 shows some of these figures.

Beside heavy metals, sewage sludge may include various organic micro-pollutants such as PAHs (polycyclic aromatic hydrocarbons), PCDDs (polychlorodibenzo-p-dioxin – Fig. 7.23), and PCDFs (polychlorodibenzofuran – Fig. 7.24).

Table 7.4 Municipal waste in both France and Turkey in the year 1993, as published by the OECD

Country	Paper and paper board	Food and garden waste	Plastics	Glass	Metals	Textile and others
France	30	25	10	12	6	17
Turkey	6	64	3	2	1	24

Table 7.5 The EC maximum permissible concentrations of heavy metals in sewage sludge amended soil (taken from Alloway and Ayres 1997)

Element	Maximum concentration (mg kg^{-1} dry solids)	Maximum rate of addition over 10 year period (kg ha^{-1})
Cd	3	0.15
Cr	400 (provisional)	15 (provisional)
Hg	50	3
Pb	300	15
Zn	300	15
Ni	50	3

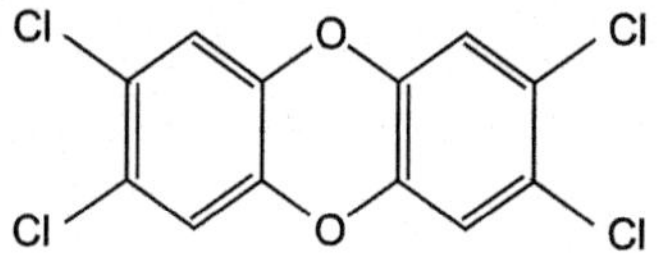

Fig. 7.23 2,3,7,8-TCDD (polychlorodibenzo-*p*-dioxin)

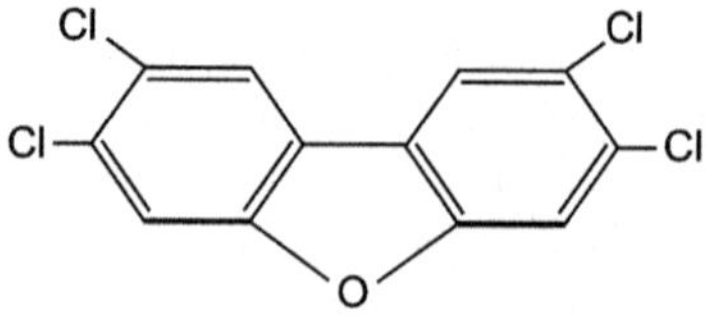

Fig. 7.24 2,3,7,8-tetrachlorodibenzofuran

PCDDs, or simply *dioxins*, are represented by over twenty isomers of a basic chlorodioxin structure and can be differentiated from each other through the number and positions of the chlorine atoms in a molecule. The most common form of dioxins is the 2,3,7,8-tetrachlorodibenzo-*p*-dioxin (Fig. 7.23). Dioxins are considered the most toxic of man-made chemicals.

PCDFs such as 2,3,7,8-tetrachlorodibenzofuran (Fig. 7.24) compare in toxicity to 2,3,7,8-tetrachlorodibenzodioxin and are both considered as examples of the most lethal synthetic chemicals.

The above-mentioned substances are synthetic chemicals and none of them have been found to form as a result of any natural process. Their main sources are the following activities:

- municipal waste incineration
- chemical industry
- coal combusting power plants
- iron and steel industry
- car traffic
- hospital ovens
- forest industry

7.3 Soil Pollution through Chemical Warfare

Use of poisonous chemicals or irritating smokes against rival troops is as old as war itself. Reports about poisoning water resources or burning sulphur to irritate the enemy are known from battles dating back to the ancient Greeks. Indeed, like a modern biological and chemical attack, the curse of Moses on the Egyptians appeared when he inflicted them with the plague of red tide (probably producing neurotoxins) that poisoned their waters and killed their fish. The Bible vividly reports on this, using the following words: "... *and the waters that were in the river were turned to blood. And the fish that were in the river died; and the river stank and the Egyptians could not drink of the water of the river*" (Exodus 7:20–21).

Yet the systematic use of lethal chemical weapons, as they are known today, is a relatively recent matter. It started and was developed by European chemists during the early stages of the First World War (1914–1918). At the beginning, the French used shells filled with ethyl bromoacetate in August 1914, and the Germans followed on October 27, 1914, at Neuve-Chappelle by using the *Ni-Schrapnell* 105 mm shell, which consisted of lead balls embedded in powdered *o*-dianisidine chlorosulfonate.

However, the turning point, which most historians consider as the starting event of modern systematic chemical warfare, came on the April 22, 1915, at 5 P.M., when the Germans discharged, 180 000 kg of chlorine gas at Ypres from 5 730 cylinders on the line between Steenstraat on the Yser Canal through Bixschoote and Langemark, to Polecappelle. The gas cloud, carried by the wind, forced the French and Algerian troops in the opposing trenches to flee after suffering heavy casualties. Professor Fritz Haber, chief of the German chemical warfare service during World War I, directed this attack, which was the first of its kind. Haber, a chemistry professor, Nobel laureate and famous for his discovery of ammonia synthesis by the combination of nitrogen and hydrogen, is often referred to as the father of modern chemical warfare.

After a second attack on April 24, 1915, against Canadian troops at Ypres, the Germans employed chlorine for the first time on May 31, 1915, on the eastern front at Bolimow, near Skierniewice, 50 km southeast of Warsaw. For this attack, they employed 12 000 cylinders, releasing 264 tons of chlorine along a 12 km line. It is assumed that during World War I, nearly 200 chemical attacks using gas released from cylinders were carried out, the largest of these occurred during October 1915 when the Germans released 550 tons of chlorine from 25 000 cylinders at Reims.

It is estimated that, beside the grievous environmental pollution caused by chemical weapons during World War I, the employment of 125 000 tons of chemical warfare agents caused about 1 296 853 casualties. A great number of people in battle regions developed serious symptoms that lasted for lengthy times after the war.

The use and advancement of chemical weapons in World War I was only a gambit for the horrific developments in this field during World War II and the subsequent years, known as the years of the cold war. During those years, chemists, armed with the experience and knowledge they had collected during World War I, discovered lethal agents that were increasingly effective in mass killing and in destroying natural resources. The development and use of herbicides and nerve agents culminated and showed its horrible face in the use of defoliation agents in Vietnam by the Americans, causing the pollution of immense forest regions and the genetic damage of a people for many generations to come.

The history of nerve agents goes back to the few years preceding World War II, to the end of 1936 when Dr. Gerhard Schrader of the *I.G. Farbenindustrie* laboratory in Leverkusen first prepared Tabun (ethyldimethylphosphoramidocyanidate, see p. 157). Tabun, a nerve poison, was very quickly identified by the Nazis as a potent warfare agent and in 1942 they started producing it on a mass scale. By the end of 1944, the Nazis had produced 12 000 tons of Tabun: 2 000 tons loaded into projectiles and 10 000 tons loaded into aircraft bombs. They stockpiled this arsenal mainly in Upper Silesia and in abandoned mineshafts in Lausitz and Saxony. The Red Army approaching Silesia in August 1944 forced the Germans to flee, abandoning the production sites and simply pouring tons of liquid nerve agents into the River Oder. It is believed that the Soviets captured both the full-scale Tabun plant and the pilot plant of another nerve poison – Sarin, which, like Tabun, is an organophosphorus compound (*o*-isopropylmethylphosphonofluridate – Fig. 7.3b). According to some reports, the Soviets resumed production at both captured plants in 1946.

The Americans were also active in developing new nerve agents during the fifties and early sixties of the last century. The main compound of these was used, under the code name *Agent Orange*, as a defoliating agent in Vietnam (Young and Reggiani 1988). Agent Orange is a mixture of herbicides, containing equal amounts of 2,4-dichlorophenoxy acetic acid (2,4-D) and 2,4,5-trichlorophenoxy acetic acid (2,4,5-T) – see Fig. 7.25. Operations of the United States Air Force against Vietnam, involving the use of Agent Orange, were stopped in May 1970, after opposition grew inside the USA.

Realising the potentially catastrophic consequences of chemical warfare for humanity, the world powers started negotiating a Convention on the *Prohibition of the De-*

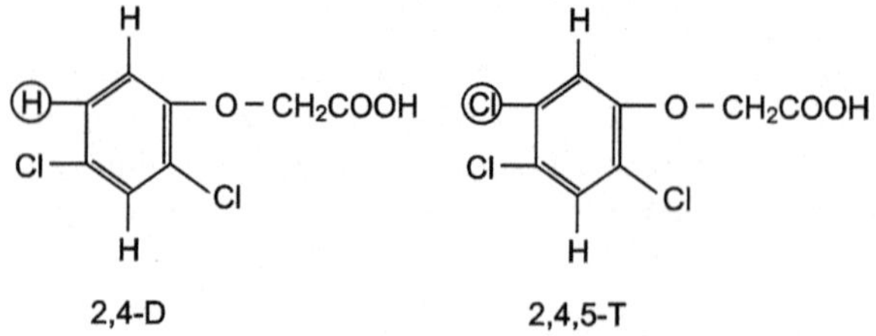

Fig. 7.25 The two main ingredients of Agent Orange

velopment, Production, Stockpiling, and Use of Chemical Weapons* and on their destruction. After twenty years of negotiations, the convention known as the *CWC* (Chemical Weapons Convention) was opened for signature in Paris, France on January 13, 1993, and entered into force on April 29, 1997.

7.3.1
Pollutants, Toxic Chemicals, and Chemical Weapons

According to the CWC, a toxic chemical is defined as: *"Any chemical, which through its chemical action on life processes can cause death, temporary incapacitation or permanent harm to humans or animals. This includes all such chemicals, regardless of their origin or of their method of production, and regardless of whether they are produced in facilities, in munitions or elsewhere."* Precursor materials used to produce toxic chemicals were also precisely defined in the CWC as: *"Any chemical reactant which takes part at any stage in the production, by whatever method, of a toxic chemical. This includes any key component of a binary or multi-component chemical system."*

In a classification of three categories (schedules), toxic or potentially toxic substances were classified according to their potentiality of being used as chemical weapons, or precursors of these. The schedules do not define the chemicals in terms of specific properties, but classify them according to general accepted usage. Where a chemical is placed in the schedules depends upon whether or not its use as a chemical weapon is legal or illegal according to the Ratified CW Convention (David R. Huff 1997):

- Schedule 1: Chemicals can be listed here, if few or no peaceful uses have yet been identified for any members of them.
- Schedule 2: Chemicals listed here are Dual-Use Chemicals of Limited Use.
- Schedule 3: Chemicals listed here are Dual-Use Chemicals of Extensive Use.

The Convention allows for future changes of listing, if required by technical or legal developments.

7.3.1.1
Schedule 1: Chemical Warfare (CW) Agents and Their Precursors

In this schedule, we can find all substances that had been produced, stockpiled, or used as chemical weapons, as defined in Article II of the convention. Furthermore, substances of similar chemical structure that may be potentially of use as chemical weapons, or can be precursors to them, are included. The following toxic chemicals and their precursors were listed on Schedule 1:

Organophosphorus Compounds
All nerve agents chemically belong to this group. They have rapid effects both when absorbed through the skin and via respiration (for a short description of these compounds see p. 139). The following organophosphorus compounds are on Schedule 1:

Fluorinated Organophosphorus Compounds
O-Alkyl ($\leq C_{10}$, incl. cycloalkyl) alkyl (Me, Et, *n*-Pr, *i*-Pr)-phosphonofluoridates.

- Example 1: *Sarin* (also written *Zarin*) – a colourless liquid, odourless in pure form and readily soluble in water and all organic solvents. It is a powerful cholinesterase inhibitor, having the chemical formula: $C_4H_{10}FO_2P$ (for structure see Fig. 7.26a).
- Example 2: *Soman* (also having the code name *GD*) – has the chemical name methylphosphonofluoridic acid, 1,2,2-trimethylpropyl ester and the molecular formula $C_7H_{16}FO_2P$ (for structure see Fig. 7.26b).

  Soman was discovered in Germany in 1944, where its laboratory testing was in progress at the end of World War II. Thus, it has never been used in combat, but was produced and stockpiled by the Soviet Union after the war. Soman, a colourless liquid when pure, has a yellow-brown colour as an industrial product (Franke 1967). The pure compound has a fruity odour, but as an industrial product, it may have a camphor-like odour, resulting from impurities.

Esters of Dimethylphosphoramidocyanidic Acid
O-Alkyl ($<C_{10}$, incl. cycloalkyl) *n,n*-dialkyl (Me, Et, *n*-Pr or *i*-Pr) phosphoramidocyanidates.

- Example: *Tabun* – a colourless to brown liquid, discovered by Schrader in 1936, Tabun is only slightly soluble in water but soluble in all organic solvents. It is an *o*-ethyl-*N,N*-dimethylphosphoramidocyanidate, having the chemical formula $C_5H_{11}N_2O_2P$ (for structure see Fig. 7.27).

Sulphonated Organophosphorus Compounds
O-alkyl (incl. cycloalkyl) S-2-dialkyl (Me, Et, *n*-Pr or *i*-Pr) aminoethyl alkyl (Me, Et, *n*-Pr or *i*-Pr)phosphonothiolates and corresponding alkylated or protonated salts

- Example: *VX* – The lethal nerve agent codenamed VX has the chemical name methylphosphonothioic acid, S-[2-[bis(1-methylethyl)amino]ethyl]-*o*-ethylester, with the molecular formula $C_{11}H_{26}NO_2PS$ (for structure see Fig. 7.28a).

  VX belongs to the phosphorylthiocholine class of compounds, which was discovered independently by Ranaji Ghosh of ICI, by Gerhard Schrader of Bayer, and by Lars-Erik Tammelin of the Swedish Institute of Defence Research in 1952–1953. As a result of intensive research by the United States at Edgewood Arsenal, VX was developed and stockpiled by the United States. A closely related compound referred to as V-gas was also manufactured and stockpiled by the Soviet Union. V-gas has the structural formula shown in Fig. 7.28b.

Mustards
Mustards are considered as the most lethal of all poisonous chemicals used during World War I; once in the soil, they remained active for several weeks.

Mustard gas (a popular name for sulphur mustards, even though they are all liquids and not gases) causes skin blistering and internal bleeding, combined with degradation of the mucous membranes. It attacks the respiratory tract and lungs, causing pulmonary edema. Effects of exposure to mustard gas may be delayed for up to 12 h, yet results normally in death several days after exposure. The burning sensation it causes upon contact with the skin is similar to that caused by oil from black mustard seeds – hence the name.

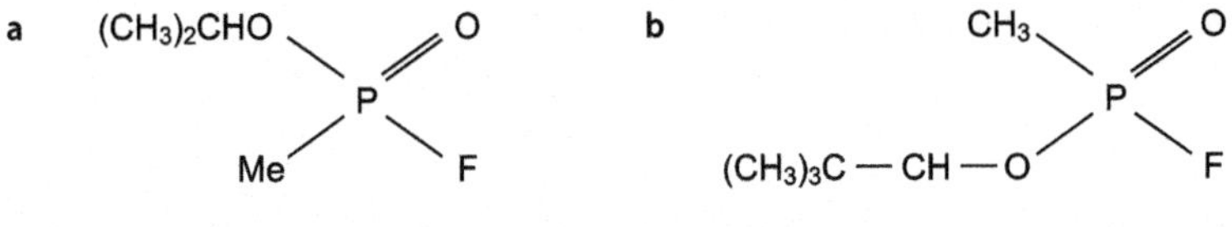

7.3 · Soil Pollution through Chemical Warfare

Fig. 7.26 Structural formula of; **a** sarin; **b** soman

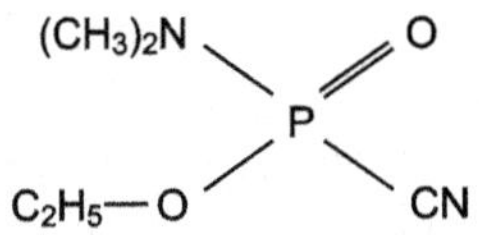

Fig. 7.27 Structural formula of Tabun

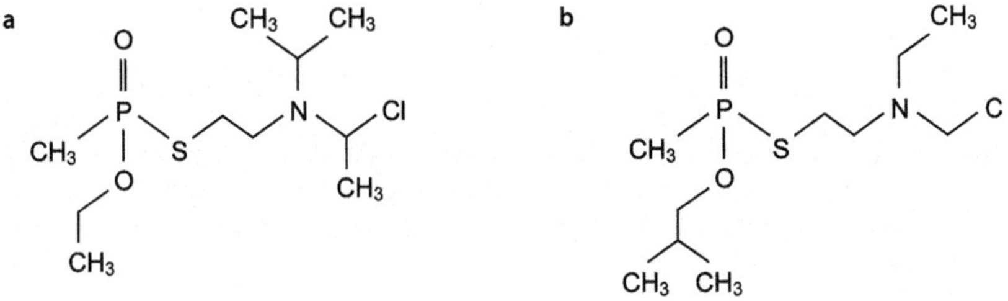

Fig. 7.28 Structural formula of; **a** VX; **b** V-gas

More recent research results show that mustard gas caused genetic damage in all systems in which it was tested. This includes bacteria, fungi, insects, cultured rodent cells, and mice exposed in vivo. It caused DNA and RNA damage in mice (ATSDR 2001).

Chemically, mustard gas is a thioether, 2,2'-dichlorodiethylsulphide, $(ClCH_2CH_2)_2S$. It can be prepared by reacting ethylene with sulphur monochloride, S_2Cl_2, or by other methods.

However, we can generally define Mustards as a group of chemicals that contain a 2-chloroethyl group attached to a sulphur atom (reactive alkyl chlorides) or to a nitrogen atom. Accordingly, one can speak of sulphur mustards and nitrogen mustards.

Sulphur Mustards

Sulphur mustards belong to the group of chlorinated thioethers. They may be regarded as derivatives of hydrogen sulphide, in which chlorinated alkyl groups replaced the two hydrogen atoms.

- Example 1: *Mustard gas* – bis(2-chloroethyl)sulphide has the chemical formula $C_4H_8Cl_2S$ (for structure see Fig. 7.29).

    Mustard gas (also known as H, yperite, sulphur mustard, *Kampfstoff Lost*) was the most important vesicant poison gas used during the First World War.

    The Germans first used it on the night of July 12–13, 1917, near Ypres in Flanders (Paxman and Harris 1982). The French followed in June 1918, while the British extended their arsenal to include it in September 1918. Documented uses of mustard in later times, according to Compton (1988), include Morocco, 1925 (by the French); Ethiopia, 1935 (by the Italians); China between 1934 and 1944 (by the Japanese) and

Fig. 7.29 Structural formula of mustard gas

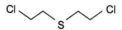

during the Iran-Iraq War (by both sides). Although no chemical warfare agents were used during World War II in Europe or in the Pacific (Franke 1967), there was a release of mustard into Bari harbour in Italy in 1943 (Compton 1988).

During the years of the cold war, the Soviet Union and the United States stockpiled considerable arsenals of this lethal weapon. Daily U.S. production of mustard gas at the end of World War I reached about 18 000 kg (40 000 lb). In 1986, about 34 million pounds (15 400 metric tons) of this terrible weapon were already stockpiled by the American Department of Defense (ATSDR 2001). According to official information, the United States no longer produces, imports, or exports mustard gas, and signed the International Chemical Weapons Convention treaty in 1997, which mandates destruction of all chemical weapons by 2007 (USCWC 2003).

Mustard gas – or *Kampfstoff Lost*, as the Germans used to call it – is a heavy liquid (b.p. 217 °C). Its vapour attacks skin and lung tissues producing blisters and burns. It can be rendered innocuous by bleaching powder. As a soil pollutant, however, mustard gas has a dangerous effect, since it can persist for decades in polluted lands (even if it is covered by water).

The following compounds belong to the sulphur mustards and are all listed on Schedule 1 of the CWC as chemical weapons:

- bis(2-chloroethylthio)methane (Fig. 7.30a)
- 2-chloroethylchloromethylsulfide (Fig. 7.30b)
- sesquimustard: 1,2-bis(2-chloroethylthio)ethane (Fig. 7.30c)
- 1,3-bis(2-chloroethylthio)-*n*-propane (Fig. 7.30d)
- 1,4-bis(2-chloroethylthio)-*n*-butane (Fig. 7.30e)
- 1,5-bis(2-chloroethylthio)-*n*-pentane (Fig. 7.30f)
- bis(2-chloroethylthiomethyl)ether (Fig. 7.30g)
- *o*-mustard: bis(2-chloroethylthioethyl)ether (Fig. 7.30h)

Nitrogen Mustards
- HN1: bis(2-chloroethyl)ethylamine (Fig. 7.31a)
- HN2: bis(2-chloroethyl)methylamine (Fig. 7.31b)
- HN3: tris(2-chloroethyl)amine (Fig. 7.31c)

All the nitrogen mustards are liquids with fish odour and are either practically insoluble in water, e.g. HN1, sparingly soluble (HN2: 0.16 g l^{-1}), or slightly soluble (HN3, 16 g l^{-1}). The accident in Bari harbour, Italy, during WWII, that killed 85 American soldiers, seems to have been due to the release of nitrogen mustards, formerly stored by the United States Navy on one of its warships.

Lewisites
Lewisite (L,2,chlorovinyldichloroarsine,2-chlorovinylarsonous dichloride) has the chemical name (2,chloroethenyl)arsonous dichloride with the molecular formula

Fig. 7.30 Structural formula of; **a** bis(2-chloroethylthio)methane; **b** 2-chloroethylchloromethylsulfide; **c** sesquimustard (1,2-bis(2-chloroethylthio)-ethane; **d** 1,3-bis(2-chloroethylthio)-n-propane; **e** 1,4-bis(2-chloroethylthio)-n-butane; **f** 1,5-bis(2-chloroethylthio)-n-pentane; **g** bis(2-chloroethylthiomethyl)ether; **h** o-mustard (bis(2-chloroethylthioethyl)ether)

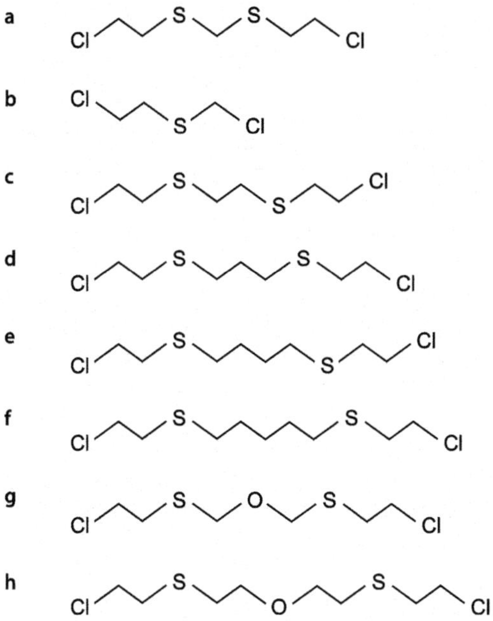

Fig. 7.31 Structural formula of; **a** HN1: bis(2-chloroethyl)ethylamine; **b** HN2: bis(2-chloroethyl)methylamine, **c** HN3: tris(2-chloroethyl)amine

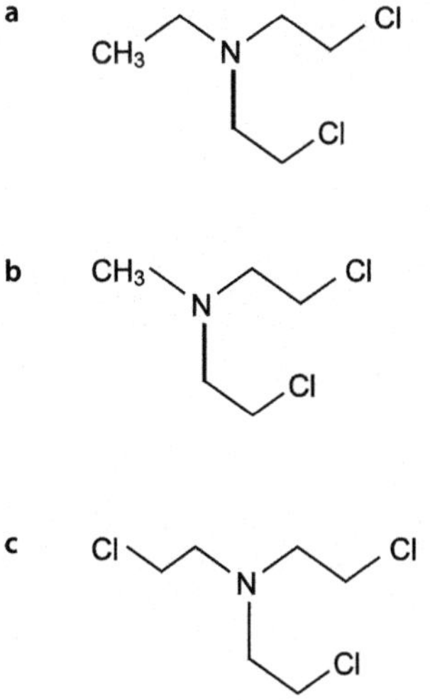

$C_2H_2AsCl_3$. Lewisite was discovered near the end of World War I by a team of Americans headed by Captain W. Lee Lewis working at the Catholic University in Washington DC (Paxman and Harris 1982). It was in fact never used in the operations of World War II. After the war, Lewisite was considered obsolete by the major powers because of the discovery that 2,3-dimercaptopropanol ("*British anti-Lewisite*") was an inexpensive and effective antidote to lewisite exposure. Crude Lewisite has a strong penetrating geranium odour; the pure compound is odourless.

Lewisite is a complex mixture of several compounds, all of which occur as *cis*- and *trans*-isomers. These are classified in three main isomers: L1, L2, and L3. In chemical agent grade Lewisite, the L1 isomer generally predominates. The three isomers have the following structures:

- L1: 2-chlorovinyldichloroarsine (Fig. 7.32a)
- L2: bis(2-chlorovinyl)chloroarsine (Fig. 7.32b)
- L3: tris (2-chlorovinyl)arsine (Fig. 7.32c)

Saxitoxin and Ricin
These are the only two naturally occurring neurotoxins listed as Schedule 1 chemical warfare agents. Saxitoxin (Fig. 7.33) is related to the deadly PSP (Paralytic Shellfish Poison), which is a potent neurotoxin produced by various algae (*Gonyaulax species*). Saxitoxin, the major component of PSP, is responsible for respiratory paralysis, which in 8% of cases results in death. It blocks the sodium ion channels in the nervous and muscle membranes. The lethal dose in humans lies between 1 to 3 mg. Numbness and respiratory arrest can occur after oral ingestion of as little as 0.5–1.0 µg of saxitoxin.

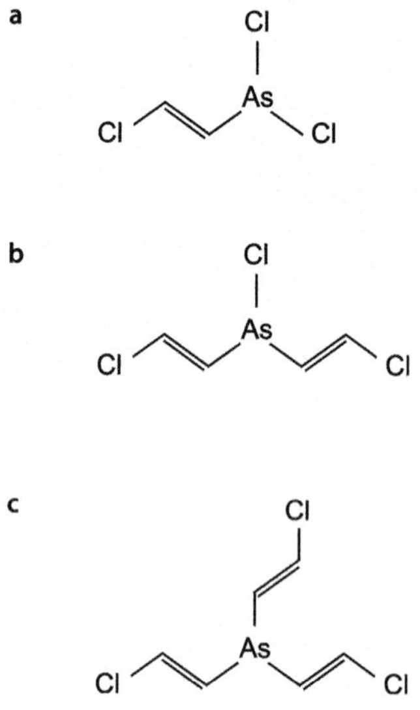

Fig. 7.32 Structural formula of; **a** L1: 2-chlorovinyl-dichloroarsine; **b** L2: bis(2-chlorovinyl)chloroarsine; **c** L3: tris(2-chlorovinyl)-arsine

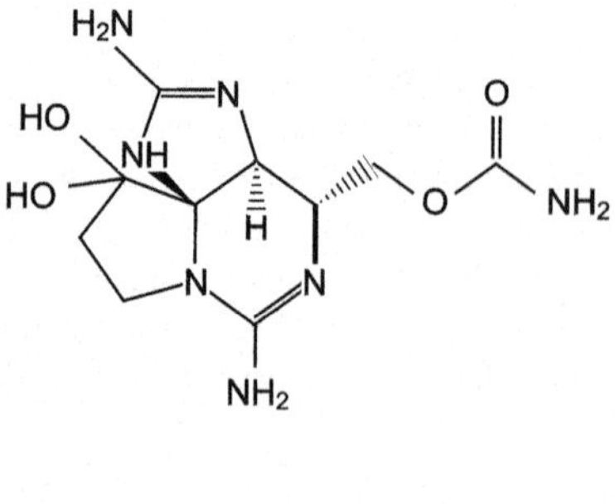

Fig. 7.33 Structural formula of saxitoxin

7.3.1.2
Schedule 2: Dual-Use Chemicals of Limited Use

The following three toxic chemicals are the leading members of Schedule 2 for chemical weapons use:

- *Amiton.* O,O-diethyl-S-[2-(diethylamino)ethyl]phosphorothiolate and corresponding alkylated or protonated salts.
    Amiton (Fig. 7.34a) is an organophosphorus insecticide that was first synthesized in the 1950s. The EPA banned its use in agriculture due to its high toxicity.
- *PFIB* (perfluoroisobutylene). 1,1,3,3,3-pentafluoro-2-(trifluoromethyl)-1-propene.
    PFIB (Fig. 7.34b) is a gas of no commercial use. It is formed as a by-product during the production of some perfluorinated polymers (e.g. Teflon), and is as toxic as phosgene ($COCl_3$) – see Schedule 3.
- *BZ.* 3-quinuclidinyl benzilate (Fig. 7.34c).

Besides toxic chemicals, known to be previously used as weapons or those which are considered as potential CW agents, Schedule 2 includes various precursors that may be used for the production of chemicals listed in Schedule 1 or Schedule 2.

7.3.1.3
Schedule 3: Dual-Use Chemicals of Extensive Use

The four toxic chemicals phosgene, cyanogen chloride, hydrogen cyanide, and chloropicrin, are the most notable among the members of Schedule 3. They may be briefly described as follows:

Phosgene: Carbonyl Dichloride
Phosgene (Fig. 7.35a) owes its name to the first method of its preparation by the action of sunlight on carbon monoxide and chlorine; the word *phosgene* means simply formed by light. It is used as a starting material for the preparation of some agrochemi-

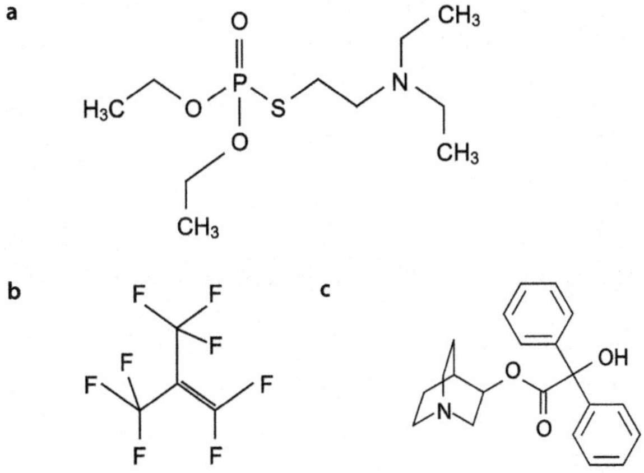

Fig. 7.34 Structural formula of; a amiton; b PFIB; c BZ (3-quinuclidinyl benzilate)

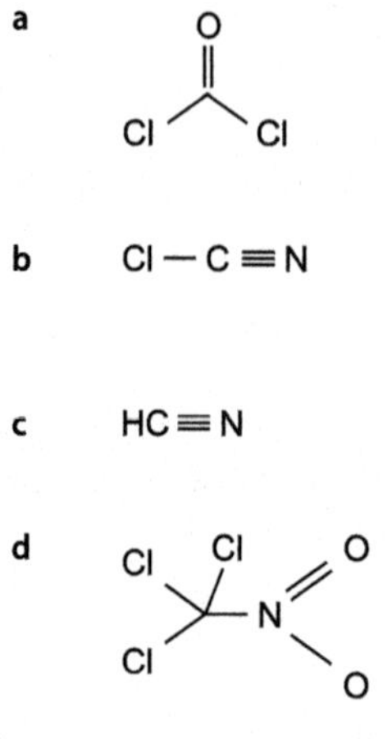

Fig. 7.35 Structural formula of; **a** phosgene; **b** cyanogen chloride; **c** hydrogen cyanide; **d** chloropicrin

a．
$$\underset{Cl\quad\quad Cl}{\overset{O}{\|}{C}}$$

b． $Cl-C\equiv N$

c． $HC\equiv N$

d．
$$\underset{Cl}{\overset{Cl\quad Cl}{\diagdown\!\diagup}}C-N\overset{\displaystyle =O}{\underset{\displaystyle \diagdown O}{}}$$

cals and in the dye chemistry. Physiologically it is a powerful respiratory poison; due to this property, it was used as a chemical weapon during the First World War.

Cyanogen Chloride
Cyanogen chloride (Fig. 7.35b) is a poisonous, lacrymatory gas, which can be condensed at 13 °C and readily trimerises to cyanuryl chloride. Cyanogen chloride acts by blocking cell respiration and causes death within a short time if high concentrations are inhaled. In the chemical industry, however, it is considered an important synthetic intermediate for the production of many organic compounds.

Hydrogen Cyanide
Like cyanogen chloride, hydrogen cyanide (Fig. 7.35c) causes death by blocking cell respiration. It is an important synthetic intermediate and was occasionally used as a pesticide.

Chloropicrin: Trichloronitromethane (Nitrochloroform: CCl_3–NO_2)
Chloropicrin (Fig. 7.35d), which can be prepared by the action of concentrated nitric acid on chloroform, is a heavy liquid, boiling at 112 °C. It is a severely irritating lacrymatory agent, which is also used as a soil sterilizer, a grain disinfectant, and an intermediate in organic syntheses.

Schedule 3 has also seven nerve agent precursors listed, which have extensive applications in the chemical industry, mostly in the production of insecticides and as chlorinating agents.

7.3.2
Soil Pollution by Military Activities During the Cold War

It is perhaps not completely true to classify pollution by military and belligerent activities as incidental sources of pollution, as we did in the foregoing section. This is because of one undeniable fact, the First and Second World Wars were, at least in Europe, only a prologue to an extended period of pollution through troop concentrations, training and stockpiling of armaments on both sides of the confrontation line between East and West. The need for armaments and munitions by less developed countries involved in regional and civil conflicts in Africa, Asia and Latin America, boosted the arms industry and increased the areas planted with land mines to horrible dimensions (e.g. Afghanistan). Due to the extensive use and handling of modern warfare agents by badly or insufficiently trained personnel in the Third World, many regions in these countries have been rendered uninhabitable due to pollution by heavy metals, fuels, oil products, explosives and various dangerous organic pollutants.

After the Second World War, Germany became the main confrontation field between the Eastern block on one side, and the NATO nations on the other. The country with a total area of 35 million ha was divided into two States – The Federal Republic of Germany measuring 25 000 000 ha in size, and the German Democratic Republic with an area of 15 000 000 ha. Until the end of the Cold War (1990), 960 000 ha (about 2.8% of the total area of the country) were used for military training by both eastern and western forces. Table 7.6 shows the contingents in ha, which were assigned to each.

According to a study made on behalf of the German Federal Agency for the Environment in Berlin (Umweltbundesamt, Berlin), the following pollution risk assessment (Table 7.7) was proposed for the whole region of Berlin, East and West. The assessment was also done for military sites formerly used by the Soviets and their allies, as well as for sites that were used by the forces of the West Alliance.

Table 7.6 Contingents assigned to eastern and western forces in Germany during the cold war

Area (ha)	Stationed troops
253 000	Federal Defence Force (Bundeswehr), West Germany
200 000	Armies of the West Alliance: USA, Britain, France, Canada, Belgium, and Holland
240 000	National People's Army (NVA), East Germany
250 000	West Group of the Soviet Armed Forces (WGT)

Table 7.7 Pollution risk assessment for abandoned military sites in Berlin (Schäfer et al. 1996)

Category	West Berlin		East Berlin		All Berlin	
	Number of sites	Area (ha)	Number of sites	Area (ha)	Number of sites	Area (ha)
A	17	300	16	300	33	1 070
B	15	280	35	220	50	500
C	21	280	57	130	78	400
Total	53	1 330	108	650	161	1 970

Table 7.8 Pollutants at abandoned military sites in Berlin (after Schäfer et al. 1996)

Class	Number of samples and detected pollutants					
	≥300	100–299	30–99	10–29	5–9	<5
Soil	Petroleum products	Xylene, Pb, Zn, Ni, BTEX	Cr, Fe, Cu, Mn, Cd, K, Mg, Na, Hg, Ca, Al, ethylbenzene, toluene, benzene, ammonium comp., nitrates, chlorides, BTEX	Nitrites, PAH	Chrysene, fluoranthene, pyrene, anthracene, benzo(a)pyrene, perylene	Chlorinated hydrocarbons, indeno-1,2,3cd-pyrene, tetrachloroethane. naphthalene, phenols, phosphates, F_2, acenaphthalene
Soil air			BTEX	Toluene, benzene, ethylbenzene	Chlorinated hydrocarbons, trichloroethane	Tetrachloroethane, AOX
Groundwater			BTEX, ethylbenzene, petroleum products, chlorinated hydrocarbons, dichloroethane, tetrachloroethane	AOX	Al, Pb, Ca, Cr, K, Cu, Mg, Mn, Na, Zn, sulphates, nitrites, PAH, tetrachloromethane	Trichloroethane, trichloromethane, nitrates, phosphates, Hg, Cl_2, Ni

BTEX = benzene, toluene, ethyl benzene, xylene, *AOX* = adsorbed organic halogens

- *Category A.* Sites of proven high contamination or sites of highly probable contamination.
- *Category B.* Sites, which according to their previous use (depots, target shooting areas, fuel stations, warehouses, car depots, parking areas, workshops, barracks, etc.) are considered as potential sites for high contamination. Further investigations are required.
- *Category C.* Low risk sites which are not initially considered as contaminated. These include: administrative buildings, transit areas, housing areas, cultural centres and guest houses. According to the same study (Schäfer et al. 1996), the following qualitative assessment for soil pollutants (Table 7.8) has been determined for representative sites.

7.4
Soil Pollution through Biological Warfare (BW)

In the foregoing section, warfare agents of pure chemical origin were discussed in some detail. However, inspection of the history of war shows that war planners have always aimed to use infectious diseases as a means of gaining the upper hand in military operations; this is known under the collective name *Biological Warfare* (BW). During the course of a biological warfare attack, infected animals or plants may be spread on the lines of the enemy, harmful biota may be introduced into his cities and shelters, or toxins extracted from biological sources may be directly used. An example from medieval history may be given by the battle of the Genoese against the Tatars in the year 1346 at the Black Sea port of Kaffa (now Feodossia, Ukraine). The Genoese used rats carrying pest-infected fleas to spread the pest among the attacking Tatar soldiers. The reaction of the Tatars supplies another classical example of a counter attack using the same strategy – they simply hurled the corpses of their casualties, infected by the pest, at the Genoese, who contracted the disease as well and had to clear the battlefield. In the 18th century, the British decimated vast numbers of American Indians by infiltrating smallpox infested blankets among them. It is also reported that plague bacilli were used in the Second World War as biological weapons. More recently (1974–1981), death-bringing mycotoxins *(trichothecene mycotoxins)* – also known as *yellow rain* – are said to have been used by the Soviet Union in South East Asia and in Afghanistan

It is not the intention to discuss here in detail the different types of BW agents or the strategies used in their production. But since they still represent, like pure chemical war agents, an important member in the chain linking human activities to the occurrence of environmental pollution, especially as they stand directly in relation to the soil system as a medium for incubation, accumulation and dispersion of these agents; a short introductory report on them is necessary.

The Devil's Tools: Biota in the Hands of War Planners

All types of parasites and disease dispersing biota, including bacteria, viruses and various low plants, were used, or are presently used, as carriers of contagious diseases or death-bringing agents by warfare planners. Important criteria for the use of harmful or disease-carrying biological agents, making them suitable for use as biological weapons, include: infectivity, virulence, toxicity, pathogenic character, incubation period, transmissibility, lethality, and stability. Other than their counterparts in chemical warfare, these agents are characterized by their ability to multiply in the body over time and actually increase their effect. Many of them may persist for very long times in the soil environment and its dwellers, as for example in different generations of rodents, bacteria or fungi. The spores of *Bacillus anthracis* (Anthrax producing bacterium – see below), for example, remain viable in the upper 6 cm of soil for periods of up to 200 years (Titball et al. 1991); it also persists for a long time in animal products (Burnett 1991; John 1996). Animals that die of anthrax release massive quantities of spores into the soil, which may remain for decades before being ingested again; that is why carcasses of animals that were infested by anthrax should be burned (not buried) to prevent long-term environmental contamination.

In the following, some of the biota abused by war planners as biological war agents, together with some description of the harm they cause to the environment, will be discussed.

7.4.1
Bacteria

As described at the beginning of this book, bacteria are small free-living organisms that reproduce by simple division. They may produce diseases, yet the disease-causing (pathogenic) species of them are only a comparatively tiny fraction of the bacteria as a whole.

Practically, bacteria occur under all environmental conditions and they may persist in soil (in dormant condition) for a very extensive time. In regard to their response to gaseous oxygen, bacteria may be classified into three groups:

- *Aerobic bacteria.* These are the species that flourish in the presence of oxygen and require it for their continued growth and existence.
- *Anaerobic bacteria.* These are the ones that cannot tolerate gaseous oxygen, such as the bacteria living in deep underwater sediments.
- *Facultative bacteria.* These are bacteria that may grow in the presence of oxygen, yet they also flourish in its absence.

7.4.1.1
Bacterial Biological Weapons and the Diseases They Transmit

Bacillus anthracis
Bacillus anthracis is a gram-positive sporulating bacillus. Its spores are resistant to heat, cold, drying, and chemical disinfections, and can persist for decades in the soil, as it was said before (Farrar 1995). It produces the fearful anthrax disease, which leads to death after a series of painful symptoms including respiratory and skin disorders.

Anthrax, a zoonotic (shared by animals and humans) disease has normally two main forms:

1. *Cutaneous anthrax.* This is developed by *Bacillus anthracis* penetrating the skin. It has an incubation period of one to seven days. It causes (in the first stages of the disease) skin ulcers that are frequently surrounded by swelling or oedema. If cuta-

neous anthrax is not treated with antibiotics, death occurs in about 10–20% of the cases. Prompt treatment, however, may reduce the mortality to less than 1%.
2. *Inhalation anthrax.* Inhalation anthrax develops when *Bacilli* are inhaled into the lungs. Its mortality rate is about 90–100%. In biological warfare, anthrax spores are delivered by an aerosol to cause inhalation anthrax and thus ensure the highest number of casualties among a population. According to an estimate by the U.S. Congress's Office of Technology Assessment, 100 kilograms of anthrax, released from a low-flying aircraft over a large city, on a clear, calm night, could kill one to three million people (McGovern et al. 1995–2001).

Natural forms of anthrax are endemic in western Asia (Afghanistan, Iran, and Turkey) and western Africa (Burnett 1991), where the disease is in more than 90% of the cases transmitted from infected animals, or their products, via skin abrasions (John 1996). It is less commonly transmitted through the ingestion or inhalation of spores (McGovern et al. 1995–2001).

Burkholderia pseudomallei (Formerly Pseudomonas)
This is a gram-negative bacillus isolated from soil and stagnant water bodies. It produces an infectious disease, known as *melioidosis*, that attacks both animals and humans. Land workers and other people who come into contact with soil may contract it, through abrasions, or if bacteria are inhaled or ingested. Melioidosis is endemic to Southeast Asia and northern Australia, but it may occur anywhere in a belt between 20° N and 20° S of the equator and is most widespread in Thailand (Handa et al. 1996). As a biological war agent, the disease is spread via aerosol; when contracted, it may appear as an acute pulmonary infection, but it may also materialise as an acute localized skin infection. Melioidosis also has the capacity of remaining latent in soil for long times. Even if treated, mortality rates due to melioidosis are over 40%. No available vaccines are known up to now (Handa et al. 1996).

Yersinia pestis
This is a gram-negative coccobacillus causing the *plague* – a fearful disease, more closely related to the rise and fall of human civilisations than any other syndrome known to man. It not only affected the current of history, but also left its traces on literature, arts and even languages, especially in western and central Europe. It is estimated that, throughout history, 200 million people were killed by the plague (Perry and Featherston 1997).

This high killing potential and effective wide spreading of the disease from person to person made *Yersinia pestis* a favoured area of research for BW-planners and hence great efforts were invested in developing technologies to aerosol them among enemy populations. However, the Japanese, during the Second World War, went in another direction, multiplying the human fleas (*Pulex iritans*), infecting them with *Y. pestis* and eventually releasing them into Chinese cities, where plague epidemics later ensued; this proved to be more deadly and subtle than the aerosol.

Francisella tularensis
Francisella tularensis is a gram-negative coccobacillus that can persist for months in upper soil, swampy water and decaying animal carcasses. It transmits a contagious

disease known as *tularemia*. This starts as a tender, red, pruritic papule that rapidly enlarges to form an ulcer with a black base. Later, the organism spreads to lymph nodes and causes serious disorders that may lead to death.

The United States Army successfully developed methods for using tularemia as a biological weapon delivered by aerosol in the 1950s and 1960s. It is fair to expect that some other nations might have achieved the same result (Franz et al. 1997).

Vibrio cholera
This is a short, curved, gram-negative bacillus that causes the dangerous diarrheal disease *cholera*. Humans acquire the disease by consuming water or food contaminated with the organism. As a BW agent, cholera will be used to contaminate water supplies; it is unlikely to be used in an aerosol form. The disease, if not treated, is fatal for many patients. It is known that cholera was responsible for decimating populations in vast areas in India and the Middle East in the 18th and 19th centuries.

Brucella melitensis, B. abortus, B. suis and B. canis
These four small gram-negative aerobic species of bacteria are responsible for the spread of *brucellosis* – a systemic zoonotic disease that resides quiescently in tissue and bone-marrow, and is extremely difficult to eradicate even with antibiotic therapy. Humans are infected when they inhale contaminated aerosols, ingest raw (unpasteurized) infected milk or meat, or have abraded skin or conjunctival surfaces that come in contact with the bacteria.

7.4.2
Viruses

Viruses are the smallest pathogens known (20–100 nm). Their reproduction mode makes them unique among all other biota, for, in order to reproduce, they require the presence of other living cells in which they replicate within the intercellular space. A virus is normally made of a protein envelope containing its genetic material. It contains only one kind of nucleic acid, either DNA or RNA, a property which is usually used for classification. Viruses have no cell structure and are not capable of producing enzymes. They are intimately dependent upon the cells of the host, which they infect. In short, they may be described as well-organised minute packages of genetic material, shaped like rods, filaments, harpoons or spheres.

Viruses may cause very serious diseases in humans – smallpox, the common cold, chickenpox, influenza, shingles, herpes, polio, rabies, Ebola, and AIDS are examples of these. It is even speculated that they are responsible for some types of cancer. Diseases caused by viruses do not respond to antibiotics. In some cases, however, they may be responsive to antiviral compounds, of which there are few available. Yet, available anti-viral agents are mostly of limited use.

When a virus comes into contact with a host cell, it inserts its genetic material into the host, literally taking over its functions, and coaxes the infected cell to produce more viral protein and genetic material instead of its usual products. At this stage, viruses may go into a dormant condition inside the cells, causing no obvious change for long periods (this stage is known as *the lysogenic phase*). If following this phase, the virus is stimulated, it passes into an active stage, known as the *lytic* one, during which formation of new viruses would commence, followed by self-assemblage, and then burst out to infect other cells.

7.4.2.1
Transport of Viruses in Soil

Viruses, when transported in soil, may reach the groundwater, transmitting disease to humans and animals. Factors controlling the transport of viruses in soil include properties of the virus itself (i.e. surface-charge, size, and morphology), the nature and properties of the soil environment (i.e. mineralogy, grain size, texture, angularity, etc.), and properties of the water (i.e. pH, ionic strength, content, and concentration of natural organic matter). A key role in these processes is also played by the properties (ionic strength, pH, chemical composition, etc.) and degree of saturation of the soil water and other fluids occupying the interstitial space. Influent virus concentrations should not be ignored in this count.

- *Soil type.* As mentioned before, the grain size of soil plays an important role in determining the capacities of soils to absorb foreign matter on its surfaces. This will also determine the availability of viruses absorbed on soil particles. Therefore, we find that clay rich soil will be in a better position to absorb viruses than coarser ones. This will result in the soil holding the viruses strongly, hindering their transport. Coarse-grained soils, on the other hand, will not hold the viruses, due to their lower absorptive capacity, thus allowing them to be transported to the groundwater.
- *Chemical conditions in the soil site.* Like bacteria, viruses have normally a negative net charge, thus they will be attracted to positively charged particles, losing the impetus for being transported. Under neutral to alkaline conditions, they will be free to move and thus get the best chance to be transported.
- *Ionic strength of the percolating water.* Experiments comparing absorption of viruses in waters of high ionic strength and at low cationic concentrations show that the soil retention of viruses increases with the ionic strength of the ambient water. Distilled water was not only a factor in reducing the absorption, according to some reports, distilled water may actually lead to desorption of viruses from the soil. This would mobilize the virus and transportation would then begin.
- *Soil organic materials (humic substances).* A competition for absorption sites on soil particles may occur between viruses and soil organic matter. In this sense, it is thought that humic and fulvic acids may reduce the potential for viral absorption on soil particles.
- *Availability of soil water (saturation).* Viruses, which are not in direct contact with the solid soil material will be less absorbed on the soil surface, thus having a higher degree of freedom for moving within the soil body. This is actually what happens when the soil is saturated with water; in this case, the viruses will move more freely and quickly within the aquatic interstitial space.

The Fate of Viruses in Soil
Beside the above-mentioned factors, which determine and control the movement and transport of viruses in soils, the survival of these organisms depends on a variety of properties, the most prominent of which were discussed in this book under the sub-

ject of soil quality and biological fertility. One can assume that the climatic and ecological conditions, such as temperature and availability of water (moisture), may play a crucial role in securing the suitable conditions required for the survival of a given virus. An important factor would also be the depth of the contaminated layer and its accessibility to sunlight, since most viruses would not be able to survive exposure to UV-rays.

7.4.2.2
Viral Biological Agents and the Diseases they Transmit

Due to the specific properties we mentioned before, and keeping their high potential of bringing death and diseases to the enemy in mind, certain viruses were selected to be used as biological weapons, to play a great role in infecting humans as well as polluting agricultural land by being adsorbed on soil particles (Lance et al. 1976; Bitton 1975). They may persist on the soil particles to repeatedly kill humans and livestock in repeating cycles of epidemics. Some of these BWs will be shortly introduced in the following sections:

Poxviridae
These viruses, also known as the *poxviruses*, are actually the largest of all viruses ($200-300 \times 10^{-9}$ m, the size of small bacteria). They belong to the group of DNA viruses, yet differ from other DNA viruses in that they replicate in the cytoplasm where they produce eosinophilic (can be easily dyed by eosin) inclusion bodies (McGovern et al. 1995-2001). They are resistant to drying and many disinfectants – a property that may add to their persistence. In this group three viral agents are used as BWs, these are *variola*, *monkey pox*, and *vaccinia*.

- *Variola*. This is the virus responsible for transmitting *smallpox* – a disease which is now considered to be eradicated and exists only in the biological war laboratories of Russia and the United States. Variola, for which the most probable route of transmission is via respiratory spread, is thought to be transmitted by monkeys.
- *Monkey pox*. This is a rare child disease, similar to smallpox, which is native in central Africa. It is acquired from monkeys or wild squirrels, but does occasionally spread from man to man in unvaccinated communities. Monkeys are thought to have antibodies against the disease, since sick animals have never been identified.
- *Vaccinia*. This is a virus strain which has been used for immunisation against smallpox. It is thought to be a genetically distinct type of pox virus which grows readily in a variety of hosts.

Rift Valley Fever Virus
This is a virus normally carried by mosquitoes in sub-Saharan Africa. In biological warfare it is most likely delivered by aerosol, and, if inhaled, would cause a febrile illness for a few days, which may in some cases (~1%) progress to a viral haemorrhagic fever syndrome; mortality in this case is roughly 50 percent. Death and later complications (encephalitis, meningitis, etc.) in all test groups have been observed.

Avoidance of mosquitoes and contact with fresh blood from dead domestic animals and respiratory protection from small particle aerosols are the most important precautions to be followed.

Congo-Crimean Haemorrhagic Fever
Congo-Crimean haemorrhagic fever (CCHF) is a viral disease transmitted by ticks, principally of the genus Hyalomma, with intermediate vertebrate hosts varying with the tick species. Occuring naturally in the Crimea, in the Middle East, the Balkans, the former USSR, and eastern China, it was first recognized in the Congo in 1965; hence the current name. CCHF would probably be delivered by aerosol if used as a BW agent.

Ebola Haemorrhagic Fever
Ebola haemorrhagic fever is a viral disease that figures very high in the discussions of BW planners. It is one of the most virulent viral diseases known by man, causing death in 50–90% of all clinically ill cases. The virus is believed to reside in the rain forest of Africa and Asia, and was first identified in 1976 in Sudan and in Zaire after significant epidemics in northern Zaire and southern Sudan. In Asia it appeared in the Philippines (near Manila) in an area reserved for monkeys intended for export.

The natural reservoir for Ebola is not yet identified; speculations include several animal groups extending from rodents to bats.

7.4.3
Rickettsiae

These are gram-negative bacteria transported by ticks, lice, fleas, mites, chiggers, and mammals. Four zoonetic pathogens in this group are known as agents that cause infections capable of disseminating in the blood to many organs. These are the genera: *Rickettsiae, Ehrlichia, Orientia,* and *Coxiella*.

Rickettsia diseases, which are used as BWs, may include, among others, diseases like *typhus, epidemic typhus* and *Q-fever*.

Q-Fever
This is a fever occurring worldwide, most often among persons coming into frequent contact with farm animals. These include farmers, veterinarians, butchers, meat packers, and seasonal workers. The disease is transmitted by an agent (*Coxiella burnetii*) through airborne transmission, via contaminated soil and dust. These infections may initially result in only mild and self-limiting influenza-like illnesses, but, if untreated, can be fatal for the infected person. Soils contaminated after BW attacks may help in spreading the disease.

7.4.4
Chlamydia

Chlamydia are intercellular parasites that like viruses are not capable of producing their own energy sources and are intimately dependent upon the host cells for reproduction. They resemble bacteria, however, in their response to broad-spectrum antibiotics. Chlamydiae transmit – among others – a disease that may be transported through sexual intercourse, causing infertility of females, if not treated.

7.4.5
Fungi

Fungi are primitive plants that are not dependent on chlorophyll, i.e. they do not use photosynthesis. Most fungi form spores and exist as free-living forms in soil. Generally fungi derive their nutrition from decaying organic matter.

Mycotoxins (toxins secreted by fungi), which are potential agents of BW, are a diverse group of small molecular weight compounds, mainly produced by members of five fungal genera: *Aspergillus*, *Penicillium*, *Fusarium*, *Alternaria*, and *Claviceps*. According to some authors (Wannemacher and Wiener 1997), there is now substantial evidence that trichothecene mycotoxins were used as BW agents in Southeast Asia between 1974 and 1981. In Laos, the attacks were described as *yellow rain* because of the yellow carrier powder that was used to spray the toxins.

7.4.6
Toxins

In the foregoing section mycotoxins were introduced together with their principal sources – algae. Toxin production, for BW-purposes, is not limited to fungi. Generally they are poisonous substances produced and derived from living plants, animals, or microorganisms; some toxins may also be produced or altered by chemical means. As it has been shown before, toxins may be very virulent BW agents and pollutants of the upper soil if they are introduced by belligerent forces. It is for this reason that some of them will be discussed in the next section.

Botulinum Toxins
This is a group of neurotoxins that block neurotransmission, causing respiratory failure due to paralysis of respiratory muscles and generally leading to death. A biological warfare attack with botulinum toxin, delivered by aerosol, would leave vast numbers of casualties. In its pure form, the toxin is a white crystalline substance that is readily dissolvable in water, but decays rapidly in the open air.

Ricin
Ricin is a glycoprotein toxin extracted from the castor plant. It destroys cells by blocking the protein synthesis through altering the RNA. Ricin is a very favourable candidate for figuring as a potent BW-agent, due to its worldwide availability and the ease of its production.

Death by ricin may be caused by ingestion of the ricin bean or by inhalation of the toxin, if introduced as an aerosol. It is most probably used this way in biological warfare.

Saxitoxin

Saxitoxin is the representative of a group of chemically related toxins, which are normally produced by marine *dinoflagellates* and blue green algae. Saxitoxin causes the disease known as *Paralytic Shellfish Poisoning* (PSP); it sometimes occurs after ingestion of bivalve molluscs that have stored dinoflagellates during filter feeding in seawater. Like in all other neurotoxins, death occurs in these cases through respiratory paralysis. Infectious bivalves may display a red colour, similar to blood smearing their meat; the ambient water in the direct vicinity of the shells may also be of the same colour. This gives the phenomenon its special name – *the red tide*. No vaccine against saxitoxin exposure has been developed for human use.

Staphylococcal Enterotoxin B (SEB)

This is one of several exotoxins produced by *Staphylococcus aureus,* causing food poisoning when ingested. A BW attack with an aerosol delivery of SEB to the respiratory tract produces a distinct syndrome causing significant morbidity and potential mortality.

Venezuelan Equine Encephalitis (VEE)

This is actually a general term for a complex of diseases produced mostly by eight distinct viruses belonging to the Venezuelan Equine Encephalitis (VEE). They primarily attack horses, donkeys and mules (Equidae). Natural infections are acquired by the bites of a wide variety of mosquitoes; Equidae serve as the viremic hosts and source of mosquito infection. A BW attack with the virus disseminated as an aerosol would cause human disease; such an attack in a region populated by Equidae and appropriate mosquito vectors could initiate an epizootic/epidemic.

Chapter 8

Pollution Mechanisms and Soil-Pollutants Interaction

The behaviour and interaction of pollutants with soil comprise of various physical, chemical, and biological processes that take place in all three components (solid, gas and liquid) of the soil medium. They generally include three main groups of processes:

1. Retention on and within the soil body
2. Infiltration, diffusion and transport by soil solutions
3. Alteration, transformation, and initiation of chemical changes within the soil

While the first two groups include mainly physical processes, by which pollutants are transported and distributed in the soil, the third group comprise of only chemical and biological processes, by which pollutants are transformed or stored as residues in the interstitial space. Figure 8.1 shows a schematic overview of the three groups of processes.

Physical processes of soil/pollutant interactions include transport and retention. They depend mainly upon the physical parameters of the medium (temperature, grain size, electric charges, etc.), while chemical processes depend largely on the type of pollutants and their chemical nature. Both groups of processes are further classified according to the mechanisms involved.

As for biological, or biologically controlled, soil pollution processes, we may include all processes of biotransformation and biodegradation, each depending on the microbial ecology, the depth, and the oxygen availability at the site of pollution. A summary of the processes involved in soil pollution may be given in Fig. 8.2.

8.1
Physical Processes and Mechanisms of Pollution

Upon encountering soil grains, pollutants will either be retained by adsorption on the surface of these grains, or be accumulated in their intergranular space, where they may form concentrations retaining their original chemical composition, or substances that have been altered by various chemical reactions. Pollutants retained thus on the soil surface, or in its interstitial space, may be organic, inorganic, or a mixture or complexes of both. They reach the soil in various physical conditions as solutes, water-immiscible liquids or suspended particles. The mechanisms of their interaction with the soil will thus depend upon physical parameters prevailing in the soil medium, such as temperature, moisture content or salinity of the soil water, as well as upon their own physical and chemical properties.

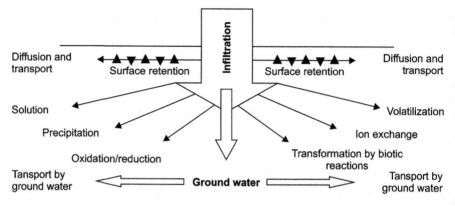

Fig. 8.1 A schematic overview of the processes representing soil-pollutant interactions

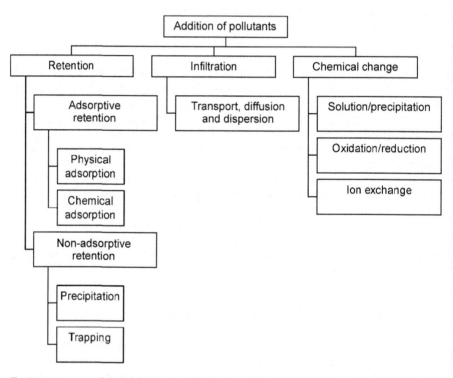

Fig. 8.2 A summary of the mechanisms involved in soil pollution

Adsorption and its accompanying phenomena are considered as the most important physical chemical mechanisms of pollutant retention on the surface of soil grains. In the following, these phenomena will be treated in some detail.

8.1.1
Adsorptive Retention

Molecules of pollutants can be retained on the surfaces of soil grains in two ways.

In *physiosorption*, or physical adsorption, molecules of pollutants will be attached to the surfaces of soil grains by *Van der Waal forces*, which are known to be weak long range forces. The amounts of energies involved in such attachments are normally of low magnitudes and are not sufficient for bond breaking. Thus, pollutant molecules sticking to the soil surface will retain their chemical identities, although they might be stretched or bent on account of their proximity to the surface.

In *chemisorption*, or chemical adsorption, the pollutants attach themselves to the grain surfaces as a result of the formation of a chemical (usually covalent) bond. In this case, the energy of attachment is very much greater than in physical adsorption. Thus, a molecule undergoing chemisorption can be torn to satisfy valency considerations arising from bond formation with the surface atoms.

Although it is very difficult to differentiate between physical and chemical adsorption, one can generally say that the amount of physically adsorbed material decreases with increasing temperature, while this relation for chemically adsorbed material is reversed.

Normally various adsorbents exist in the soil medium; some examples of these are given by clay minerals, zeolites, iron and manganese hydrated oxides, aluminium hydroxide, humic substances, bacterial mucous substances, and plant debris. Many rock-forming minerals such as micas, feldspars, some pyroxenes, and some amphiboles are also considered as good adsorbents of pollutant molecules.

The capability of clay minerals and colloids in general in adsorbing foreign molecules on their surfaces is attributed partially to their high surface energy, and partially to the existence of a net surface charge (σ_s), which may be caused by some functional groups (e.g. M–OH). These normally possess charges that are dependant in sign and magnitude upon the composition of the ambient liquid phase, as well as on the nature of the surface that they are bound to. Such net charges attract ionic pollutants to the surfaces of the adsorbent material. However, non-ionic pollutants can also be adsorbed by soil grains; this occurs principally through electrostatic forces.

8.1.1.1
The Theory of Diffuse Double Layer (DDL)

To explain the adsorption of charged particles on the surfaces of solids, Helmholtz (1879) assumed an electrical double layer of positive and negative charges at the surface of separation, between the colloidal particle and the dispersion medium. The double layer, according to Helmholtz, consists of one layer firmly attached to the surface of the particle and a second oppositely charged layer at a monomolecular distance from the particle in the surrounding medium (Fig. 8.3a). Guoy and Chapman (1910, cited in Atkins (1978)) and further workers, in the following years, modified this theory to a form that culminated into the *diffuse double layer model*. According to this model, the charges on the surface of the solid are not balanced by a single movable layer in the surrounding phase, but by a layer more diffuse in character, that ex-

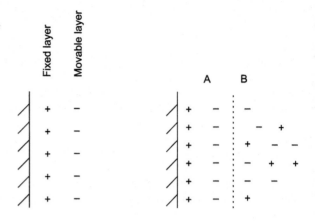

Fig. 8.3 a Helmholtz double layer; **b** Gouy-Chapman diffuse double layer

tends into the ambient phase as shown in Fig. 8.3b. One part of the double layer (A in Fig. 8.3b) is firmly attached to the surface of the colloid substance and thus becomes an integrated part of it. The second part (B in Fig. 8.3b) lies in the surrounding phase. Thus the potential drop between the solid surface and the surrounding phase is formed of two parts: (*i*) a part between the solid surface and the firmly attached layer A; (*ii*) and another part between layer A and the bulk of the surrounding phase. This second potential drop is called the *electro kinetic* or *zeta potential* (represented by the Greek letter ξ).

The theory explains the formation of the double layer through the argument that the surfaces of colloids are considered as planar surfaces, upon which electrical charges are uniformly distributed. The colloid surface with its layer of net charge, on encountering the front of the approaching liquid or gaseous phase, will be faced by a layer of equal, but opposite charges, formed by the ions, to be absorbed on the surface of the colloid. This will cause the approaching ions (if possessing an opposite charge) to be attached to the colloid surface. Following the electrostatic attachment of the oppositely charged ions on the colloidal surface, another process will take place, viz.: the repulsion of the similarly charged ions, which in this case will be drifted by diffusion to form a layer further from the surface of the colloid. Thus, two layers will surround the colloid surface – one of them is the layer of attracted opposite charges (called the layer of counter ions), and the other is a diffusive layer of repelled ions, migrating towards the ambient liquid or gaseous phase. The double layer theory is suitable to explain adsorption of ionic pollutants on the surface of soil particles, especially in connection with ion exchange mechanisms. This will be explained below in some detail.

8.1.1.2
Chemical Adsorption or Chemisorption

As it was stated before, the formation of covalent bonds during chemical adsorption makes the energy of attachment very much greater than in the case of physical adsorption. A molecule undergoing chemisorption can be torn to satisfy valency considerations arising from bond formation with the surface atoms.

Generally, the amount of adsorbed material in physical adsorption is inversely proportional to temperature. This relation is reversed in chemical adsorption. In chemisorption, molecules undergoing this process normally lose their identities as the atoms are rearranged (Hassett and Banwart 1989).

8.1.1.3
The Extent of Adsorption (Adsorption Isotherms)

The extent of adsorption depends upon the exposed surface area of the adsorbent, as well as upon the concentration of the sorbate in the soil solution (partial pressure in case of gases), and the temperature of the medium. As it will be later shown, the adsorption process arrives at equilibrium when the number of molecules adsorbed will be equal to the free ones in the surrounding medium. If measured, then the adsorption data are plotted against the concentration values of the adsorbate in the surrounding medium; a graph known as the *adsorption isotherm* can be obtained.

The simplest isotherm is the *linear distribution coefficient*, K_d (also called *linear partition coefficient*), which is widely used to describe adsorption in soil and near surface aquatic environments. According to this equation, the amount of contaminant adsorbed is directly proportional to the concentration of the adsorbate in the ambient solution. It has the form:

$$S = K_d C$$

where S = amount adsorbed (μg/g solid), C = concentration of substance to be adsorbed in the ambient solution (μg/ml), and K_d = distribution coefficient.

Another coefficient, also widely used in problems of soil pollution, is the *organic carbon partition coefficient* (K_{oc}). It is derived by dividing the distribution coefficient (K_d) by the percentage of the organic carbon present in the system (Hamaker and Thompson 1972).

Accordingly

$$K_{oc} = K_d / (\% \text{ organic carbon})$$

where K_d is the distribution coefficient.

Readers interested in the application of K_d to explain adsorption processes of organic pollutants in soil environment, are referred to Karickhoff (1984).

The Langmuir Isotherm
This isotherm was originally developed to describe adsorption of gases on homogeneous surfaces and can be derived as follows:

For an adsorbent surrounded by a gaseous phase, gas molecules crashing on the exposed surface will be trapped only if they quickly dissipate their energy into the vibrations of the underlying lattice. Otherwise, they will be bounced back to the surrounding phase. The rate of the collisions with the surface, which will successfully lead to adsorption, is called the *sticking probability* and is given by the relation

$$s = \frac{\text{Rate of adsorption of molecules by the surface}}{\text{Rate of collision of molecules with the surface}}$$

The sticking probability depends upon the exposed surface area – it drops to smaller values as the surface sites get filled. The extent of the surface coverage is given by the relation:

$$\theta = \frac{\text{Number of adsorption sites filled}}{\text{Number of adsorption sites available}}$$

At the moment equilibrium is attained between the molecules sticking on the surface and the molecules free in the gas phase, θ will depend on the pressure of the gas. This relation between θ and the pressure at a given temperature delivers what we called above the *adsorption isotherm*.

Assuming that every adsorption site is equivalent, and that the ability of a molecule to attach itself to the surface is independent of whether the neighbouring sites are occupied or not, we may represent the dynamic equilibrium between the adsorbed molecules and the free ones by the equation:

$$A(g) + M(\text{surface}) \underset{k_d}{\overset{k_a}{\rightleftharpoons}} AM$$

where k_a and k_d are the rate coefficients for adsorption and desorption respectively.

If the fractional coverage is denoted by $(1-\theta)$, then the number of vacant sites will be given by $N(1-\theta)$, where N is the total number of sites.

Since the rate of adsorption is proportional to the pressure of A (P_A), as well as to the number of vacant sites, we may represent it by the equation:

Rate of adsorption = $k_a P_A N(1-\theta)$

The rate of desorption, however, is proportional to the number of adsorbed species $N\theta$, so that it can be represented by the equation:

Rate of desorption = $k_d N\theta$

At equilibrium, the rate of adsorption = rate of desorption, i.e.

$k_a P_A N(1-\theta) = k_d N\theta$

solving for θ, we get the Langmuir isotherm:

$\theta = KP_A / (1 + KP_A)$

The Langmuir isotherm was originally developed to describe the adsorption of gases on solids. However, in dealing with solutes in soil water, P_A is replaced by C_A, so that we get the Langmuir isotherm:

$\theta = KC_A / (1 + KC_A)$

where θ stands for the amount adsorbed per unit mass of adsorbent and C_A for the concentration of adsorbate in solution.

At low degrees of surface coverage, the graphical relation between θ and the concentration of adsorbate is given by a straight line. However, this was found to deviate gradually as the adsorption sites are increasingly occupied by the molecules of adsorbate (Fig. 8.4). Such deviation hints to the fact that the following assumptions underlying the derivation of the Langmuir isotherm are quite unrealistic:

- The energy of adsorption is equal for all sites and is independent of the degree of surface coverage.
- The adsorbed entities are attached to the surface at definite homogeneous localised sites.
- Forming a monolayer with no interaction between adjoining adsorbed molecules.
- The energy of adsorption is independent of temperature.

These considerations and the observation that the energy of adsorption logarithmically decreases with increasing coverage of the adsorbent surface led to the empirical derivation of the *Freundlich equation*, that modifies Langmuir isotherm to fit more realistic conditions.

This is given by $\theta = KC^{1/n}$ where K and n are empirical constants.

The Brunauer, Emmet and Teller (BET) Isotherm

This equation was developed to allow for multilayer adsorption, which is charateristic for phenomena of physiosorption. It has the form:

$$\frac{P}{V(P_0 - P)} = \frac{1}{V_m C_h} + \frac{(C_h - 1)P}{V_m C_h P_0}$$

- P stands for the equilibrium pressure at which a volume V of a gas is adsorbed.
- P_0 is the saturation pressure of the gas.
- V_m is the volume of gas corresponding to a mono-molecular layer.
- C_h is a constant related to the heat of adsorption of the gas on the adsorbent.

Fig. 8.4 The Langmuir isotherm for several values of K

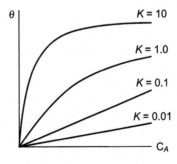

8.1.1.4
Adsorption of Ionic Pollutants

According to the Diffuse Double Layer model (DDL), soil grains, surrounded by a gaseous or liquid environment, will be faced by a front of one or more layers of counter ions (ions of opposite charge) or co-ions (ions of similar charge). Many soil components (e.g. clay minerals) have a marked tendency of replacing some of their ions with similar species from the ambient medium (solution or gaseous phase). When the species lost or gained cations, the phenomenon will be described as cation exchange, otherwise we speak of anion exchange. Cation exchange plays a dominant role in the soil environment, while anion exchange processes are very rare. This is because anions, as acid radicals, may in the presence of hydrogen, lead to dissociation of adsorbent materials such as the clay minerals.

A cation exchange process between an adsorbent and the surrounding soil solution is a reversible process. It can be represented, for an adsorbent A, by a simple reversible equation, as shown by the following example:

$$Ca^{2+}A + 2Na^+ \rightleftarrows 2Na^+A + Ca^{2+}$$

Classic examples of soil components showing marked cation-exchange behaviour are the clay minerals; especially the montmorillonites and illites. They possess a negatively charged repetitive structural framework, having well defined, negatively charged sites, occupied by singly or doubly charged cations. A great variety of materials, other than the clay minerals, exhibit the same behaviour. Among these, include most silicate minerals, silicate glasses, arsenates, vanadates, molybdates and related species.

The CEC of an ion-exchanger is normally defined as the equivalent mass (in milliequivalents) of exchangeable cation per 100 g of exchanger at pH = 7.

In Table 8.1, the CECs of some soil components are given. These are clay minerals and zeolites in the first place. Zeolites may occur in soils formed on volcanic substrates; they exhibit high CECs, and normally cause the enrichment of soil water in $NaHCO_3$, due to their tendency for sodium replacement with other cations. Montmorillonite, which occurs in various soils, is normally considered as the most important cation exchanger in the weathering zone.

8.1.1.5
Ion Selectivity

Natural cation exchangers do not attract all ions with the same intensity. This preference, or selectivity, depends principally on the cationic concentration in the solution, the cationic dimensions, as well as on the structural properties of the exchange surface. It has been generally postulated that cations of higher valency, and those held tightly in their crystal lattice, are preferred by soil components. Hydrogen, however, forms an exception to this rule; it behaves in the course of ion exchange as if it were of higher valence (II or III). The general series of ion-preference may be represented as follows:

$$H^+ > Rb^+ > Ba^{2+} > Sr^{2+} > Ca^{2+} > Mg^{2+} > K^+ > Na^+ > Li^+$$
$\longleftarrow$ stronger   ATTACHMENT   weaker $\longrightarrow$

Table 8.1 Cation exchange capacities of some soil components (after Caroll 1959 and Grim 1968)

Rock forming mineral or sediment	Cation exchange capacity (meq/100 g at pH 7)
Kaolinite	3 – 15
Halloysite ($2H_2O$)	5 – 10
Halloysite ($4H_2O$)	40 – 50
Illite (hydrous mica)	10 – 40
Chlorite	10 – 40
Glaukonite	11 – 20
Palygorskite	20 – 30
Allophane	25 – 50
Montmorillonite	80 – 150
Silicagel	80 – 150
Vermiculite	100 – 150
Zeolite	100 – 130
Organic substance in soil and recent sediments	150 – 500

Among monovalent cations, the preference takes place according to the following series:

$$Cs^+ > Rb^+ > K^+ > NH_4^+ > Na^+ > Li^+$$

On soil organic compounds, multivalent cations are generally preferred to monovalent cations, and transitional group metals to the strongly basic metals.

A quantitative measure for the selectivity of an exchanger towards a pair of monovalent cations (or its tendency to bond one of them more strongly than the other), can be derived using the Law of Mass Action as follows:

Assuming the exchange reaction to be:

$$AX + B^+ \rightleftarrows BX + A^+$$

at chemical equilibrium, we have

$$\frac{[A^+][BX]}{[B^+][AX]} = K_{AB}$$

Rearranging we get:

$$\frac{[A^+]}{[B^+]} = K_{AB} \frac{[AX]}{[BX]} \tag{8.1}$$

where $[A^+]/[B^+]$ is the ratio of ion activities in the solution; [AX] and [BX] refer to the concentration of A^+ and B^+ in the exchanger, in moles per unit weight of exchanger (they may also be expressed in mole fraction), and K_{AB} is the selectivity constant, which expresses the inequality of the activity ratios of the cationic pair. The following example illustrates the use of the relation.

a The ratio of ionic activities of A^+ and B^+ in a solution was found to be 1. The mole fraction of both species in the exchanger is also equal to unity. What is the selectivity of the exchanger?

In this case, according to the Eq. 8.1, $K_{AB} = 1$, the exchanger has no selectivity and both A^+ and B^+ are bonded to it with equal strength.

b If in the same example $K_{AB} = 10$ and the ratio of ion activities in solution $[A^+]/[B^+] = 1$, which ion will be represented by higher occupation on the exchanger?

In this case, $[AX]/[BX] = 0.1$, i.e. the exchanger is predominantly occupied by B^+.

c Under what condition can equal occupation occur in the case (b)?

This can only happen if $[A^+]/[B^+] = 10$, while $[AX]/[BX]$ will be equal to unity.

J. F. Walton, after studying the results of various experimental works, replaced Eq. 8.1 with the following empirical relation which is basically the same, except for the empirical exponent n.

$$\frac{[A^+]}{[B^+]} = K'_{AB}\left(\frac{[AX]}{[BX]}\right)^n$$

where $[A^+]$ and $[B^+]$ are the activities of cations in solution, [AX] and [BX] are the concentrations of ions in the exchanger, n is an empirical exponent and K'_{AB} is the exchange constant.

For a monovalent-divalent ion exchange, the reaction can be written:

$$2AX + B^{++} \rightleftarrows BX_2 + 2A^+$$

Accordingly, the empirical equation takes the form:

$$\frac{[A^+]^2}{[B^{++}]} = K'_{AB}\left(\frac{[A_2X_2]}{[BX_2]}\right)^n$$

For exchange involving more than two cations, the following relation applies:

$$\frac{[A^+]^2}{[B^+][C^+]} = K'_{AB} K'_{AC}\left(\frac{[AX]}{[BX][CX]}\right)^{n+m}$$

8.1.1.6
Factors Affecting Adsorption

The intensity of adsorption depends upon several factors, including physical and chemical properties of the pollutants themselves, as well as the soil matrix, composition, and surface properties. It is generally possible to summarise all these factors as follows:

- Mineralogical composition of the soil;
- Grain size distribution in the soil;
- The content and distribution of humic substances in soil;
- Chemical and physical properties of the soil solution;
- CEC of organic and mineral components;
- The pollutants, their nature, and chemical constitution;
- We may add to the above external conditions such as climatic conditions and agricultural practices.

In the following, each of these groups and collective factors will be briefly discussed.

Mineralogical Composition of the Soil

As mentioned before, clay minerals are the most important adsorbents in the soil environment, followed by some silicates and organic components. Accordingly, the intensity of adsorbance in soils will largely depend on the clay content of the soil, as well as on the proportion of other silicates in the mineralogical composition. The negative framework of the clays consists essentially of sheet structures of aluminium silicates, in which the exchangeable cations occupy interlayer positions or are located adjacent to the particle surfaces (Sect. 2.1.3). The structural properties of individual clay minerals normally play the principal role in determining selectivity, intensity, and mechanism of adsorption on these substances. In this respect we can identify the following types of adsorption on clay minerals:

- Adsorption on planar external surfaces (as in kaolinite) – here the tetrahedral layers are strongly held by hydrogen bonds, leaving only the external surfaces as available sites for ion exchange.
- Exchange in the interlayer space – here the ability of the layers to swell on hydration will contribute to the feasibility of ion exchange in the interlayer space, as in montmorillonite. In addition, the bonding of adjacent layers by cations, such as in vermiculite, will lead to cation exchange, if size conditions are fulfilled.

The CECs of clays, however, can considerably increase in the presence of other mineral matter such as aluminum and iron hydroxides. Terce and Calvet (1977) found that the two hydroxides increase the adsorptive capacity of montmorillonite.

Soil Matrix (Grain Size Distribution)

It has generally been observed that the rate of adsorption is higher on finer sediments than on coarser ones. Kennedy and Brown (1965) found that the content of total calcium and sodium in a sandy sediment was represented by about 90% in the grain size fraction of 0.12–0.20 mm, while the coarser fraction of 0.2–0.50 mm delivered only 10% of the whole content. Malcolm and Kennedy (1970) interpreted this behaviour as being due to the slow diffusion rates in coarse fractions compared to the fine ones. Despite the fact that concentration of certain cations in finer sediment fractions is also known from other sediments (e.g. carbonates), the following possibilities should be taken into consideration when interpreting this phenomenon for silicates. In the case of sandy sediments, where calcium and sodium are concerned, the relative hardness of certain silicate minerals (e.g. feldspars) and their lower resistance to abrasion, relative to harder silica compounds with low calcium and sodium content (e.g. quartz), may control the distribution of such cations between the different grain size fractions in the same sandy layer. To attain better bases for interpretation, the mineralogical composition of the different grain size fractions must be determined using X-ray diffraction methods. However, one should also bear in mind that the high surface area, and hence high surface energy, of fine sediments, supplies an excellent interpretation for the higher rates of ion adsorption on them.

Humic Substances and Their Distribution in the Soil
Humic substances containing carboxyl and phenolic hydroxyl functional groups increase the CEC of the soil. In general, the presence of active functional groups (e.g. carboxyl, hydroxyl, carbonyl, methoxy and amino groups) is thought to be a positive influence on the CEC of a soil.

Chemical and Physical Properties of the Soil Solution
A major part of pollutants, however, passes in solution, or in particulate form, to the vadose, or even saturated groundwater zone. In the presence of clays, water molecules are adsorbed on their surfaces to form hydration shells; these provide adsorption sites for pollutant molecules. Water adsorbed on clay molecules generally has higher rates of dissociation, providing surfaces with an acidic character (Yaron et al. 1996) which may increase the exchange capacity of the soil. Mechanisms by which pollutants are transported to deeper horizons of the soil are collectively called *infiltration*. They form the optimum environment for the spreading of pollutants, as reported by Calvet in 1989. He discovered that pesticides are transported to the adsorbing surfaces by water. Descending contaminated fluids, that might end in joining the vadose or the saturated zones of the groundwater, are generally known as *leachates*. In the vadose zone, leachates spread horizontally in the direction of the groundwater flow. Such movements are controlled by the laws governing transport phenomena in groundwater, and will be discussed later.

The Pollutants, Their Nature and Chemical Constitution
The composition and nature of contaminants control to a considerable extent not only the solution and diffusion processes, but also adsorption on the soil grains. Such control may be explained by the fact that ion exchange and hydrolysis reactions are particularly sensitive to the parameters (pH, Eh) of the chemical environment, created by the contaminants in their direct vicinity. An example of this may be provided by the adsorption of organophosphorus pesticides on clay surfaces. As mentioned before (p. 139), organophosphorus pesticides are members of the phosphoric acid ester group,

Fig. 8.5 General formula of organophosphorus compounds

$$\begin{array}{c} \quad O(S) \\ RO \diagdown \; \| \\ \quad P - O - X \\ RO \diagup \end{array}$$

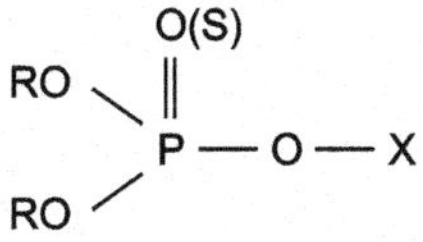

having the general formula shown in Fig. 8.5, where the two alkyl groups (R) may be methyl or ethyl, but they are the same in any given molecule. X (the leaving group) is generally a complex aliphatic cyclic group. It was found that adsorption of organophosphorus compounds on clay surfaces is influenced by the nature of the constituent group X (Yaron 1978). This is due to the fact that such esters are stable at pH-values ≤ 7, i.e. at neutral or acidic media, but they are susceptible to hydrolysis under alkaline conditions, where the P–O–X ester bond breaks down. The rate of this process is related to the nature of the group X.

8.1.2
Nonadsorptive Retention

8.1.2.1
Trapping

The entrapment of solid particles and large dissolved molecules in the pore space of the soil forms one of the major mechanisms in the retention of pollutants in the soil. This type of retention occurs following the three mechanisms shown in Fig. 8.6, drawn after J. R. Boulding (1995). The figure illustrates the three mechanisms, which may be concisely described as follows:

1. *Caking.* This may occur physically when the pollutant particles are larger than the soil pores. In this case, the entrapped particles form a layer (*cake*) on the surface where the pore sizes become too small. Caking may also result from biological activities through which particles cluster in bigger lumps, clogging the soil pores.
2. *Straining.* Straining occurs when pollutant particles are about the size of the soil pores. They move down the pores until they are entrapped at the entrance to a pore which is too small to pass through.
3. *Physical-chemical trapping.* The limitation of the flow, through the clogging of the pore space, may occur because of physical or chemical transformation, such as the production, by chemical reactions, of new products having molecular sizes that exceed that of the soil pores.

An example is the flocculation of colloidal material resulting from the precipitation of iron and manganese oxides.

8.1.2.2
Precipitation

The retention of contaminants in the soil may often occur through the passing of contaminants from a dissolved form to an insoluble form, in the course of geochemical

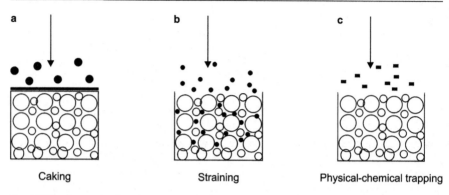

Fig. 8.6 Trapping mechanisms in porous media (based on Boulding 1995, after Palmer and Johnson 1989)

reactions taking place within the soil pores. Precipitation reactions are controlled by acid-base equilibria and redox conditions. They are reversible and may lead to the dissolution of formerly precipitated compounds, if conditions are changed. A further discussion of these types of reactions will be given later in the section dealing with transformations.

8.1.2.3
Infiltration

This is perhaps the most common mechanism of contamination of soil solutions in the vadose zone, as well as deeper regions of the saturated zones of the groundwater. As fluids move downward under the influence of gravity, they dissolve materials to form leachates that contain inorganic and organic constituents. As they reach the saturated zone of the groundwater, the contaminants spread horizontally and vertically by joining the main cycles of the geochemical flows. Fortescue (1980), following the pioneering work of Kozlovskiy (1972), classified the patterns of material flow in landscapes into three main categories which may be described as follows (see also Fig. 8.7):

1. *Main migrational cycle (MMC)*. This type of flow resembles the one familiar in geochemical cycles, i.e. the chemical substances are predominantly transported in a vertical direction upward from the soil to plants and animals, and then downward from plants and animals to the soil, approaching a steady state (Fig. 8.7a).
2. *Landscape geochemical flow (LGF)*. This involves a progressive transport of material parallel to the soil surface (see Fig. 8.7b). It takes place within a prism (Landscape prism: Fortescue 1980), including portions of the atmosphere, the pedosphere, and the lithosphere, as shown in Fig. 8.7. An example of chemically active air migrants in the LGF is carbon dioxide and other gases that would dissolve in soil water, causing a shift in its chemical constitution.
3. *Extra landscape flow (ELF)*. A third type of material flow in landscapes is ELF. Applying this to soils as a portion of the landscape prism, we may define it as the flow of chemical substances into the soil where they would be accumulated (+ve flow), or out of it (−ve flow), see Fig. 8.7c.

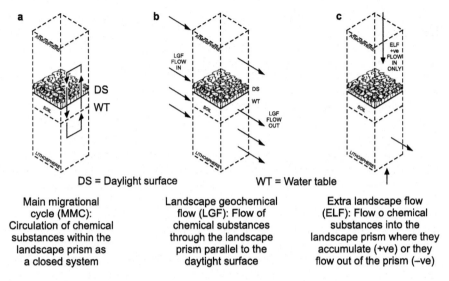

Fig. 8.7 Illustration of the three flow patterns according to Fortescue (1980)

8.2
Contaminants Transport

The spreading and transport of contaminants during any of these cycles of geochemical flows occurs according to two principle transport mechanisms: (1) advection, movement caused by the flow of groundwater; (2) dispersion, movement caused by the irregular mixing of fluids during advection.

Advection
This is the mechanism controlling fluid flows in the soil and underlying earth layers. It is quantified by Darcy's law:

$$Q = \frac{-K\rho A(h_2 - h_1)}{\eta l}$$

where Q is the total discharge of fluid per unit time (cm³ s⁻¹), A is the cross-sectional area of flow path (cm²), l is the length of the flow path, ρ is the density of fluid (g cm⁻³), η is the dynamic fluid viscosity (mPa s), h_2-h_1 is the hydraulic head or pressure drop across the flow path (g cm⁻²), K is the permeability constant in *Darcies*.

Dispersion
Besides advection, contaminants may be transported in the soil by hydrodynamic dispersion, which is defined as the net effect of a variety of microscopic, macroscopic,

and regional conditions that influence the spread of a solute concentration front through an aquifer[2].

Figure 8.8 shows in a schematic way the different mechanisms involved in spread processes of contaminants through dispersion.

8.2.1
Microscopic Dispersion: Molecular Diffusion

On a microscopic scale, dispersion may occur due to: (*a*) molecular diffusion along concentration gradients, (*b*) transport along viscosity or density gradients, or (*c*) transport by forces arising from change of pore geometry. The transport process through diffusion occurs in gases, liquids and solids. A contaminant dissolved in soil water may diffuse along concentration gradients in the fluid to attain uniform concentration in a given portion of the pore space, which will again contribute to the geochemical gradient in neighbouring regions. Gases in the soil air, resulting from volatile components of contaminants, such as fuel spills, will diffuse as well throughout the pore system. The rate of diffusion (matter transport) was found to be proportional to the concentration gradient. This relation finds its mathematical expression in *Fick's First Law of Diffusion*:

$$J_z \text{ (matter)} = -D(dN/dz)$$

It states that the flow of matter along an axis z is proportional to the concentration gradient along the axis (see Fig. 8.9).

Density and viscosity changes in soil fluids control the transport of contaminants in the pore space. This follows from the fact that the mobility in the pore space is related to other physical parameters such as temperature, density, and viscosity. The diffusion constant is related to viscosity via *Stokes-Einstein relation*:

$$D = kT / 6\pi\eta a;$$

where D = diffusion coefficient, T = absolute temperature, η is the viscosity and a is the radius of flow cylinder.

Kaufman and Mckenzie (1975) reported that the apparent hydraulic conductivity of an injection zone in the Floridan aquifer receiving hot organic wastes increased about 2.5 times due to temperature differences. Oberlander (1989) also reported that density variations might cause errors in estimations of flow directions.

Transport by Forces Arising from Changes of Pore Geometry

In Darcy's law (see above), K (the permeability constant) describes permeability as the property by which fluids are allowed to pass through a medium without changes in the structure of the medium or displacement of its parts. It depends largely upon

---

[2] Op. cit.

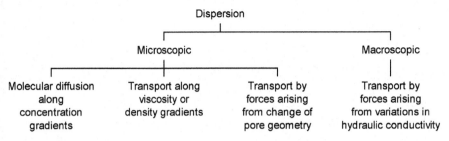

Fig. 8.8 Dispersion processes in soil

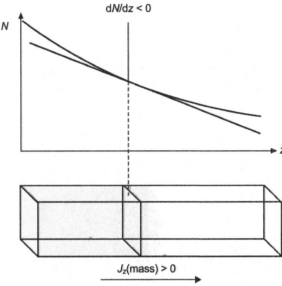

Fig. 8.9 Flux of particles down a concentration gradient (Atkins 1978)

soil texture and the geometry of its pores. Generally, it is related to porosity by the following theoretical relation:

$$\phi = a + b \log k$$

Where ϕ is the porosity and k is permeability. However, permeability may change without an alteration of porosity, due to properties inherent in the geometrical organisation of the pores, such as the small-scale roughness of the pore walls or the path length the fluid must follow during transport. The latter is commonly known as *the tortuosity factor*. In calculating the mass flux of vapour in soils, Jury and Fluhler (1992) make allowances for reduced cross-sectional areas and increased path length of gas molecules in soil, by introducing a tortuosity factor in Fick's first law, so that it takes the following form:

$$J_g = \xi_g(a) D_g^a \frac{\partial C_g}{\partial z}$$

where J_g is the gas flux, D_g^a the binary diffusion coefficient of the vapour in air, and $\zeta_g(a)$ is the tortuosity factor.

Porosity, however, may change in a way that enhances transport or triggers off higher transport rates. Such cases occur when the soil is in some parts fractured or has dissolution cavities. In such cases, forces arising from the changes of pore geometry may lead to an enhancement of the transport processes. Some examples of pore geometry leading to an increase in fluid transport are shown in Figs. 8.10–8.13.

8.2.2
Macroscopic Dispersion

The classical advection/dispersion model for contaminant transport in soil is only valid for dispersion on a micro scale, i.e. so long as the pore system of the soil is taken into consideration. A different physical situation is encountered in soils where, due to particle aggregation or development of cracks, the hydraulic conductivity is considerably changed. An example of this is the situation in clay rich soils, when consecutive cycles of wetting and drying produce shrinkage cracks in the soil body, that eventually serve

Fig. 8.10 Pore geometry in a homogenous fine sediment (SEM) (*P* stands for pore). Observe roughness of the pore walls (tortuosity)

as preferential routes for fluid transport. During transport through cracks and large pores, retention on soil surfaces is reduced to a minimum because in this case only a small portion of the soil surface encounters the fluid.

Fig. 8.11 Pore geometry in a fine sediment of different grain sizes (SEM)

Fig. 8.12 Sheet pores in calcite aggregates (SEM)

Fig. 8.13 Pore geometry in skeletal crystals (dolomite aggregates – SEM)

8.3
Behaviour of Non-Aqueous Phase Liquids (NAPLs) in Soils

Synthetic organic solvents, which are insoluble or slightly soluble in water, are grouped into one category of contaminants known collectively as *non-aqueous phase liquids* (NAPLs). In fact, one uses the term NAPL for all immiscible pure chemicals that may contaminate the soil. The behaviour of this category in soils depends on various factors, such as the degree of saturation of the soil, the density and viscosity of the NAPL relative to water, and the volume or dimension of spill introduced into the landscape. As in the case of the aqueous phase liquids, contaminants of this category may be retained on the soil surface, spread in the vadose zone, or, if added in great volumes, may infiltrate the soil to reach the groundwater table at the saturated zone. Transport of NAPLs follows principally the same hydrologic principles controlling permeability in porous media. Absolute permeability is independent of the nature of the fluid; it depends only on the medium which is described through its coefficient of permeability k, given by the equation:

$$k = Nl^2$$

where N is a dimensionless number depending on pore space characteristics, such as grain shape and packing (it may be constant for a given soil), and l is the length of the pore structure of the soil (a factor related to grain size).

If different phase fluids are present, such as gas, an organic solvent, or water, the fluids create complex mutual interferences; a so-called effective permeability for each phase (k_g, k_o, k_w) would control material transport. Absolute permeability of the me-

8.3 · Behaviour of Non-Aqueous Phase Liquids (NAPLs) in Soils

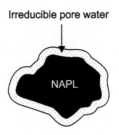

Fig. 8.14 Non-aqueous phase liquid (NAPL) displacing water in pore space

dium, as given by the above-mentioned equation, ceases to be the only control factor for transport in the medium. In such a case, $k_g + k_o + k_w$ may be lower than the observed permeability k_{abs} due to the fact that mutual interferences are retardative rather than enhancing. This clearly shows that the validity of Darcy's law is restricted to single phase homogeneous or laminar fluid flow.

In cases of NAPL spills on water-saturated soils, water and NAPL would be competing for flow within the pore system, as mentioned before. In such cases, the NAPL drives the pore-water into finer and finer spaces where capillary forces hold it. Soil air and any existing gaseous phases are also driven out of the pore-space in the course of this process, so that, at the end, the central portion of the pore fills with non-aqueous phase liquids, while irreducible pore-water, held by capillary forces, may form a thin layer lining the pore (see Fig. 8.14). At this point, the fluid saturation, with respect to water, decreases to almost zero, while the fluid saturation with respect to NAPL reaches its maximum. The fluid saturation is defined as the fluid volume expressed as a fraction of the total pore space. The velocity of the transport of a given fluid is directly proportional to its fluid saturation in the medium. In cases of small spills on a soil not fully saturated with water, the volume of NAPL would not be enough to expel all the water out of the pores. This leads to a distribution of the fluid saturation of the soil between the NAPL and the pore water, followed by a depression of the transport velocities of both to a level lower than the expected one, if either of them is dominant in the pore space. A certain fraction of the NAPL adheres to soil particles in the vadose zone – forming the so-called *residual saturation*. This may be later transported by dissolution or be volatilised in the pore space.

The behaviour and transport patterns of NAPLs depend on two important factors. These are the relative density of the NAPL with respect to water (lighter or denser than water) and the size of spill. In the following, both classes will be separately discussed.

8.3.1
NAPLs Lighter than Water (LNAPLs)

- *Small spills.* LNAPLs may, in the case of small spills, be retained on grain surfaces within the vadose (unsaturated) zone. Further penetration can only follow if the LNAPL, retained as residual saturation, is dissolved by penetrating waters and carried in solution to deeper parts of the contaminated site. This is often the case with contaminants such as benzene, toluene, or xylene.

    A further path of LNAPL's residual saturations to deeper horizons of a soil may be provided by evaporation and diffusion within the soil pore space.

- *Large spills.* Larger spills of LNAPLs will normally be followed by dispersion both in vertical and horizontal directions. At the beginning, contaminants reaching the water table may change the wetting properties of water (i.e. changing capillary pressure and/or viscosity at the interfaces in the system mineral-water-NAPL), leading to a collapse of the capillary fringe (see Fig. 10.13 in Sect. 10.4). Such a dramatic change will normally be followed by a depression of the water table. However, when the discharge of NAPL stops, the contaminant flows preferentially in a horizontal direction in the vadose zone until residual saturation is reached. This leads to a relief in the upper horizons, making it possible for the water table to rebound to its original level (Palmer and Johnson 1989).

8.3.2
NAPLs Denser than Water (DNAPLs)

- *Small spills.* Denser non-aqueous phase liquids would, due to their denser nature, displace water on their way to deeper parts of the vadose zone, yet the confrontation between the NAPL with water which is more viscous than the contaminant results in an unstable liquid-water boundary, forming viscous fingers penetrating the vadose zone until residual saturation is reached. Penetrating waters and dense vapours affecting the capillary fringe may help in forming a contaminant plume. At places where the chemicals are held in place between the soil grains, aggregates of NAPL may persist in forming local concentrations, known as *ganglia*.
- *Large spills.* Depending on the dimensions of the spill, the contaminant may either be dispersed around or near the water table, until residual saturation is reached, or penetrate deeper into the saturated zone, forming pools at the surface of impermeable layers.

The surface tension between NAPLs and the water wetting the surfaces of soil grains make it difficult for both phases to mix, and largely reduce their miscibility. This, together with the low solubility of NAPLs, makes these pollutants dissolve very slowly in the groundwater. A variety of processes, including the following, have been proposed to increase their dissolution rates.

- *Flushing with hot water.* The reasoning behind this process is the possibility of increasing the chemical solubility of the NAPLs through decreasing water viscosity. However, Imhoff et al. (1995a) found that this process has very little effect on the dissolution rate of NAPLs at contaminated sites.
- *Steam injection.* Hunt et al. (1988) demonstrated through experimental work that steam injection in porous media (sand) was very effective in removing immiscible pure phase liquids such as trichloroethylene, toluene, and gasoline. This method is very effective, yet it is connected to high energy-costs. The energy from 6.8 l of fuel oil would be necessary to clean one cubic meter of contaminated aquifer.
- *Flushing with solvents.* Some solvents, e.g. methanol, are capable of decreasing the interfacial tension between NAPLs and water, leading to the mixing of the two phases and hence increasing the chemical solubility of the pure phase. Imhoff et al. (1995a) used different concentrations of methanol – water solution and found out that us-

ing a 60% methanol solution decreased the surface tension to 25% of its original value, thus largely increasing the chemical solubility of the pure phase contaminant in water.
- *Flushing with surfactants.* Like in the above-mentioned methods, the reasoning here depends upon changing the wetting properties of water, in particular decreasing the interfacial tension between water and the pure contaminant phase. Surfactants can largely decrease the interfacial pressure, through partitioning of the pure phase into surfactants micelles. This may be followed by an increase of the chemical solubility of the NAPL.

Chapter 9
Pollutants' Alteration, Transformation, and Initiation of Chemical Changes within the Soil

In a highly complicated system of different phases, such as soil subsurface, penetrating substances will go through a myriad of chemical, physical, and biological processes that will determine their fate, besides controlling as well the degree of their toxicity to the environment. Such subsurface processes are broadly classified into the following groups:

- Physical processes (processes related to chemical mobility)
- Chemical processes
- Biological processes

9.1 Processes Related to Chemical Mobility

These include processes in which no net chemical change occurs. They normally affect physical conditions that control phase distribution of the substance, i.e. its association with aqueous or solid phases under given environmental conditions. It is due to this that these processes are collectively known as *distribution processes.*

Distribution processes include processes such as advection, dispersion and volatilisation. Besides the processes discussed in Chap. 8, they include all those processes that may affect the mobility of a substance in the subsurface environment. Of these, the following are most prominent:

- Immiscible phase separation
- Acid-base equilibrium
- Precipitation-dissolution reactions

9.1.1 Immiscible Phase Separation

Depending on their miscibility in water, fluids or gases will separate forming an independent layer in a multi-component system. An example may be given by NAPLs separating to form a floating layer on the surface of groundwater (LNAPL), or sinking to form a deposit on an impermeable bed at the base of the aquifer (DNAPL).

9.1.2
Acid-Base Equilibrium

The Brønstead-Lawry theory defines an acid as any substance that can give or donate a proton. Likewise, the theory defines a base as a substance ready to accept a proton. The following examples illustrate both definitions:

- Ionisation of hydrochloric acid:

$$HCl^0 \text{ (neutral)} \rightleftarrows H^+ + Cl^- \tag{9.1}$$

- Ionisation of sulphuric acid:

$$H_2SO_4^0 \text{ (neutral)} \rightleftarrows 2H^+ + SO_4^{2-} \tag{9.2}$$

- Ionisation of carbonic acid:

$$H_2CO_3^0 \text{ (neutral)} \rightleftarrows H^+ + HCO_3^- \tag{9.3}$$

All three reactions are reversible. In Reaction 9.1 (ionisation of hydrochloric acid), the acid completely dissociates to give a proton (H^+) and an anion (Cl^-), which is in the reverse reaction ready to recombine with the proton, restoring the original structure of the acid. In this case, Cl^- may be called a base according to the definition. In Reaction 9.2 (sulphuric acid), the same happens as in Reaction 9.1 and here also SO_4^{2-} (the sulphate group) plays its role as a base that may restore the acid on recombining with the released protons. Reaction 9.3 (carbonic acid), though principally the same as the others, shows a basic difference in that the acid does not completely dissociate like in the two other cases. It partially dissociates so that on reaching equilibrium, a fragment of the original neutral species (HCO_3^-) remains unionised in the aqueous solution. Such acids, incapable of complete dissociation in one step, are generally known as *weak acids*, while acids completely dissociating in one stage are referred to as strong acids. Aqueous solutions of weak acids have pH-values ranging from 4–6, while those of strong acids possess pH-values lower than 4. Bases may also be classified, on the same grounds, into weak and strong bases. Thus, bases like NaOH or CaOH, capable of complete dissociation to give a cation and an anion (see Eq. 9.4), are also known as *strong bases*.

$$NaOH^0 \text{ (neutral)} \rightleftarrows Na^+ + OH^- \tag{9.4}$$

The hydroxyl ion in the last equation might also combine with a proton to form water, thus emphasising its role as a base. In comparison to sodium, potassium, or calcium hydroxides, bases like $Fe(OH)_3$ and $Al(OH)_3$ are described as weak bases because they do not ionise further in water. This may be illustrated by the dissociation of the mineral jarosite, which generally dissolves in water to form ferric oxyhydroxide and K^+, producing a considerable acidity, according to the following equation:

$$KFe_3(SO_4)_2 + 3H_2O \rightleftarrows K^+ + 3Fe(OH)_3 \text{ (ppt)} + 2SO_4^{2-} + 3H^+$$

Thus, the mineral seen here is a salt of a strong acid (H_2SO_4) and a very weak base $Fe(OH)_3$, which forms a precipitate that almost does not ionise.

Acid-base equilibria in a subsurface environment control the prevailing pH-values and hence the stability and solubility of the substances present. This plays a principle role in determining mobility, fate, and toxicity of the penetrating pollutants.

Buffering Capacity

If a strong acid or base is added to a solution, the pH of the solution changes according to whether H^+ is removed or released in the solution. Some solutions, however, resist changes in their pH from the addition of small amounts of acid or alkali. Such solutions are called *buffer solutions*, or simply *buffers*, and the degree of resistance to change is called the buffering capacity.

To understand this, let us consider the following example of equilibrium between a weak acid (acetic acid) and its sodium salt (sodium acetate):

$$CH_3COOH \rightleftarrows CH_3COO^- + H^+ \tag{9.5}$$

$$CH_3COONa \rightleftarrows CH_3COO^- + Na^+ \tag{9.6}$$

If to a solution of acetic acid and sodium acetate a slight amount of hydrochloric acid is added, the hydrogen ions from the hydrochloric acid will combine with a portion of the acetate ions to form unionised acetic acid, which is a weak acid, i.e. slightly dissociates. The removal of the added hydrogen ions to form acetic acid means that they will not affect, or change, the hydrogen ion concentration of the solution, and thus will not change its pH.

Similarly, if OH^- ions are introduced by the addition of a small amount of an alkali, the OH^- will combine with the hydrogen ions resulting from the dissociation of acetic acid to form water. Removal of H^+ in this way will disturb the equilibrium in Eq. 9.5, enhancing the acetic acid to increasing dissociation, in order to restore equilibrium. Thus, the pH of the solution will practically remain unchanged. Other weak acids and their salts (e.g. carbonic acid) will also display the same pattern of behaviour by retaining their pH after the addition of small amounts of strong acids or bases. The best examples of these are seawater and other brines containing carbonic acid and its salts, which may resist changes in its pH on being exposed to small spills of strong acids or bases. The reaction in this case would be controlled by the following equilibrium:

$$CO_2(aq) + H_2O \rightarrow \underset{\text{carbonic acid}}{H_2CO_3} \rightleftarrows \underset{\text{bicarbonate}}{HCO_3^-} + H^+ \rightleftarrows \underset{\text{carbonate}}{CO_3^{2-}} + 2H^+$$

On attaining equilibrium, the concentration of H^+ remains constant. If the solution is exposed to a strong acid, the concentration of H^+ increases, leading to a shift in equilibrium to the left, forming unionised carbonic acid, upon which the excess H^+ ions will be removed, keeping the pH of the solution practically unchanged. If OH^- ions are added to the solution by the introduction of a strong base, they combine with H^+ to produce water, keeping the pH of the solution at its initial value. The buffer capacity, however, depends on the initial concentration of carbonates and bicarbonates, since an exhaustion of these will stop the formation of carbonic acid, thus allowing an increase in the hydrogen ion concentration, and consequently leading to a lower pH-value.

9.1.3
Dissolution-Precipitation Reactions

Solubility and precipitation are perhaps the most characteristic phase distribution processes that take place in the soil environment. For in dissolution there would be a transition from a gaseous or a solid phase into an aquatic one, and, if at any time precipitation occurs, it will follow just the opposite way. Thus, dissolution obviously induces an increase in mobility, while precipitation acts as an inhibitor.

The solubility of a substance in the soil environment depends on the nature of the substance, as well as on physical parameters such as temperature, pressure, pH, and Eh (redox potential). As for the nature of the contaminant substance, we find that organic toxic substances are less soluble than inorganic salts. This is naturally to be expected, due to the hydrophobic character of non-polar substances. Another factor determining solubility in water would be the already existing concentration of the same substance in solution, and how far it is from the equilibrium concentration at a given temperature. This is mainly controlled by the solubility product, which is an expression of the maximum amount of a substance that will dissolve in a solution at a given temperature and pressure. Precipitation occurs when the value of the solubility product is exceeded. In aquatic systems, this may occur due to a change of equilibrium conditions such as temperature, pressure, pH, or Eh, making the boundary separating distribution processes from processes encompassing chemical change to fade.

9.2
Chemical Transformation Processes

In fact, chemical transformation processes and distribution arrangements that may affect the chemical mobility of a substance are complementary in nature. They normally go hand in hand with one or the other, implementing the advent of its successor. Nonetheless, the two groups of processes are completely different in their mode of action. While, as said before, distribution processes mainly affect the mode of association of a given substance, chemical transformation will in the first place change the chemical structure of the substance, bringing about a net chemical change. Both may happen parallel to one another, and it is rarely observed in the soil environment that one of them occurs without the other following on its heels. This may be illustrated by the case of precipitation, which, as mentioned above, is one of the most typical examples of phase distribution. Perelman (1967) classified processes of metal precipitation from natural waters into the following types, which include a great deal of transformation going hand in hand with the processes affecting mobility and phase distribution:

- *Oxidation type.* An example of this is the precipitation of iron and manganese oxides by the oxidation of reducing waters.

- *Reducing type.* Examples are given by the precipitation of uranium, vanadium, copper, selenium, and silver, as metals, or lower valency oxides, by the reduction of oxidising waters. This is usually caused by encountering organic matter or by mixing with reducing waters or gases.
- *Reducing sulphide type.* Sulphate waters carrying ions of copper, silver, zinc, lead, mercury, nickel, cobalt, arsenic, or molybdenum, may be reduced to precipitate sulphides of these metals. This occurs usually by the action of sulphate-reducing bacteria, or through an encounter with organic matter.
- *Sulphate and carbonate type.* Alkali metals such as barium, strontium, and calcium, may be precipitated as carbonates following a shift in equilibrium relations. Griffith et al. (1976) reported lead-precipitation from landfill leachates as carbonates.
- *Alkaline type.* Percolation of acidic solutions into carbonates and silicates, as well as their encounter with alkaline solutions, lead to precipitation of metals like calcium, magnesium, strontium, manganese, iron, copper, zinc, lead, and cadmium.
- *Adsorption type.* This type encompasses all transition metals which are susceptible to adsorption on clays and other particulate substances.
- *Oxidation – reduction type.* Mobility of trace metals in aquatic solutions is largely influenced by the redox status of their environment, even though they generally are not directly involved in oxidation-reduction reactions. Figure 9.1 shows the stability relations in the system $Zn + S + CO_2 + H_2O$. In this system, three solid phases (precipitates) are possible: the sulphide, carbonate, and hydroxide of zinc. Under reducing conditions, sulphide and hydroxide are the stable phases at high pH-values. At lower pH-values, however, the hydroxide dissolves and only the sulphide is precipitated.

  Oxidising conditions induce the precipitation of amorphous iron and manganese oxihydrides, which later, as adsorbents, affect tremendously the mobility of trace metals in the solution. Reducing conditions induce the reduction of Mn^{4+} to Mn^{2+} and Fe^{3+} to Fe^{2+}, thereby solubilising their associated and adsorbed trace metals. Figure 9.2 summarizes the redox chemistry of iron and manganese.
- *Complex formation and chelation.* Most trace elements exist in water as hydrated ions rather than as free ones. In the course of hydration, water, due to its polar character and the unsatisfied charges on both hydrogen and oxygen, orients its molecules so that oxygen points to the cation while hydrogen points away from it (see Fig. 9.3). The water molecules, thus connecting themselves to the cation, form a hydration shell around the cation, producing a species known as *cation aquocomplex* (copper aquocomplex in Fig. 9.3). Each of the water molecules will be called a *ligand* and, as it is seen here, ligands are not attached to the cation through electron sharing, but rather due to electrostatic forces arising from pairs of electrons that exclusively belong to the cation. The so formed species may further be associated to anions that would form an additional sphere at the outer surface of the water shell, forming what is known as an *outer sphere complex.* Anions in the outer sphere complex are associated to the cation by long-range electrostatic forces; their association is not strong enough to displace any of the water molecules in immediate contact with the cation.

  A classical example of the formation of outer sphere complexes is given by the formation of complex cobalt chlorides. Cobalt forms a simple ionic chloride – $CoCl_3$, in which three electrons are transferred from a cobalt atom to chlorine atoms. In the presence of ammonia, $NH_3(aq)$, however, this cobalt(III) chloride can form a series

Fig. 9.1 Stability relations in the system $Zn + S + CO_2 + H_2O$

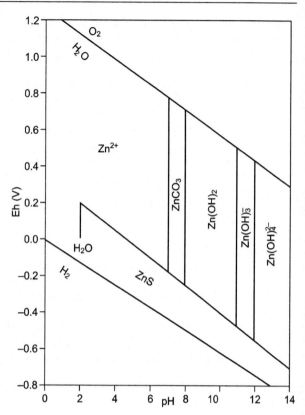

Fig. 9.2 Schematic representation of the redox chemistry of iron and manganese

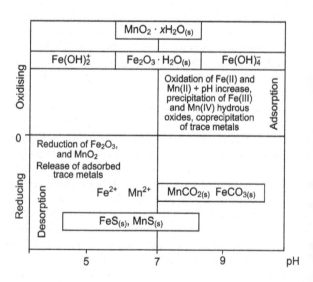

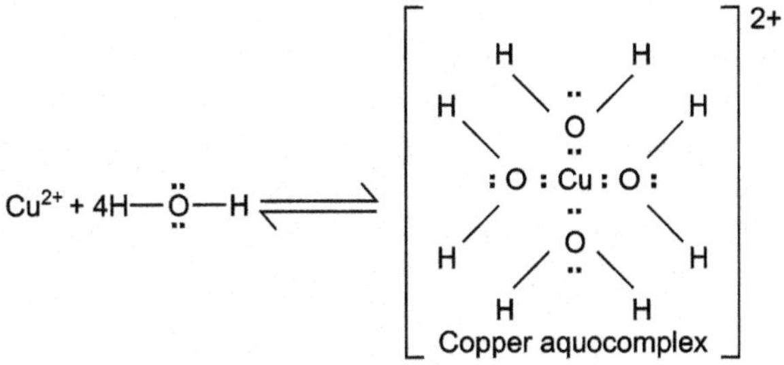

Fig. 9.3 Formation of copper aquocomlex

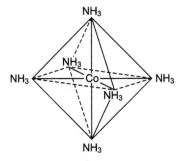

Fig. 9.4 Octahedral coordination of cobalt with ammonia to form $[Co(NH_3)_6]^{3+}$

of distinct compounds having up to six ammonia molecules as ligands surrounding the cobalt cation $[Co(NH_3)_6]Cl_3$, as shown in Fig. 9.4.

A ligand is defined as a species capable of donating an electron pair to a central metal ion at a particular site in a geometrical structure. Therefore, a ligand is essentially considered as a Lewis base and the metal ion as a Lewis acid. If a ligand is capable of donating a single electron pair, as in ammonia, it is called a *unidentate* (single-toothed) ligand. Some ligands, however, are capable of donating more than a single electron pair to different sites in the geometrical structure of a complex ion – these are called *multidentate ligands*. Most inorganic ligands are unidentate, while multidentate ligands are normally organic ones.

Humic acids are multidentate ligands, forming cage structures around metal ions when they associate with their complexes (see Fig. 3.5). When bonding between a metal and a multidentate ligand results in such ring or cage structures, the process is called *chelation*, the species produced is a chelate and the multidentate ligand is a chelating agent. The expression *chelate* is derived from the Greek word *Chela* meaning a crab's claw.

Complex formation in aquatic systems renders metals less bioavailable for organisms. It also influences the adsorption of metals on colloid substances and may increase the solubility of minerals. Solutions with high ionic strengths are favourable media for the formation of complexes, the stability of which will be directly proportional to the cation charge and inversely proportional to its radius.

9.2.1
Hydrolysis

As discussed in Sect. 1.2 under chemical weathering, hydrolysis is a pure chemical process during which a proper chemical reaction takes place between water and another substance, to produce, or consume, a proton (H^+) or an electron (OH^-). A typical example was given by the reaction of the mineral albite with slightly acidic water to produce kaolinite (see Eq. 1.2).

Organic pollutants percolating through the soil environment may be chemically transformed by hydrolysis, either through water addition to the molecule, or though replacement of some functional groups by water. An example of replacement reactions is the replacement of halide ions in alkyl halides to form alcohol:

$$RX + H_2O \rightleftharpoons R\text{-}OH + HX$$

where R = alkyl group and X = halide.

Hydrolysis reactions taking place by the addition of water molecules normally proceed more readily than those taking place by replacement. An example of these may be given by the addition of water to alkenes to form alcohol (Fig. 9.5).

9.3
Biodegradation and Biologically Supported Transformations

As mentioned before (see Sect. 3.2), organisms (soil biota) form a very important integrative constituent of soil, and as such they play a decisive role in determining the fate of foreign substances added deliberately or accidentally to the soil body. They normally respond to the addition of xenobiotics by initiating two main types of reactions:

- *Primary metabolic reactions* (also known as phase I biotransformation), during which the foreign substance is rendered more soluble in water by the addition or exposure of functional groups on it.
- *Secondary metabolic reactions* (phase II biotransformation), through which the products of the primary reactions are conjugated with endogenous groups to facilitate their excretion. On passing into solution, the foreign substance will be capable of penetrating the organism with rates which are specific for organisms and their anatomical and biological peculiarities. In case of high rates of penetration, i.e. if a foreign substance enters the organism more quickly than it can be eliminated, it accumulates in some of its organs, and, if the substance is toxic, this goes on until a toxic concentration is reached. At normal rates of penetration, secondary metabolic transformations will lead, by conjugation with endogenous compounds, to the formation of substances that may be used as energy sources by the organism, or to ones that are easily eliminated by excretion. Figure 9.6 explains this in a schematic way.

From this schematic representation it becomes clear that the level of metabolism is a principal factor in determining the persistence, or degradation, of the foreign sub-

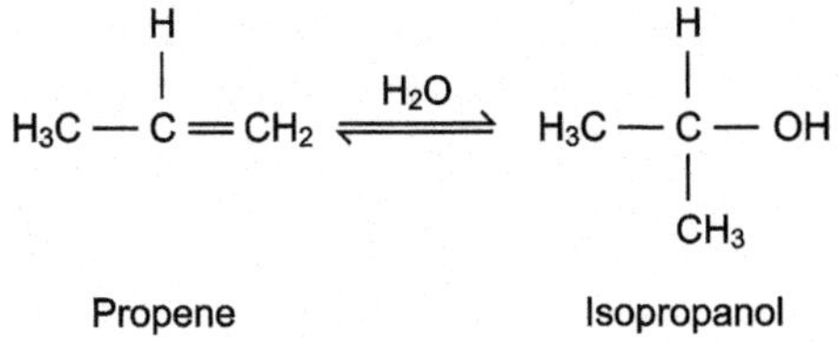

Fig. 9.5 Addition of water to alkenes to form alcohol

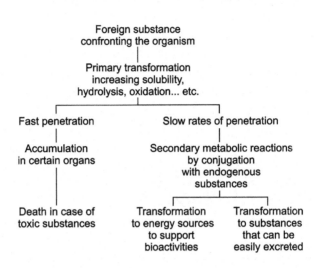

Fig. 9.6 Schematic representation of the response of soil biota to foreign substances

stance in the soil, thus forming one of the most important conditions under which biodegradation succeeds.

As it may be expected, the activation of these reactions requires a high energy demand, or at least a source of catalysis, to secure their advancement. This catalysis is supplied in organisms by *enzymes*. These catalytically active high molecular weight proteins enable the activation of biological transformations of substances into energetic sources or easily eliminated chemicals. In order to understand the fundamental action of enzymes in biologically assisted transformations in soil, an outline of the chemical processes involved in enzymatic actions will be shortly introduced.

9.4 Enzymatic Transformations: A Primer on Enzymes, Their Types and Mode of Action

Enzymes take their specific names from that of the substrate, the substance they help to change, by adding the suffix "-ase" to the name of this substance, e.g. proteinase, lipase, etc. According to the type of chemical reactions in which they are usually involved, enzymes are classified into six main groups, namely, the *hydrolases*, those that assist hydrolysis; the *transferases*, those that help transfer a certain group to another substrate, not usually water; the *oxidoreductases*, those that transfer hydrogen or electrons between two substrates; the *lyases*, those that remove groups from their substrates; the *ligases (synthases)*, those that catalyse the joining of two molecules (i.e. synthesise a C–C bond) at the cost of chemical energy; and the *isomerases*, which are enzymes that catalyse intramolecular rearrangements. Each of the six groups is further classified into subgroups, as seen in Fig. 9.7.

An enzyme generally consists of two fragments: a protein portion, forming a colloidal carrier, and a non-protein fragment, made of a simple, well defined compound, which, unlike the protein portion, can be dialysed and is more resistant to heat. The

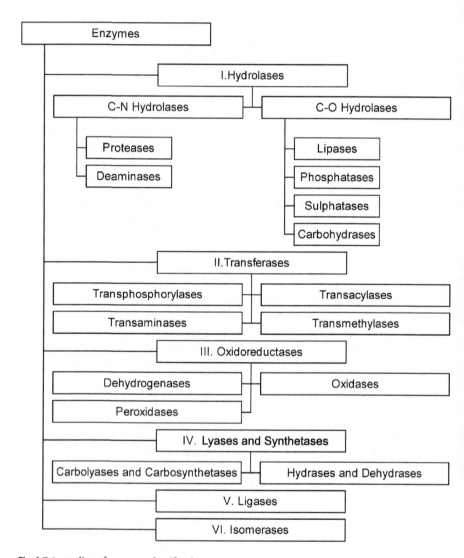

Fig. 9.7 An outline of enzymes classification

non-protein fragment is known as *the prosthetic group* – if tightly attached to the colloidal carrier; or as *the coenzyme* – if it is loosely attached to the same.

The catalytic activity of enzymes, and specifically the special groups upon which this activity depends, are subject to intensive research, for in many cases these groups are not completely well known. In some cases –SH groups play the major role; in other cases, metals bound to the protein may be the principal factors of catalysis. However, for most enzymes, the catalytic activity may be traced to the non-protein fragment and may be dependent on certain substances, without the presence of which the catalytic activity cannot function. These are called *cofactors*.

9.4.1
The Hydrolases

These form about one third of the known enzymes and act mainly on peptides, ester, glycosidic, amide, and similar bonds. They catalyse the hydrolysis of their substrates and may further be classified into C–N and C–O hydrolases according to the type of bond which may be attacked under their support.

9.4.1.1
C–N Hydrolases

Specific for the hydrolysis of C–N bond, the C–N hydrolases may be classified according to the C–N bond they hydrolyse into *proteases* and *deaminases*.

Proteases
These are those hydrolases that help degrade proteins (see Fig. 2.25) by hydrolysing the internal peptide bonds along the chain linking the individual amino acids in the protein molecule:

$$-CO-NH- + H_2O \rightarrow -COOH + H_2N$$

Normally, proteases are differentiated into the *proteinases* that cleave high molecular-weight proteins into simpler compounds (*polypeptides*), and the *peptidases* that further degrade the polypeptides resulting from protein degradation by proteinases (see Fig. 2.25).

$$\text{Proteins} \xrightarrow{\text{Proteinases}} \text{Polypeptides} \xrightarrow{\text{Peptidases}} \text{Simpler compounds}$$

Deaminases
These C–N Hydrolases catalyse the hydrolysis of certain types of carbon-nitrogen bond. The most important of them are *urease* and *arginase*. Urease assists the hydrolysis of urea into ammonia and carbon dioxide (see Fig. 9.8).

It occurs in soya beans, watermelons, moulds, and bacteria. Arginase is principally found in animal liver. It helps in splitting arginine (a compound of proteins) into ornithine (α,δ-diaminovaleric acid) and urea. Ornithine in the liver of birds removes the toxic benzoic acid in the form of its dibenzyl derivative.

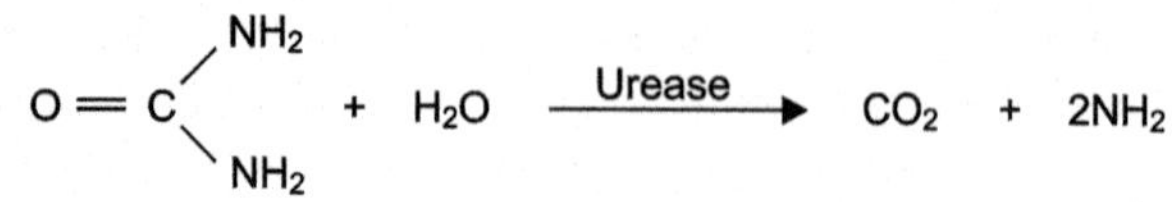

Fig. 9.8 Hydrolysis of urea by urease

$$O=C\begin{array}{c}NH_2\\NH_2\end{array} + H_2O \xrightarrow{\text{Urease}} CO_2 + 2NH_3$$

9.4.1.2
C–O Hydrolases

C–O hydrolases catalyse the hydrolysis of natural esters. Depending upon the type of their substrates, they may be classified into various groups, the most important of which are:

- *Lipases*. These are capable of hydrolysing fats. Lipases of vegetable origin are known as *phytolipases*. An example of them is castor lipase which is found in the seeds of *recinus communis*.
- *Phosphatases*. These C-O hydrolases are capable of catalysing the hydrolysis of phosphoric ester groups. They occur in almost all living cells and their catalytic activity is pH-dependent.
- *Sulphatases*. These are capable of hydrolysing natural sulphuric esters, e.g. phenyl sulphuric acid.
- *Carbohydrases*. As it is clear from their name, this group of hydrolases is capable of degrading simple sugars as well as transforming polysaccharides into simple ones. A prominent example of them is *cellulase*, which degrades cellulose and is only found in bacteria, fungi, and in the digestive juices of certain snails and worms.

9.4.1.3
The Role of Hydrolases in Pesticide Degradation

Hydrolases, whether they are specialised in the C–N or the C–O bond, are capable, with their various groups and individual enzymes, of degrading many pesticides containing ester, amide or phosphate linkages, as explained in the foregoing section. Examples of such pesticides are various; to mention a few of them, one may consider the group of organophosphorus pesticides, the carbamate insecticides, or the urea and carbamate herbicides.

Detoxification of malathion (see Table 7.1 and Fig. 9.9), by the action of a C–O hydrolase (carboxylesterase), provides an example for the degradation of an organophosphorus pesticide through cleaving a carboxylethyl linkage.

Hydrolysis of the carboxylethyl linkage follows the following reaction path:

$$R-COO-C_2H_5 \xrightarrow[H_2O]{\text{Hydrolase (carboxylesterase)}} R-COOH + C_2H_5OH$$

The low toxicity of malathion for some higher animals may be due to the efficiency of this process (Hassall 1982).

Organochlorine herbicides such as esters of 2,4-D (Fig. 7.9b) are also degraded through the catalytic action of carboxylesterase. After penetrating into the weed, the esters are hydrolysed to release the biologically active dichlorophenoxyacetic acid.

Fig. 9.9 Malathion

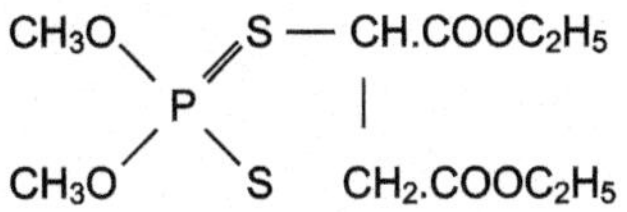

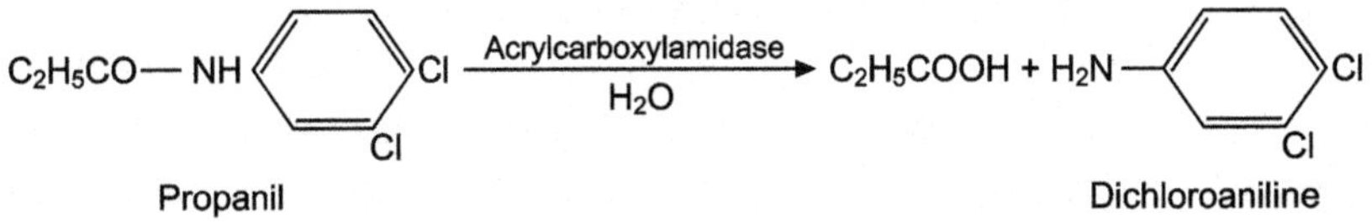

$$CH_3O\diagdown \underset{CH_3O\diagup}{\overset{S}{P}} \diagup \underset{S}{\diagdown} \begin{array}{l} S-CH.COOC_2H_5 \\ | \\ CH_2.COOC_2H_5 \end{array}$$

$$C_2H_5CO-NH-\underset{Cl}{\underset{Propanil}{\bigcirc}}-Cl \xrightarrow[H_2O]{Acrylcarboxylamidase} C_2H_5COOH + H_2N-\underset{Cl}{\underset{Dichloroaniline}{\bigcirc}}-Cl$$

Fig. 9.10 Hydrolysis of propanil by acrylcarboxylamidase

Propanil – an organochlorine herbicide, derived from aniline (Fig. 7.10a), was found to be hydrolysed by the action of a C–N hydrolase (acrylcarboxylamidase), found in rice plants (Yih et al. 1968) to produce dichloroaniline (Fig. 9.10).

These were just few examples of the enzymatic degradation of some insecticides and herbicides by the action of hydrolases. The interested reader is referred to Hassall (1982) for a detailed treatment of this theme.

9.4.2
The Transferases

Transferases derive their individual names from that of the transferred group, either with the prefix "trans-" and suffix "-ase", or followed by "transferase". So one finds names like: *trans-phosphoryl-ase*, *trans-acyl-ase*, or *dihydroxyacetone transferase*. The following groups are the most important among the transferases:

9.4.2.1
Transphosphorylases

As mentioned before, some enzymes need the assistance of a third substance, normally called the *cofactor*, to fulfil their action of transforming their substrates into other chemical constitution. Cofactors are usually phosphate esters of sugars or of compounds between sugars and nitrogen-containing bases. A good example of such cofactors is adenosine triphosphate (ATP) – formed from the base adenine (Fig. 9.11a), the sugar ribose (Fig. 9.11b), and three phosphate groups.

Transphosphorylases, being one of those enzymes that need a cofactor to carry out their task, which is adding phosphate groups to their substrate, use ATP as a source for the required phosphate groups.

9.4.2.2
Transacylases

Acyl groups (from *acidum* = acid) are groups of general formula RCO, where R is an alkyl group. They may be viewed as fatty acids (R–COOH), lacking a hydroxyl group (OH$^-$). Examples are acetyl (CH$_3$CO), formyl (H–CO), etc. Enzymes capable of trans-

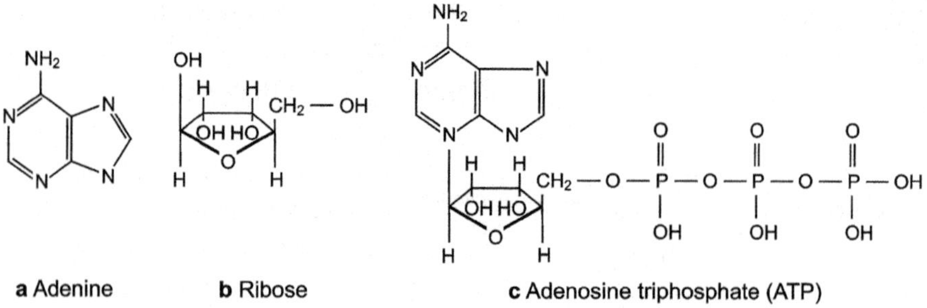

Fig. 9.11 Constitution of adenosine triphosphate (ATP)

a Adenine   b Ribose   c Adenosine triphosphate (ATP)

ferring an acyl group from a donor to an acceptor molecule are known as *transacylases*. These have a loosely attached non-protein fragment functioning as a coenzyme that cleaves the acyl bond in an intermediate step, followed by its transfer to the acceptor group. An example can be given by the transfer of an acetyl group (CH_3CO) attached to a molecule X (donor) to another molecule Y (acceptor). Such a process proceeds along the following reaction path:

Acetyl–X + CoĀ–SH → Acetyl–S–CoĀ + XH

Acetyl–S–CoĀ + YH → Acyl–Y + CoĀ–SH

(CoĀ = coenzyme A without –SH group)

Coenzyme A – a coenzyme of transacylases, made of adenine, ribose, phosphate, pantothenic acid, and cysteine – plays a decisive role in all biochemical acylations, especially the oxidative degradation of carbohydrates and fats. It also forms a key substance in the conversion of carbohydrates into fatty acids.

9.4.2.3
Transaminases

These enzymes effect a reciprocal process of amination-deamination between keto-acids, mainly pyruvic acid (Fig. 9.21) and glutamic acid (Fig. 9.12), without incurring the presence of free ammonia. In this reciprocal reaction, known as *transamination*, L-glutamic acid functions as a nitrogen-carrier in the synthesis of amino acids in animal organisms.

In plants, the same role is played by L-aspartic acid. Meanwhile, pyruvic acid, which plays a key role in transamination, is sometimes produced by the degradation of aliphatic herbicides such as dalapon (Fig. 7.15).

9.4.2.4
Transmethylases

Like the other transferases, transmethylases transfer a group (in this case methyl) from a donor molecule to an acceptor.

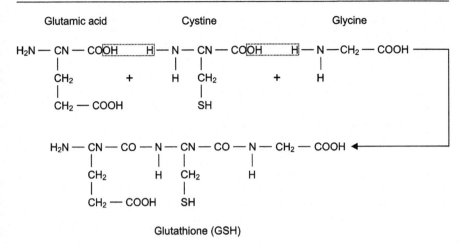

Fig. 9.12 Formation of glutathione

9.4.2.5
The Role of Transferases in the Degradation of Pesticides and other Contaminants

Among the transferases, the system known as *glutathione-S-transferase* (GST) plays an important role in the detoxification of contaminants, by catalysing the conjugation of the endogenous substance (GSH) to an electrophilic site on the intruding xenobiotic material. Glutathione (GSH) is a tripeptide, formed by a combination of the three amino acids: glycine, cysteine, and glutamic acid, as shown in Fig. 9.12.

Catalysis of glutathione conjugation on xenobiotics by GST generally takes place on substrates sharing three common features: they must be hydrophobic to some degree, they must contain an electrophilic carbon atom, and they must react nonenzymatically with glutathione at some measurable rate (T. J. Rees 1993).

Of the several known glutathione-S-transferases, the following three groups were considered by Hassall (1982) as being among the most important.

Glutathione-S-Epoxide Transferases
Alkylene oxides or epoxides are groups formed by the oxidation of olefins (see Fig. 9.13).

Epoxides may be formed during phase I of biotransformation as intermediate stages in the degradation of numerous unsaturated compounds. A subsequent opening of the ring, by conjugation of an endogenous substance on the epoxide, leads, in most cases, to the formation of substances less harmful to life. This may be illustrated by the GST catalysed glutathione conjugation on the intermediate epoxide of allyl phenyl ether (a substance used in polymer synthesis – see Fig. 9.14). In this reaction, glutathione conjugation on the substrate leads to an opening of the epoxide ring and the formation of a glutathione complex with the detoxified original substance as shown in Fig. 9.14.

With benzene rings this reaction precedes more readily if the ring contains chlorine atoms, this is why they may be of great help in degrading organochlorine epoxides used as pesticides, such as dieldrin (Fig. 7.5b) and heptachlor epoxide (Fig. 7.5c).

Fig. 9.13 Formation of alkylen oxide or epoxid

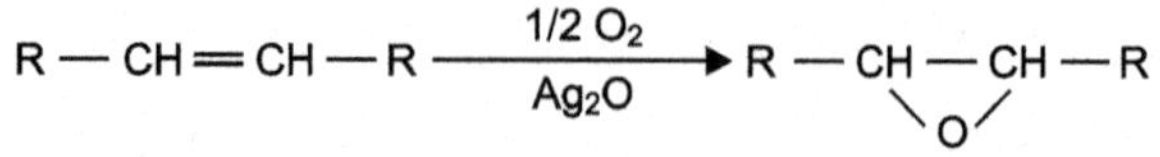

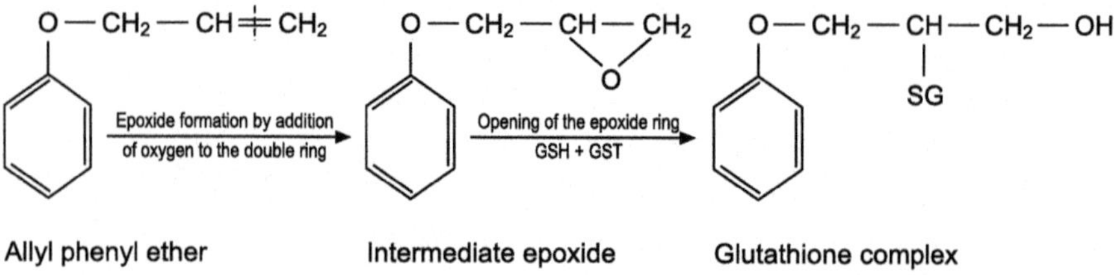

$$R-CH=CH-R \xrightarrow[Ag_2O]{1/2\ O_2} R-CH-CH-R$$
$$\diagdown O \diagup$$

Allyl phenyl ether — O—CH$_2$—CH=CH$_2$ attached to phenyl ring

Epoxide formation by addition of oxygen to the double ring →

Intermediate epoxide — O—CH$_2$—CH—CH$_2$ with epoxide O, attached to phenyl ring

Opening of the epoxide ring GSH + GST →

Glutathione complex — O—CH$_2$—CH(SG)—CH$_2$—OH attached to phenyl ring

Fig. 9.14 Degradation of allyl phenyl ether by glutathione (GSH)

Glutathione-S-Aryl Transferases

This group effects mainly the elimination of hydrogen halides from their substrates. They are most effective in the detoxification of organochlorine pesticides (e.g. triazine derivatives). The chemical mechanism, according to which such reaction occurs, may be illustrated by the following example (Fig. 9.15) of eliminating a hydrogen halide group from a molecule of the herbicide atrazine (Fig. 7.13b).

Various organophosphorus pesticides are also degraded by chemical reactions comprising glutathione conjugation. An example of this is the degradation of the pesticide diazinon which is strongly adsorbed onto soils. Diazinon is used to control pests that have become resistant to organochlorine compounds. Its cleavage by GST follows the pathway shown in Fig. 9.16 (Shishido et al. 1972).

Being capable of reacting with alkyl halides, these enzymes were found effective in removing methyl groups from organophosphorus insecticides that contain the CH$_3$–O–P group. However, they are less effective in removing ethyl and larger alkyl groups. This may be illustrated by the demethylation of parathion (Table 7.1) as shown in Fig. 9.17.

Non-Enzymatic Glutathione Conjugation

Substrates susceptible to glutathione conjugation, as stated before, must be in a position to react nonenzymatically with glutathione (GSH) at some measurable rate. It seems that even in absence of GTS this type of reaction plays an important role in the detoxification of some aromatic pesticides. An example may be given by the degradation of the herbicide propachlor, during which the glutathione is conjugated to an electrophilic site with the elimination of a hydrogen halide to form a complex that subsequently decomposes to an N-acetyl cysteine derivative known as *mercapturic acid*. The polarity induced by the insertion of COOH group makes such a compound more soluble and hence easy to be excreted in urine or faeces (Fig. 9.18 shows the pathway of such a process). Glutathione levels in soil biota are accordingly used sometimes as biomarkers in soils, i.e. substances that may indicate the level of pollution by relating this to the degree of stress incurred in the organisms. In addition, metabolites such as mercapturic acid may be used for the same purpose.

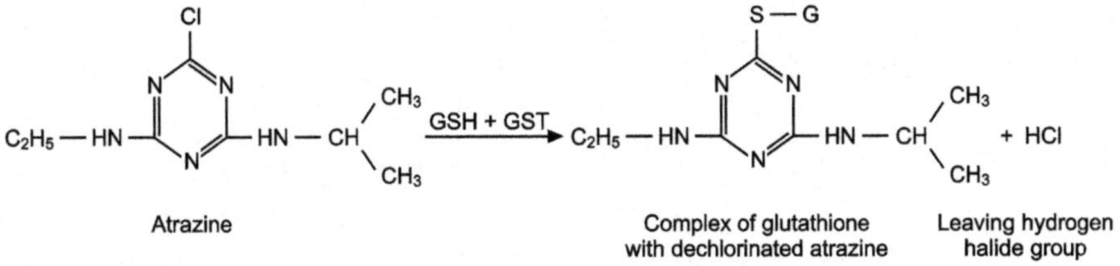

9.4 · Enzymatic Transformations: A Primer on Enzymes, Their Types and Mode of Action

Fig. 9.15 Elimination of a hydrogen halide from atrazine by GSH

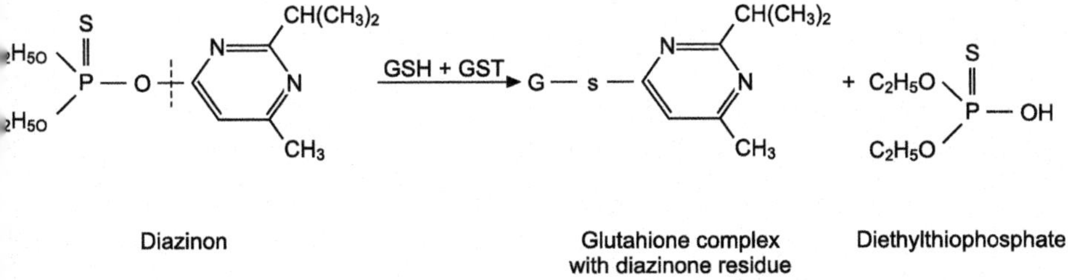

Fig. 9.16 Cleavage of diazinon by GSH

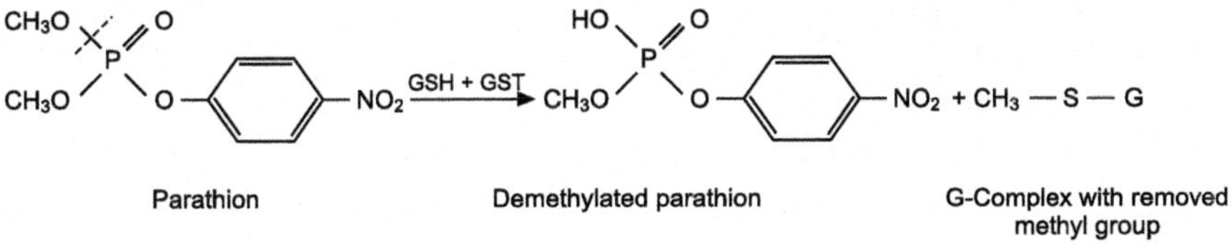

Fig. 9.17 Demethylation of parathion by GSH

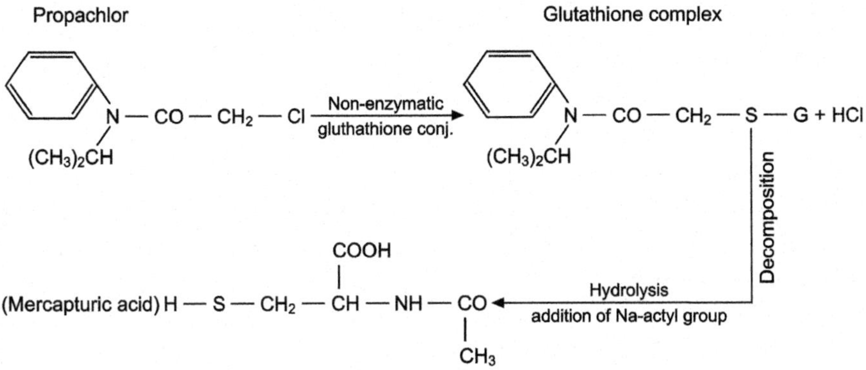

Fig. 9.18 Nonenzymatic propachlor degradation by glutathione

9.4.3
The Oxidoreductases

These are enzymes capable of transferring hydrogen or electrons between substrates. They are of three types: the *dehydrogenases*, the *oxidases*, and the *peroxidases*. While dehydrogenases mainly transport hydrogen or electrons to acceptor enzymes or to oxygen to produce H_2O_2, the oxidases react directly with oxygen to give water. The peroxidases, however, catalyse the decomposition of hydrogen peroxide, which has a toxic effect, into oxygen and water.

9.4.3.1
Dehydrogenases

The catalytic activity of dehydrogenases is characterised by the transfer of two hydrogen atoms from one organic donor to an organic acceptor. This may be schematically represented as shown in the following two equations in which the enzyme is represented by the box with the letter D.

$$XH_2 + \boxed{D} \rightarrow X + \boxed{D}\text{-}H_2$$

$$\boxed{D}\text{-}H_2 + Y \rightarrow \boxed{D} + YH_2$$

As shown by the equations, two atoms of hydrogen were transferred from the donor X to the acceptor Y, while the enzyme as a catalyst was recovered unchanged at the end of the process.

9.4.3.2
Oxidases

This group of enzymes, also known as *aerobic electron transferases*, helps complete the reduction of oxygen into water. They thus differ from the majority of dehydrogenases, which carry the reduction of oxygen only to the peroxide stage. They are metalloprotein compounds carrying heavy metals and are accordingly classified into cytochrome oxidases (iron proteins), and cuprotein oxidases.

9.4.3.3
Peroxidases

These are, as mentioned before, enzymes that attack hydrogen peroxide; they are present in animal organs and most plants and are made of iron porphyrin compounds. Peroxidases catalyse reactions between hydrogen peroxide and other substances according to the equation:

$$XH_2 + H_2O_2 \xrightarrow{\text{Peroxidase}} X + 2H_2O$$

A similar catalytic function is exerted by *catalases*, which are also found in almost all animal organs, cells and tissue fluids, as well as plant tissues. They effect the decomposition of the cell toxin (hydrogen peroxide) into water and oxygen according to the equation:

$$2H_2O_2 \xrightarrow{Catalase} 2H_2O + O_2$$

9.4.3.4
The Role of Oxidoreductases in Soil Formation and Biodegradation of Contaminants

The decay of plant debris and other natural organic matter forms one of the main sources of soil humus; among other processes involved in this process, biodegradation of lignin holds a central position. This degradation process is initiated by several oxidoreductases excreted by white rot fungi. Examples of such enzymes are haem-containing peroxidases, lignin peroxidase (LP), manganese-dependant peroxidase (MnP), as well as copper-containing phenol oxidase, laccase (Camarero et al. 1999a,b).

Further research revealed that lignin-degrading enzymes, excreted by white rot fungi, are also capable of oxidising high-molecular-weight polycyclic aromatic hydrocarbons (PAH). This discovery directed attention to them as potential agents for bioremediation of contaminated soils (Kotterman et al. 1998). Various oxidases use nicotinamide adenine dinucleotide (NAD) or its phosphate derivative nicotinamide adenine dinucleotide phosphate (NADP) as coenzyme. NADP-dependent reactions were found, as it will be shown in the following examples, to be important for the biochemical degradation of many insecticides.

Hydroxylation of Carbaryl
Carbaryl (Fig. 9.19), a carbamate insecticide used to control pests mainly on maize and soybeans, may suffer ring hydroxylation by NADP assisted oxidases to give a mixture of 4-hydroxy and 5-hydroxy carbaryl, as shown in Fig. 9.15.

O-Dealkylation of Methoxychlor
Methoxychlor (Fig. 9.20) is a minor organochlorine insecticide similar to DDT, yet differing in that it has p,p'-dimethoxy groups instead of chlorine atoms. It can be easily dealkylated by NADP-dependent enzymatic reactions to produce the polar compound demethyl methoxychlor that can undergo further conjugations, facilitating its removal from animal bodies.

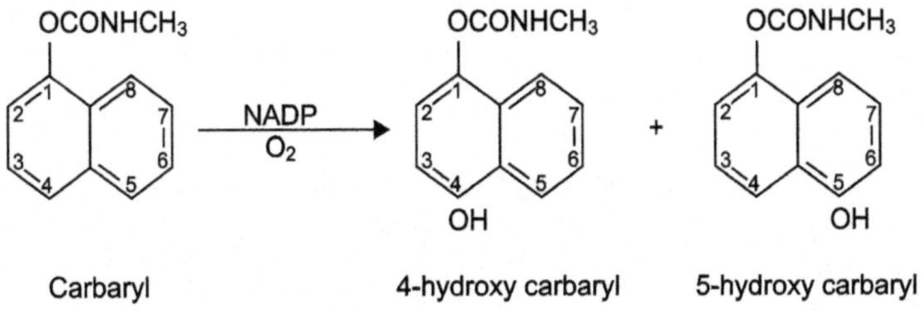

Fig. 9.19 Ring hydroxylation of carbaryl

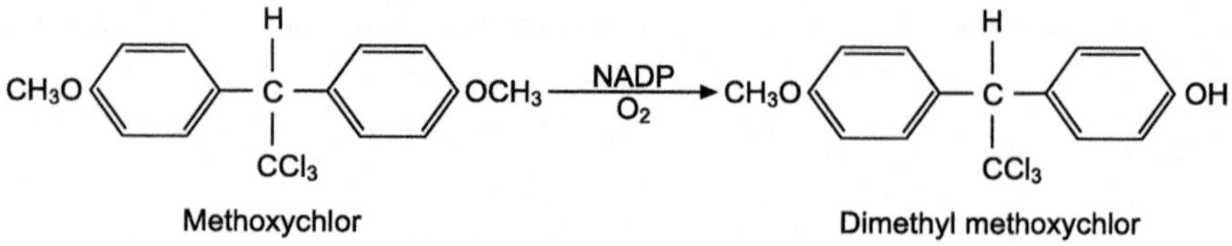

$$CH_3O-\underset{\underset{CCl_3}{|}}{\overset{\overset{H}{|}}{C}}-\text{C}_6H_4-OCH_3 \xrightarrow[O_2]{NADP} CH_3O-\underset{\underset{CCl_3}{|}}{\overset{\overset{H}{|}}{C}}-\text{C}_6H_4-OH$$

Methoxychlor → Dimethyl methoxychlor

Fig. 9.20 O-dealkylation of methoxychlor

In his book on the chemistry of pesticides, Hassall (1982) quotes the eminent toxicologist Barnes with the words that *"if methoxychlor had been marketed instead of the slightly cheaper DDT, the persistence of organochlorine insecticides may never have been regarded as a factor of major ecological importance."*

9.4.4
The Lyases

This group helps mainly in the cleavage of a C–C bond. They are subdivided, according to the bond they split, into the carbolyases and the dehydrases.

9.4.4.1
Carbolyases

An important member of this group is α-carboxylase (pyruvate decarboxylase). It splits pyruvic acid into acetaldehyde and carbon dioxide (see Fig. 9.21).

Another important member of the group is the amino acid decarboxylase that occurs in microorganisms, in higher animals, and in many plant tissues. It plays a key role in the putrefaction of proteins by catalysing the anaerobic decarboxylation of amino acids into amines and carbon dioxide (see Fig. 9.22).

9.4.4.2
Dehydrases

This group catalyses the elimination of water. It plays an important role in the citric acid cycle. Its function may be illustrated by the action of the enzyme *aconitase* (citric-isocitric isomerase) that helps change citric acid into *cis*-aconitic acid (see Fig. 9.23).

9.4.5
The Ligases

Ligases (synthases) catalyse the linking of two molecules; they are sometimes classified according to the type of bond formed under their catalytic action. Thus, we may have carbon-sulphur bond forming, carbon-oxygen bond forming, or carbon-carbon forming ligases.

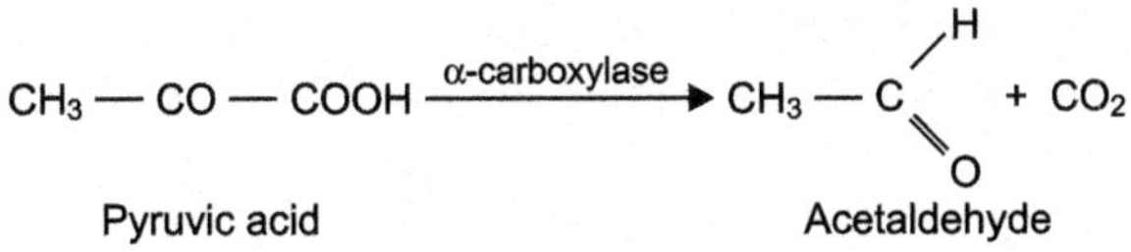

Fig. 9.21 Funktion of α-carboxylase

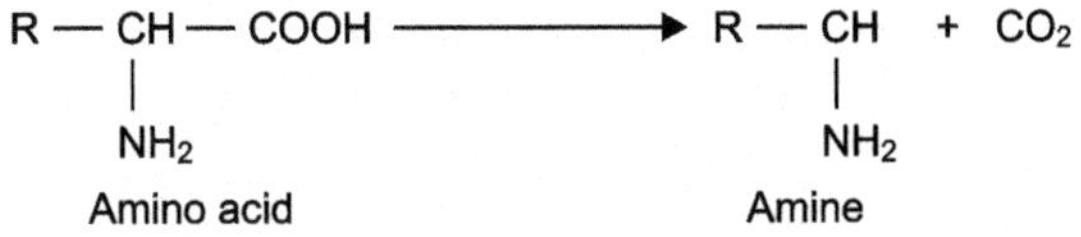

$$CH_3-CO-COOH \xrightarrow{\alpha\text{-carboxylase}} CH_3-C{\overset{H}{\underset{O}{\diagdown}}} + CO_2$$

Pyruvic acid → Acetaldehyde

Fig. 9.22 Anaerobic decarboxylation of amino acids

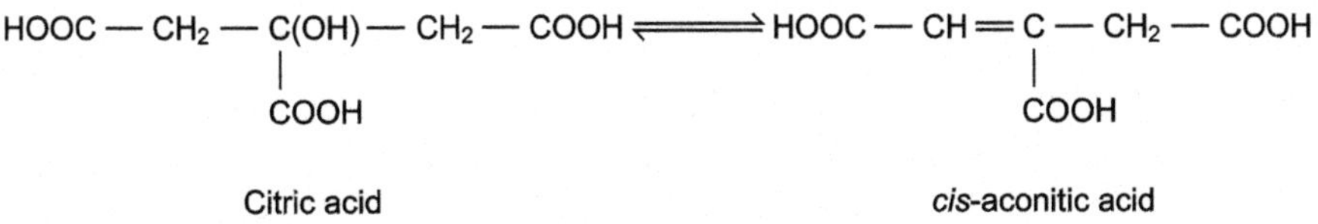

$$\underset{\substack{|\\ NH_2\\ \text{Amino acid}}}{R-CH-COOH} \longrightarrow \underset{\substack{|\\ NH_2\\ \text{Amine}}}{R-CH} + CO_2$$

$$\underset{\text{Citric acid}}{\underset{\substack{|\\ COOH}}{HOOC-CH_2-C(OH)-CH_2-COOH}} \rightleftharpoons \underset{\textit{cis}\text{-aconitic acid}}{\underset{\substack{|\\ COOH}}{HOOC-CH=C-CH_2-COOH}}$$

Fig. 9.23 Function of aconitase

9.4.6
The Isomerases

This class of enzymes catalyse intramolecular rearrangements. An example may be given by D-arabinose isomerase, an enzyme that catalyses the intramolecular rearrangement of D-arabinose to give D-ribulose.

9.5
Transformations Assisted by Bacterial Action

In the foregoing sections it was shown that soil organisms, assisted by their natural metabolic processes, create an environment in the soil that allows the whole system to develop and to avoid conditions that would bring it to collapse. It should, however, be reminded that the main target of all metabolic processes is to secure the energy required for an organism to continue its life activities. Energy production in plants is reached through a complex of light-assisted biochemical processes known collectively as *photosynthesis*. In animals, cleavage (mostly oxidation) of complex organic material fulfils the same purpose – the whole process, in this case, is known under the collective name *chemosynthesis*.

Bacteria and other low organisms use both processes, yet some bacteria are also capable of energy production through chemical transformation of inorganic material and thus play a role in determining the fate of inorganic pollutants in soil.

Changes in the soil environment by bacterial action are brought about mainly by organisms known as *the lithotrophs* (rock eaters). The Russian microbiologist Sergei Winogradsky who first described them in 1885 gave them this name. As the name indicates, they are capable of covering their need of energy by oxidising soil inorganic compounds. This process may be pure chemotrophic, as explained before, or it may be a phototrophic one, using visible light as a source of energy (Fig. 9.24 summarises these relations).

The major groups of lithotrophic bacteria that play principle roles in the modification of the soil environment are the following:

- Sulphur bacteria
- Nitrifying bacteria
- Iron oxidising bacteria
- Methane oxidising bacteria
- Hydrogen bacteria

9.5.1
Sulphur Bacteria

The natural ecological community of sulphur bacteria is generally known under the collective term *sulfuretum*, coined by Baas-Becking in 1925.

A sulfuretum exists normally in oxygen-deficient environments and may even flourish under extreme conditions, as in the case of the highly alkaline sediments of Wadi Natrun in northwestern Egypt (Trüper 1982). Organisms living in such a community must not essentially follow the same trophic mode – some may be chemotrophic, while others could be phototrophic. As examples of both types, we may mention the following:

Photolithotrophic Sulphur Oxidising Bacteria
A good example of these is provided by the so-called *thiospirills* – a group belonging to the family Rhodospirillaceae. It embraces two subgroups, using different types of proton-donors in photosynthesis. The first subgroup, known as *thiorhodaceans*, uses H_2S or S, while the other subgroup, *athiorhodaceans*, uses H_2, derived from the breakdown of organic substances such as fatty acids.

Chemolithotrophic Sulphur Oxidising Bacteria
Two prominent examples represent this group – *Beggiatoa* sp., and the frequently mentioned *Thiobacillus denitrificans*. *Beggiatoa* has a peculiar thread-like appearance and is always in a continuous state of worm-like motion (see Fig. 9.25).

Sulphur organisms of the beggiatoa species use both hydrogen sulphide and elemental sulphur (S^0) for their chemotrophic oxidation processes, according to the following equations:

$$H_2S + 2O_2 \rightarrow SO_4^{2-} + 2H^+$$

$$S^0 + 2H_2O \rightarrow SO_4^{2-} + 4H^+$$

Thiobacillus denitrificans is capable of oxidizing hydrogen sulphide using nitrate (NO_3^-) as an oxidizing agent instead of oxygen. As a result of this process, molecular nitrogen (N_2) and sulphuric acid (H_2SO_4) are released.

Fig. 9.24 Classification of lithotrophic bacteria according to their energy sources

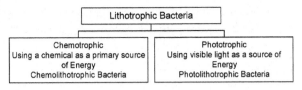

Fig. 9.25 *Beggiatoa* sp.

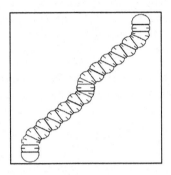

Fig. 9.26 The combined work of *Nitrosomonas* and *Nitrobacter*

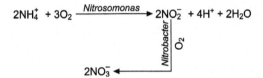

Some bacteria species are also capable of reducing sulphates to produce sulphides under anaerobic conditions, of these we may mention *Desulfovibrio* sp. and *Desulfotomaculum* sp. Sulphite reduction to sulphide is carried out by the anaerobic bacterium *Clostridium pasteurianum* (Trüper 1982).

9.5.2
Nitrifying Bacteria

These widely distributed bacteria, represented by two groups living in a symbiotic community – the nitrite bacteria (e.g. *Nitrosomonas europaea*) and the nitrate bacteria (e.g. *Nitrobacter agile, Nitrobacter winogradskyi*) – are very important soil dwellers that cause important chemical changes in the soil environment. The first group changes NH_4^+ (ammonium) into NO_2^- (nitrite), while the second group oxidises nitrite further into nitrate (NO_3^-) (see Fig. 9.26).

Nitrosomonas europaea was first discovered and described by Winogradsky in 1890. It flourishes under aerobic conditions in arable soils, where ammonia (NH_3), resulting from the breakdown of organic substances, is continuously oxidised to nitrite (NO_2^-) according to the equation:

$$2NH_3 + 3O_2 \rightarrow 2HNO_2 + 2H_2O$$

The energy set free in this reaction supplies the major part of energetic requirements for the chemosynthesis of carbon dioxide by this group of bacteria.

9.5.3
Iron Oxidising Bacteria

Many species belonging to the order *Caulobacterials* possess the capacity of oxidising Fe(II) into Fe(III), whereby the oxidation product will be stored in their mucous cells giving them the characteristic rusty brown colour of Fe(III)-OH. This may in some cases supply an explanation for the appearance of rusty horizons (patches) in soils and marine sediments, inhabited by bacterial colonies such as *Gallionella ferroginea* or *Sidercapsa treubii*. The first of these flourishes near iron rich springs and may also grow in shallow marine environments.

9.5.4
Methane Oxidising Bacteria

It seems that anaerobic methane oxidation goes hand in hand with sulphate reduction, for it was proposed by many authors that methanogenic bacteria (e.g. *Scarina methanica*) may be producing and oxidising methane at the same time. Scarina methanica uses carbon dioxide as a proton acceptor to change alcohols to methane as the equation demonstrates:

$$2CH_3CH_2OH + CO_2 \rightarrow CH_4 + 2CH_3COOH$$

The acetate produced in this reaction allows and supports the growth of sulphate reducing bacteria, which may, in turn, assist the anaerobic oxidation of methane into carbon dioxide and water.

9.5.5
Hydrogen Bacteria

This group, assisted by the catalytic effect of the enzyme hydrogenase, is capable of oxidising molecular hydrogen into water. *Hydrogenomonas* – a species belonging to the family Nitrobacteriaceae – is capable of oxidising molecular hydrogen to produce water according to the equation:

$$2H_2 + O_2 \rightarrow 2H_2O$$

Part III
Monitoring Soil Pollution

Monitoring is the regular surveillance and quantification of the amount of pollution present in a given location of soil. It should be carried out in a way that enables the detection of spatial as well as time variations in the concentration of pollutants at the site of investigation. Monitoring should provide information on the following:

- Nature of the pollutants, their quantities, sources and distribution;
- Effect of the pollutants;
- Concentration patterns, pedological changes and their causes;
- Possibility and feasibility of remediation.

It may be carried out using physical, chemical, or biological methods.

Chapter 10
Monitoring and Monitoring Plans

Before setting up any monitoring plans, complete information about the investigation site should be carefully collected and analysed. This saves a great deal of effort and financial costs, and allows a suitable selection of the monitoring technical installations. Important information about the site should include geological, pedological, hydrogeological, and historical data about land use in the area. Geological and hydrogeological information can be taken from geological maps or from information supplied by the local geological survey. For pedological information, soil maps or individual field investigations are normally used. The main step after collecting and analysing the available information is to determine the main objectives of monitoring, providing careful answers for the following questions:

- Is the main objective of the monitoring project to detect the presence or absence of a given contaminant? If yes, this will be a detecting monitoring (see below).
- Is it known that the site is already contaminated with the alleged pollutant, where in this case the main objective of the plan would be to determine the extent of contamination? If yes, this will be an assessment monitoring (see below).
- Is the objective of the plan to evaluate the feasibility and required financial burden for the remediation of a pre-investigated area? If yes, then one speaks here of a performance monitoring.
- Will the monitoring process be a part of a follow-up plan to evaluate the success of remediation efforts? If yes, a research monitoring (see below) will be the main activity.

Each of the above-mentioned questions delivers an answer that helps in selecting the method of monitoring and consequently the technical installations required for the project. This allows a proper planning and assessment of the financial burden in advance. In fact, each of the above-mentioned criteria serves to limit the choice of the monitoring technique planned for the site under investigation. All techniques and possibilities have in common the way in which the preliminary steps follow a general strategy, starting with site characterisation, followed by determining the objectives of monitoring, and finally determining the techniques and field measurements suitable for the objectives and type of monitoring to be conducted – in short, the proposal of a sampling and monitoring plan.

An integrated monitoring plan comprises generally of the following essential parts:

- Site characterisation
- Data acquisition
- Data quality control
- Interpretation
- Reporting

10.1
Site Characterisation

Collecting all available data about the site of interest should precede all fieldwork in a monitoring project; this includes reviewing all published information about geomorphic and pedological characteristics of the area. If no maps are available, the area should be mapped using any of the known standard techniques, whereby the following types of maps should be obtained:

- *Base map*. A base map helps characterise the morphology of the site and shows the run-off conditions, including water bodies, which might be influencing hydrologic flows in the site. In a later stage, all sampling stations should be plotted on the base map. If biological monitoring is also planned, the stands of all trees and plant communities used for measurements are to be marked.
- *Geologic map*. Since soil types are normally related to the bedrock and the weathering processes that produced them, a geologic map of the area, detailing the main rock types, is either to be prepared or obtained from the local geological survey authority. Such maps provide valuable information about the soil chemistry and the background concentration of heavy metals in the unsaturated zone of groundwater. It also helps in determining the potential for flow, adsorption, and retainment of pollutants in the soil material at the site of investigation.
- *Hydrologic map*. Based on the interpretation of all geologic information about the area, they supply information on groundwater relations such as availability of groundwater, the depth to the water table and the direction of groundwater flow.

  Normally, a geologist is in a position to read all required information about subsurface flow characteristics in the area using such maps. Hydrologic maps can be obtained from the local geological survey, otherwise test bore holes, coupled with careful observation of available wells in the locality, may help in delineating the hydrologic conditions in the site of investigation to a fair degree.
- *Overburden map* (soil material). Pedological maps or soil maps provide information on soil types prevailing in the area according to one of the pedological classification systems mentioned in Part 1 of this book. They provide a means of interpreting any differences in the soil chemistry of the area and help in understanding any inconsistency in chemical characteristics at different parts of the investigation site. Such inconsistency may result from the existence of different soil types at the site of investigation. Soil maps obtained from the local survey authorities normally include useful information on soil texture, soil depth, as well as soil chemistry data such as heavy metals, pH, and CEC-values.

Beside geographic and geologic characterisation of the site, any available information about historical conditions of the site may be of great help in later evaluation of the data obtained. This includes photographs, maps, and any published or unpublished data about earlier land use activities.

10.2
Data Acquisition

Following the preliminary preparation and characterisation of the area, the principal step in a monitoring plan, comprising of data acquisition and generation, should be started. This stage includes the preparation of a sampling plan and determination of the chemical and physical characterisation of the problem. Figure 10.1 shows in a schematic way the relations connecting all those processes required for data acquisition and generation. Two main steps are essential for this aim – the preparation of a sampling plan, and the planning of all chemical and physical investigations that will form the framework of data collection.

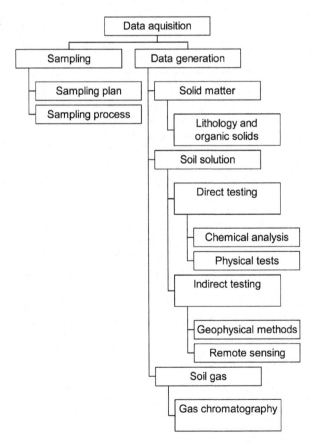

Fig. 10.1 The main steps in soil monitoring

10.2.1
Sampling-Planning and Realisation

In all the types of monitoring mentioned above, sampling forms the principal chain in all processes needed for the delineation and characterisation of the environmental quality parameters of the area under investigation. It helps locate the site(s) of contamination and provides information on time and spatial patterns related to the distribution of pollutants.

Location of Sample Points

In designing a sampling plan, the location of sampling points will largely depend upon the purpose of monitoring, as well as on the topography and geologic conditions of the area. A detection monitoring plan, or an assessment monitoring plan, for example, imply a dense network of sampling stations that are more or less systematically distributed over the whole area of investigation, while evaluation monitoring will require a concentration of sampling stations in the neighbourhood of the pollution sites. It must also be emphasised that a good knowledge of the geology and hydrology of the area is very decisive for the selection of the sampling locations.

Spatial patterns of sampling follow two fundamental models that form the basis for all types mentioned by different authors (see for example Boulding 1995). These are the simple rectilinear grid type and the traverse type, both of which were originally developed and used by exploration geochemists.

In the first type (grid pattern), a rectilinear grid of samples is taken at equal intervals along evenly spaced lines (systematic, Fig. 10.2a). Another variation of this type is to take random samples within every block of the grid (random, Fig. 10.2b).

In traverse line sampling, points are selected on a traverse following topographic, geologic or geophysical data. The environmental and geochemical data known about the area is decisive for the direction and density of the sample points. Samples along a traverse may also be taken systematically (at equal intervals) or at random distances. However, rectilinear grid patterns of sampling are more popular because of the ease in laying out the fieldwork and in plotting the data. The reliability of samples, as representatives for the environmental conditions in the investigated area, depends principally on factors like frequency of sampling, technical procedures of sample collection, objective and technical errors of the operator, as well as storing, handling and treatment of samples in the course of their collection and transport to the laboratory.

Fig. 10.2 Rectilinear grid patterns of sampling; **a** systematic grid pattern; **b** random block pattern

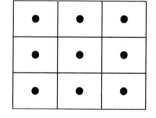
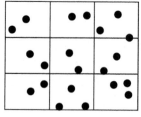

Systematic grid pattern   Random block pattern

Generally, however, sampling points should be concentrated around hot spots, i.e. around spots where contamination is expected, or near areas that are suspected of being contaminated.

10.2.2
Sampling Procedures

10.2.2.1
Sampling Solid Soil Matter

Soil samples are taken either at fixed depths or separately from each pedogenic horizon, if detailed work is intended. The following depths are recommended by the UN/ECE ICP Forests programme (UN/ECE ICP Forests 1994): 0–5 cm, 5–10 cm, 10–20 cm and 40–80 cm. Samples from shallow depths (up to 30 or 60 cm) are collected from small pits. Deeper samples are normally collected using a soil auger. Simple soil augers (Fig. 10.3) were found satisfactory in sampling depths from 1 to 2 metres.

Simple soil augers are normally used to sample loamy soil. For stony soils, especially if deep samples are to be collected, light power augers are recommended. Peat samples are collected using box-type samplers or by means of special augers, such as the Hillier peat borer (Fig. 10.4).

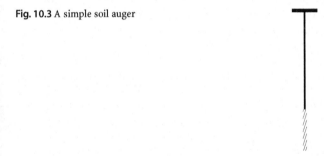

Fig. 10.3 A simple soil auger

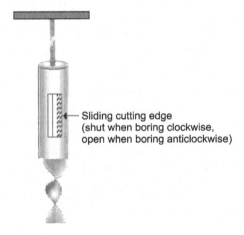

Fig. 10.4 Hilliers peat borer

Modern peat borers include the so called *Russian peat borer*, developed by aquatic research (www.aquaticresearch.com) and identified by EPA as a new technology for hazardous waste sampling.

As shown in Fig. 10.5, the ARI Russian Peat Borer is a manually driven core sampler, designed to consistently collect uncompressed samples of bog and marsh sediments and may also be used to collect samples at shallow water sites. Undisturbed soil samples can be obtained by using a piston sampler or may be taken from a soil pit, dug near to the plot. For more details on this borer the reader is referred to EPA/600/F-99/008.

10.2.2.2
Sampling Soil Solution

The sampling of solution, for the sake of soil monitoring as well as other factors leading to stress in the soil environment, may be carried out using different techniques, some of these are destructive, while others are non-destructive. The selection of methods will naturally depend on the strategy and aim of the collection. One can tabulate the methods of soil water collection as shown in Fig. 10.6.

Non-Destructive Methods of Sampling Soil Solution
Non-destructive methods (mainly lysimetric methods) depend upon two physical phenomena of water flow in the soil, either the fluids flow under normal pressure in the soil, using no other means (zero tension); or vacuum pumps are used to collect the samples at vessels located at or near the ground surface. Collection vessels, generally known as *lysimeters*, consist of two main types:

1. *Zero-tension lysimeters* (ZTL) (Fig. 10.9) are normally used to sample soil water as it moves through saturated soil. A lysimeter of this type consists of a cylinder placed under an undisturbed core of soil, which has been originally removed by a piston sampler. Two flexible connections are attached to tubes at the soil surface. One of the connected tubes serves as an air let tube, while the other may be attached to a pump for collecting the samples. ZTLs may be of different shapes and types, the best-known being the plate-shaped type and the funnel type (see Figs. 10.7 and 10.8).

    ZTLs were found useful in monitoring soil water during and after in situ remediation. This helps to monitor water variability and make a comparison of water properties during the whole process of remediation. It should, however, be observed that plate type ZTLs can only be installed close to the soil surface in order not to cause greater disturbance to the soil profile (Derome et al. 2002).
2. *Suction (tension) lysimeter* (Fig. 10.10). This is a device used to sample soil water in unsaturated soils. Acting as a continuous vacuum, it draws the water into the lysimeter through a porous membrane. Water samples are brought to the surface by suction through a tube dipping deep into the device.

    Experience shows that tension lysimetry is the technique most widely used for sampling solution. This is due to the fact that in zero-tension flow, the water collected represents the end result of preceding chemical and physical changes taking place in the soil environment (e.g. buffering, neutralisation, etc.). To reduce the costs of

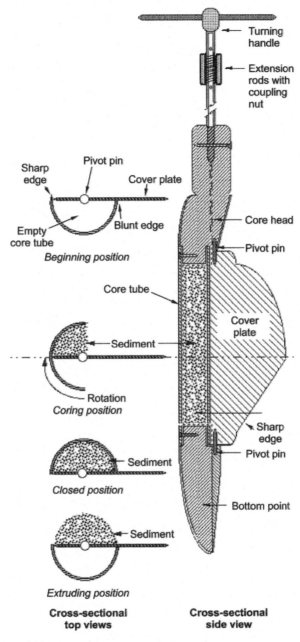

Fig. 10.5 ARI-Russian peat borer according to EPA-SITE, Program Demonstration Bulletin (EPA/600/F-99/008). (Reproduced by kind permission of Aquatic Research Corp.)

tension lysimeters (with a constant operating vacuum pumps), one can use pumps that are capable of constantly keeping the vacuum at the desired level (normally, 300–600 hp).

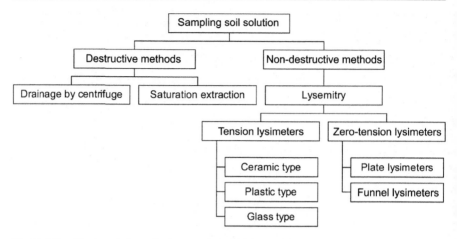

Fig. 10.6 Classification of the methods normally used for soil solution sampling

Fig. 10.7 Plate type lysimeter

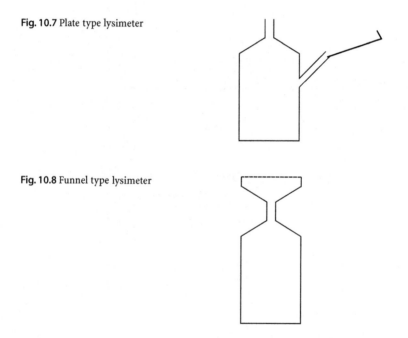

Fig. 10.8 Funnel type lysimeter

Installation Depths for Lysimeters
John Derome et al. (2002) strongly recommend installing lysimeters at least at two depths, i.e. within the rooting zone (10–20 cm) and below the rooting zone (40–80 cm), in order to allow collecting samples that may deliver information on both nutrition and toxic elements within this zone.

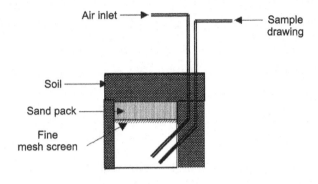

Fig. 10.9 Structure of a simple zero-tension lysimeter (after Thompson and Scharf 1994)

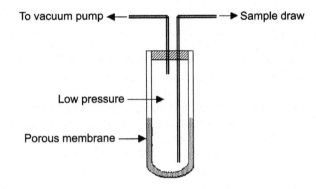

Fig. 10.10 A suction lysimeter

Destructive Methods of Sampling Soil Solution

In these methods, intact cores or mixed composite samples are transferred to the laboratory, where soil water may be extracted for investigation either by centrifuge drainage or by saturation extract.

- *Centrifuge drainage.* In this method, bulk samples are centrifuged in two piece cups (see Fig. 10.11), where bulk samples are placed in the upper part, on a perforated plate. On applying centrifugal forces (about $10\,000 \times g$ for about 30 minutes), soil solution samples can be collected in the lower part of the tube.
- *Saturation extract method.* In this method, water is added to change the concentration of some anions, such as chloride and nitrate. The original concentrations can be recalculated if the water content of the soil sample and the amount of water added is known. This method is not very popular, yet it is used for some soils where other methods are not successful, such as stony and clay soils.

10.2.2.3
Sampling Soil Air

Although soil gas investigation may be helpful in monitoring various environmental parameters, it is best used to investigate pollution by highly volatile hydrocarbons such

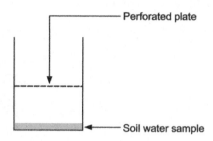

Fig. 10.11 Two-piece cups used for collecting soil water samples by centrifuging

as benzene and toluene. It is also useful for detecting contamination by chlorinated solvents and other organic pollutants.

Soil gas sampling can either be done by simple methods, such as air extraction from holes bored with an auger using a manual suction pump, or by relatively advanced methods, where soil air is sucked through a probe, rammed in the soil to a depth of about 1m, by pumps mounted on a vehicle equipped with proper analytical facilities. If manual pumps are used, the collected samples should be stored in airtight stainless steel or glass containers before being sent to the laboratory.

Another method of soil gas sampling, which is relatively new, is the so-called *passive sampling*. This technique, which is principally used for detecting volatile organic matter, uses an activated charcoal rod buried in the soil as an in situ adsorbent for VOCs. After a few days or weeks the charcoal rod is retrieved and analysed by gas chromatography or any other suitable analytical method.

10.3
Field and Laboratory Investigations

Test methods of the collected samples, whether solids, liquids or gases, will depend upon the purpose of monitoring as well as the degree of precision desired. Screening or field methods will some times suffice for a first assessment of the degree of contamination. In other cases, however, detailed laboratory tests are necessary to determine the next steps required for the project.

10.3.1
Investigation of Solid Matter

The sampling personnel normally screen the lithology of soil in the field by visual assessment. This is a relatively easy task that can be carried out by workers who have some training in geology. In the case that more detailed information is required and if grains cannot be identified using a pocket lens, a microscopic examination in the laboratory may be carried out by a geologist. XRD (X-ray Diffraction) analysis may in some cases be required. Through visual assessment, the presence of certain contaminants can also be ascertained in the field. Examples of theses contaminants are oil, tar and other coloured materials. For oil, however, the method of UV-fluorescence that has been successfully used in oil exploration is now in wide use.

If field assessments for organic material do not supply sufficient information, laboratory investigations using IR Spectroscopy (Infrared), GC (Gas Chromatography) or any other method are carried out.

Bulk chemical analysis of the soil matter is very efficient in determining pollution by heavy metals. This may be carried out by AAS (Atomic Absorption Spectroscopy) or by any other suitable method.

10.3.2
Investigation of Soil Solution

- *Direct methods.* Samples collected in the field by the previously mentioned methods should be analysed immediately after being brought to the laboratory. The time between sample collection and investigation should always be kept at minimum. Physical parameters, such as pH, Eh and salinity, are to be determined at the beginning. Determination of metal contents is normally preceded by adding acid to the samples (0.5 ml conc. HNO_3) to ensure desorption of metals from the walls of the storage bottles. If samples are collected by lysimeters, filtering might not be needed because the porous membranes, or the sand pack in the case of ZTL, act as filters themselves.

  The concentrations detected may however be misleading, due to the fact that the samples collected by the previous sampling methods normally originate from the coarse and medium pores of the soil, which often have concentrations deviating from those in the finer pores. Also the possibility that the samples may be collected from depths where the soil solution is not so much affected by the plume can form a source of error. In order to ascertain better quality of data, soil samples from different depths and locations may be centrifuged in the laboratory. The collected soil solution can be analysed to supply a control for the data gained from analysing lysimeter samples.

- *Indirect methods.* Indirect methods that may provide data about soil solution without disturbing the surface of the soil include surface geophysical methods as well as remote sensing techniques. Both make use of measuring variations in physical parameters such as gravity, electrical conductance or the Earth's magnetic field. Remote sensing methods depend upon measuring the response to specific parts in the electro magnetic spectra, such as gamma rays, visible light or microwaves.

10.3.2.1
Surface Geophysical Methods

Inorganic plumes are often detected and mapped using the fact that ions present in leachates are capable of increasing the specific conductivity of soil solution. The classical methods, which have been successfully used in this category, are those depending on measurements of resistivity or self-potential properties of the soil. However, electrical geophysical methods used in environmental research depend upon measuring variations of two parameters which are crucial for the conductance of electric current within earth materials. These are the resistivity and the conductivity.

Resistivity, which is the reciprocal of conductivity of any material, is defined as the electrical resistance of a cylinder of this material, having a unit length and a cross-sectional area of unity. It controls the amount of electric current that can pass through the material and can be determined by the following equation:

$$R = \frac{\rho S}{l}$$

where R = the resistance, l = the length, S = the cross-sectional area and ρ = the resistivity. The unit of resistivity is the ohmmeter or ohm-centimetre. Conductivity for a continuous medium is given by $1/\rho$ and its unit is Siemens/m or mho/m.

In earth materials, the porosity and the chemical constitution of the fluids filling the pores are the primary factors determining the resistivity of the rock or soil body. Due to this, the salinity of soil fluids plays the principal role in determining the conductivity of the electric current within the soil at shallow depths. Making use of this important fact, the distribution of resistivity at shallow depths of the soil is determined by surface measurements and the so gained results are used to locate contaminant plumes at the site of investigation. This is done in the field by the transmission of a direct electric current through the soil, followed by measurement of the induced potential. Two pairs of electrodes are used at the surface. One pair of electrodes (current electrodes, A and B in Fig. 10.12) is used for introducing current into the soil, while the other pair (potential electrodes, C and D in Fig. 10.12) is for measuring the potential associated with the current.

The potential at point C is given by:

$$V_C = \frac{I\rho}{2\pi}\left(\frac{1}{r_1} - \frac{1}{r_2}\right)$$

where r_1 is the distance between potential electrode C and current electrode A, and r_2 is the distance between B and C.

The potential at point D is given by the equation:

$$V_D = \frac{I\rho}{2\pi}\left(\frac{1}{R_1} - \frac{1}{R_2}\right)$$

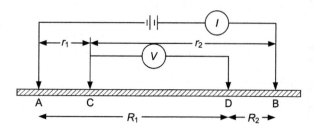

Fig. 10.12 Arrangement of the electrodes for resistivity measurements in the field (after Dobrin et al. 1988)

where R_1 is the distance from D to A, whereas R_2 is the distance from D to B. In both cases the resistivity ρ can be calculated from the equation, since V is known from measurements.

10.3.2.2
Remote Sensing Methods

Remote sensing is the term used to describe methods of obtaining information about an object with a sensor that is physically separated from the object. Sensors may be mounted on a variety of platforms, ranging from ground-based tripods to aircrafts and balloons. Generally though, aircrafts equipped with photographic devices are used.

The majority of remote sensing methods make use of measuring the response to specific parts of the electromagnetic spectrum such as gamma rays, IR, visible light, microwaves or radio waves. They also make use of measuring changes in potential fields that may indicate anomalies of the subsurface, such as gravity and the Earth's magnetic field. Very good results have been achieved in locating buried waste and determining the depth to groundwater through Ground Penetrating Radar waves (GPR). This technique (GPR) has also been successfully used for the measurement of organic pollutants in the groundwater. It measures the depression of the capillary zone, resulting from the presence of hydrocarbon films on the surface of water table – a property, which was already mentioned on p. 196 of this book (for example see Olhoeft 1986).

10.4
Monitoring of Groundwater Flows

As it was mentioned before, pollutants infiltrating the soil may be transported or absorbed on grain surfaces in the vadose zone. They may also penetrate the substrate to join the saturated zone where they are further transported and spread on a wide scale. Study of the dynamic properties of groundwater and the mechanisms of flow may be very important in the planning and designing of remediation projects. A monitoring of the velocities and directions of flow is thus an essential part of the whole monitoring plan. To understand the dynamics of groundwater, a characterisation of the levels of water concentrations and the extent of change in their dynamic properties according to the prevailing geologic conditions will be shortly explained.

10.4.1
The Different Zones of Groundwater

Groundwater is concentrated under the surface in two main zones representing two different concentration (or pressure) conditions. Starting from the surface, the vadose zone is that column reaching down to the water table. It is also known as the aerated zone or the unsaturated zone, despite the fact that it may, under some conditions, be intermittently saturated. Its description as aerated refers to the fact that the hygroscopic waters that collect here create an environment with a pressure less than that of the atmospheric air. At the base of the vadose zone there is a region where the water rising by capillary pressure from the water table forms a fringe, the height of which depends on the pore sizes of the sediments. Following the principles of capillary rise, the fringe reaches maximum width at lowest pore size in the aquifer. It may, however, collapse through contamination with non-aqueous phase liquids, as already mentioned in Chap. 8. The capillary fringe forms a transition zone between the permanently saturated phreatic zone and the vadose zone. The water table (the upper layer of the saturated zone) forms the base of the capillary fringe. Figure 10.13 shows, in a schematic way, the distribution of groundwater zones in a hypothetical aquifer.

10.4.2
Monitoring Flow Directions

Flow directions may be roughly deduced from the topography of the site or by consulting a hydrogeologic map; however, for accuracy on a local scale, this is done by systematic measurements of the water table height above a standard level (normally sea level) and plotting the results on a base map depicting the distribution of hydrologic heads. For this purpose, observation wells reaching below the water table are drilled at several (as many as possible) locations of the investigation site. They should be cased with tubes of equal diameters (~4"). Measurements of the depth to the water table are systematically performed at equal periods and the height above sea level is calculated. Location of the test wells depends on two conditions, without which erroneous results could be achieved. These are:

i   All test holes should be sunk to a depth that does not go beyond the same aquifer.
ii They should possess a good hydraulic conductance relative to each other.

Fulfilment of the first condition requires a fundamental knowledge of the subsurface geology in the area, so that all observation holes should be drilled at depths and positions that remain within the boundaries of the same aquifer.

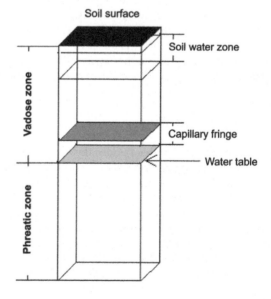

Fig. 10.13 Schematic representation of groundwater zones in a hypothetical profile (not to scale)

However, to be sure of a good hydraulic conductance among the test wells, pumping experiments are required. Hölting (1980) quotes Nattermann (1962) with an empirical rule for experiments needed to determine the suitability of a site as a systematic measuring station for water level monitoring. According to this rule, good hydraulic conductance is proven in a test well if the following relation has a numerical value higher than 0.0115:

$$\varepsilon = \frac{2}{t}\left(\frac{h_1 - h_2}{h_1 + h_2}\right)$$

where:

- h_1 = drop of the water table level by pumping relative to the original stand.
- h_2 = recovery of the water level after time t.
- t = time in minutes.

To illustrate this, the following example may be considered:
In a pump experiment, the water table level dropped by 45 cm. After 5 minutes, however, back flow caused the level to rebound to a height 28 cm higher than the lowest level reached by pumping. Is the site suitable for being chosen as a water table monitoring station?

h_1 (level drop) = 45 cm

h_2 (recovery) = 28 cm in t = 5 minutes

Therefore, $\varepsilon = (2/5)[(45 - 28)/(45 + 28)] = 0.093$, which is higher than 0.0115, i.e. the site is suitable as a monitoring station for water height measurements. It possesses a good hydrologic conductance within the aquifer. Data collected over a long time are later used to plot water table maps, upon which potentiometric contours are drawn to depict the directions of groundwater flows. These are generally perpendicular to the contour lines. In Chap. 12, techniques for the construction of potentiometric maps will be introduced.

10.4.3
Monitoring Hydraulic Heads

Measuring the height of the water table in different observation wells is not only useful to determine directions of flow, but can also supply information to be used in Darcy's equation to determine many other parameters, such as the velocity of groundwater flow, the transport of contaminants in groundwater, or the time required for a given amount of contaminants to reach a certain point in the area.

Practically, the hydraulic head in an observation well is measured by subtracting the depth to the water table at the observation well from the topographic height of its casing top above datum level (normally sea level). So (in Fig. 10.14) if the top of the casing of an observation well lies at a height of 87.0 m above sea level and the depth

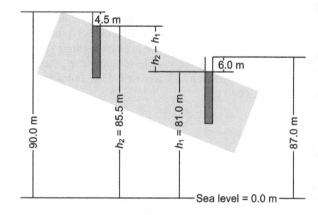

Fig. 10.14 Monitoring of hydraulic heads in two wells

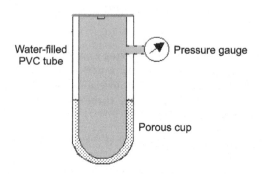

Fig. 10.15 Tensiometer

to the water table measured from the top of the casing is 6.0 m, its hydraulic head would have a height $h_1 = 87.0 - 6.0 = 81.0$ m. In Chap. 12 the use of such data will be elaborated on.

10.4.4
Measuring Hydraulic Heads in the Vadose Zone

Hydraulic heads in the vadose zone are measured using a tensiometer. This is made of a porous cup attached at its upper end to a PVC tube, on which a transducer or a common pressure gauge is attached (see Fig. 10.15). The cup and PVC tubing, which is sealed at the top, are filled with degassed water and placed into the soil at the depth required. Through the porous cup, hydraulic contact between the soil water and the tensiometer is established. It reaches dynamic equilibrium when flows in both directions have attained equilibrium.

Tensiometer readings give the so-called *tensiometer potential* (ψ_t), which is related to the degree of saturation of the soil. Its value at full soil saturation (where no exchange between the soil and the instrument takes place) is equal to 0, i.e. zero flux. Normally tensiometers of different lengths are buried at specific depths, where potentiometer potential gradients can be read.

Chapter 11
Biological Monitoring

Every plant and every living organism inhabiting a part of the terrestrial environment is, in a way or another, a product of this environment. It reflects the prevailing natural conditions, whether in a chemical or a physical sense. Plants especially have, in environmental geochemistry as well as in exploration geochemistry, proved to be excellent indicators of the chemical conditions prevailing in their substrates – a property which is now finding wide application in soil monitoring. One should, however, mention that observations associating plant species to chemical conditions in landscapes are not just modern pieces of wisdom that grew with the wide development of environmental science. Old miners in Europe and elsewhere have always known of the indicative character of certain plant species for buried ore deposits. Brooks (1979) reports about the use of *Lychnis alpina* by Scandinavian miners searching for copper in medieval times. Also, in South Tunisia and in the Sinai (Egypt) nomadic tribes have always been using plants as indicators of saline substrates. In middle Anatolia (Turkey) inhabitants identify saline substrates through small shrubs of *saltwort* – *Salsola nitraria*, known there as the *red kursalik*, due to its beautiful pink colour. These observations are now classified and critically investigated in the discipline of geobotanics.

Martin and Coughtrey (1982) classify indicator plant species into two main groups: universal indicators, which exclusively grow on soils having high concentrations of the metal indicated by them; and local indicators, which are associated with metal-bearing substrates in certain geographical areas, but which also grow in non-mineralised areas. This means that the universal indicators are the most useful types of these. However, this classification is not as rigid as it may appear on the face of it. Some universal indicators may occasionally grow in areas having low concentrations of the metal characterising their typical substrates. An example of this may be given by *Crotalaria cobaltica*, which is otherwise considered as a universal indicator for cobalt. In their book published in 1982, Martin and Coughtrey include a list of plant indicators together with their specific metals.

Plants indicate the condition of their substrates through two properties: geographical distribution or formation of certain assemblages; and the appearance of abnormal morphological changes, due to stress or metabolic problems resulting from metal uptake. Some shrubs growing on raised reefs in North Luzon (Philippines) are miniature forms of their species growing elsewhere. This is a typical symptom known as *dwarfism*. Other symptoms of disturbed metabolism may include gigantism, chlorosis of the leaves, distortion of fruits, or change of flower colours.

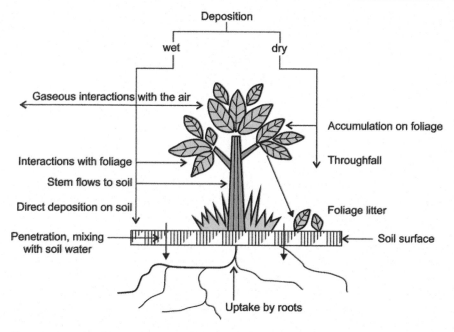

Fig. 11.1 Interactions between vegetation, soil and atmosphere

For environmental studies, however, all parts of a plant are considered in the monitoring process. This is especially the case with leaves, which are capable of supplying wide information not only about the chemical constitution of the soil, but also about any atmospheric wet or dry deposits that might be affecting the ecosystem (Fig. 11.1). Dry deposits may accumulate on the surfaces of leaves, leading to interactions with the tissues, or simply to changes of the chemical environment on these surfaces; they may fall through the canopy to join the soil environment. Wet deposits on the leaves may disturb their gaseous interactions with the atmosphere (Fig. 11.1) or even stop it, thus leading to changes that may appear as pathological symptoms. Wet depositions may also directly attack the surfaces of the leaves causing serious changes (acidic rain). A part of the wet deposits, however, may be transported to the soil by stem flows or by falling directly onto the soil surface.

The directly deposited material on the soil surface (*throughfall*), together with the material formerly accumulated on leaf surfaces, contribute the most to the soil environment. This occurs through a washing of the litter material by percolating surface waters and the penetration of the resulting solutions into the vadose zone.

11.1
Planning and Implementation of Biological Monitoring

Planning of biological monitoring should start at an early stage of the monitoring project. It forms an integrated part of the plan starting as early as the stage of site characterisation. Preparation of site maps should include, besides soil, geologic and hydrologic maps, separate maps showing the distribution of plant communities and tree stands, and maps showing the distribution of vegetation indicating any peculiar morphologic changes. Field and laboratory works should include the following.

11.2
Foliage Sampling and Investigation

Foliage should be sampled, for monitoring purposes, yearly at fixed periods. For example, foliage of deciduous species is sampled in late summer, while that of evergreens is normally sampled at the dormant times. In selecting sample trees, one should take those trees that are in the vicinity of the locations where soil samples were taken (provided that their roots were not damaged during soil sampling). The sample trees should represent the dominant species on the plot. Their number must cover and represent the area of investigation. It should not be less than 10% of the number of the predominant species.
For all foliage samples, the leaves should represent the current year, or the last three years in case of pine. They should be collected from the upper third of the tree. The uppermost leaves, however, are normally not selected.

11.3
Chemical Investigation of Foliage

After the samples are collected and brought to the laboratory they should be prepared for chemical analysis. Normally it is not necessary to wash them, but, in the case they were collected in an area of high air pollution, a systematic washing may be recommended. Later, they should be dried in an oven at 80 °C for 24 hours and then ground to obtain a homogeneous powder. For the chemical analysis itself, samples of the powder are digested either by wet or dry digestion techniques. If wet digestion methods are selected, the samples would be digested by hot mineral acids, or by a mixture of acid and an oxidising agent such as hydrogen peroxide. In dry digestion methods, dry ashing of the samples takes place in an oven at a high temperature (500–600 °C) followed by dissolution in water or dilute acids.

The analytical techniques may depend upon the elements to be determined and measured; yet, for most metals, atomic absorption methods were found very useful.

It was also found that X-ray fluorescence methods are helpful in determining the concentrations of metals and non-metals down to fluorine. This method has the benefit of not requiring digestion, since the measurements are done on compacted vegetal powder.

11.4
Sampling and Investigation of Litterfall

The sampling and analysis of shed foliage (*litterfall*) can provide useful information about the concentrations of nutrients and pollutants in the vegetation, brought about by uptake from the soil or resulting from atmospheric deposition.

Litterfall sampling is carried out monthly or at fixed periods to ensure registration of any temporal changes. This is done by distributing litter sacks on the plot at a depth of about 0.5 m under the trees, in order to secure the collection of the litter before it is blown away by wind. Litterfall sacks should be made of inert material. Their placement in the ditches should be in such a position that they do not touch the ground. This ensures that soil humidity would not mix with the collected material – a process which might accelerate decomposition.

After drying the samples at 40 °C, chemical analyses are carried out similar to the analysis of foliage material. Results from both processes are later compared to deduce any conclusions about the nutrition status and transport of contaminants.

Throughfall samples can also be collected in association with litter samples and analysed in the same way.

Part IV
Modelling of Soil Pollution

"Nature is essentially simple; therefore we should not introduce more hypotheses than are sufficient and necessary for the explanation of observed facts." These words, which were written by Sir Isaac Newton in his *Principia* about 275 years ago, represent, according to the science theoretician Gerald Holton (1988), a rule of simplicity and *verae causae* (real causes) that should never be forgotten when treating natural phenomena in a scientific way. Newton and Holton are not alone in propagating this principle of simplicity; Einstein himself was a great adherent of it. If we recall his equation: $E = mc^2$. How elegant and how simple it appears, and yet how deep and pervasive its influence has been on the development of modern science!

Scientists call such an equation a mathematical model of the relation between energy and matter. It summarises the results of centuries of observations and experimentation, and simplifies them in this general mathematical expression. So do all models, whether mathematical as this equation or of any of the other types, which will be explained below. They are necessarily simplifications of reality.

Chapter 12

Models and Their Construction

A model is formally defined as an object or concept designed according to a structural, functional or logical analogy to a corresponding origin in the real world. It is mainly used to solve a problem or find an answer, when, due to certain conditions the direct operating on the origin, is rendered unfeasible, difficult or impossible. Mathematical models, like the previously mentioned equation of Einstein, are logical constructions based on the essential quantitative or geometrical relations connecting the basic parameters of the original substrate, so that it allows an access to general information, explanation or prediction of behavioural or evolutionary patterns of the investigated object or phenomenon. Such patterns may serve as bases for work hypotheses, theories or laws, describing or controlling the investigated problem. Thus, we may conclude that models are abstractions of real systems. They may be mathematical or descriptive (geometrical); they may, or may not, be accurate or inaccurate, depending on the information used in their construction.

Soil, in the course of material addition or subtraction, can be considered as a system reacting to perturbation. This system is a composite one, embracing many closely related subsystems, each of which is aware of the perturbation and reacts in a way to resist any changes resulting thereof. Such reactions can be modelled and the results may, if accurate, be used for predictions of behavioural patterns that may be present under similar conditions. Take for example the soil solution, the goal of which is to uphold the chemical equilibrium and thus maintain the conditions required for a balanced soil environment. In this system, serious changes of hydrogen ion concentrations (pH) may lead to far reaching disturbances that could lead to a collapse of the system performance as a whole. Following the onset of such concentration changes, the soil solution system resists the perturbation by rearranging its chemical environment to restore the original pH-values. Such rearrangements (generally known as *buffer-reactions*), which were already explained in Chap. 8, can be mathematically modelled so that future developments, under similar conditions, may be predicted. Another example can be given by the behaviour of soil organisms towards xenobiotic substances. As it was mentioned before in Chap. 9, organisms try at the beginning to digest or detoxify the added material in order to keep the system at balanced conditions. This may go up to a certain degree, yet overworking the system by more perturbations may eventually lead to a complete collapse in the system's performance and only external efforts in the form of remediation measures can bring it back to its original conditions. Experimental studies on this phenomenon, followed by mathematical treatment of the results, may lead to the development of generalised formulations that enable predictions and facilitate the planning of remediation measures. Such formulations, based on field observations or experimental work, are nothing but abstract models, treating individual parts of the whole system. From this, we may conclude that modelling in the soil environment can encompass many aspects, such as reaction kinetics, biochemistry of the soil environment, or the physical and chemical properties of mineral matter. Each of these modelling studies provides vital information not only for understanding the conditions under which the system may collapse (e.g. strong pollution), but also for its remediation, if required.

12.1
Types of Models

It is very difficult to suggest a general classification for models which may be valid in all cases, yet for our purpose, i.e. the classification of models describing relations pertaining to soil pollution, we may adopt a simple scheme of model classification. According to this scheme, we may, for problems related to soil pollution, use either natural analogue models or mathematical models, describing in mathematical terms the dynamic conditions pertaining to the whole soil system. The latter may be deterministic (analytical as in the examples mentioned above) or they may be stochastic, i.e. using parameters of probabilistic nature.

12.1.1
Space Analogue Models

It goes without saying that the most important models in this category are maps and other graphical representations, as those mentioned in Sect.10.1. Maps (topographical, geologic, pedological, or hydrological) are space analogues of the site under investigation. They are not only important for site characterization but they also supply the information required for prediction through mathematical analysis (mathematical modelling). They are of great importance in planning sample collection, understanding flow relations, or later designing remediation measures. The base (topographic) map provides the fundamental construction for all subsequent models. Through contour lines (lines of equal elevations) a model representing a two dimensional view of the topography is constructed. It is, however, useful to see this within the framework of the whole region of which the study area forms a small part. This enables an understanding of the physical conditions of the area in an integrated view of the topographic, geologic and hydrologic conditions prevailing in the region. Figure 12.2 shows the relation between the contour map and the topography of a hilly area characterized by a ridge extending in a North-South direction (Fig. 12.1). The study area lies at the foot of the northern end of the ridge. Out of this block view of the whole region, a detailed view (close up) of the study area is constructed to illustrate the local topography (Fig. 12.3).

To complete the model, a profile representing a stratigraphic section (a diagram, showing the spatial distribution of the different lithologic units in the subsurface) is constructed. This allows a better understanding of the results of soil chemical analyses, since the soil characterising the area is a product of their weathering (see Figs. 12.4 and 12.5).

12.1 · Types of Models

Fig. 12.1 Topographic (contour) map of the region where the area of investigation lies

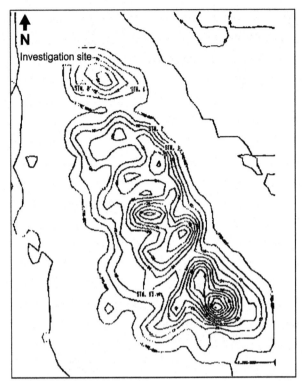

Fig. 12.2 The relation between contours and real topography of the region mentioned in the above example

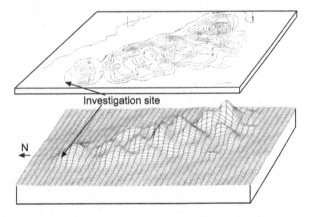

Fig. 12.3 Local topography of the site

Fig. 12.4 A block diagram showing a close up of the topography at the streambed together with the original contour map to show the relation between both

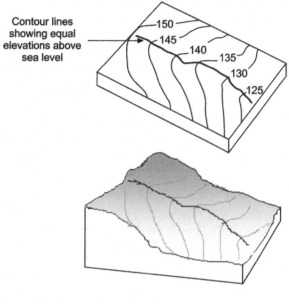

Fig. 12.5 A stratigraphic profile showing the main lithologic units forming the substrate

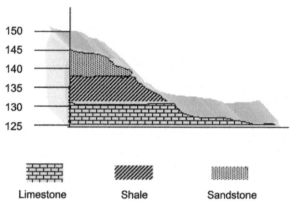

Modelling of Fluid Transport in Soil

One of the rules for constructing useful models is to define the objects of modelling together with the characterising parameters, in order to deliver a clear-cut imagery of the problem to be solved. So, in order to speak about fluid transport in this section, we require a definition of the fluids meant and the parameters defining its dynamic evolution and flowage in the soil medium. Soil fluids may include natural fluids (groundwater) or any contaminants infiltrating the soil, either as separate fluids or as solutions in the groundwater. In general, however, a model dealing with fluid transport in soil should have the following objectives:

12.1 · Types of Models

- Depicting the direction of flow in a spatial sense;
- Depicting the feasibility and velocity of flow in distance per unit time;
- Depicting the amount of flow in quantity per unit time.

Determining the Direction of Flow and Construction of Potentiometric Maps

Water level measurements, as explained in Sect. 10.4.3, deliver the basic material for constructing water table isopotentiometric maps, out of which the general direction of groundwater flow may be determined. Basically, an isopotentiometric map is constructed by plotting the measurements from monitoring wells on the base map, and connecting points with equal values of water level heights above sea level (or any standard datum) to form contours (isopotential lines) of the same. In an aquifer of isotropic nature, the direction of groundwater flow is perpendicular to the isopotential lines.

Figure 12.6 gives an example of such maps and shows the relation between the data registered at the observation wells and the general direction of groundwater flow. The map in Fig. 12.6b represents a graphical model of the flow relations in the area, which can be considered as a spatial analogue model. In constructing the potentiometric map, various hypothetical observation wells (Fig. 12.6a) supply enough data for the geometrical construction of the model, but what about in real life when considerably lower numbers of observation wells are available? This leads to the question about the minimum number of observation wells required to draw a potentiometric map. In fact one can use as low a number of wells as three, out of which a rough delineation of the isopotential lines by geometrical extrapolation may be done. The extrapolated relations should always be compared with the subsurface geology and corrected, if required, to suite the real conditions prevailing in the area. This method of extrapola-

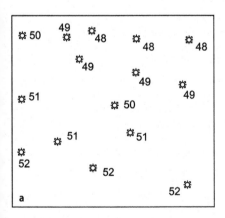 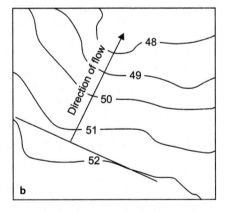

Fig. 12.6 Data collected from level measurements plotted on a base map (*a*), together with the corresponding potentiometric map (*b*); **a** base map showing positions of observation wells and the levels of groundwater measured in them; **b** potentiometric map: the contour lines connect observation wells of similar groundwater levels

Fig. 12.7 Extrapolation of the contours for determining the flow direction using data from the three observation wells – hydrologic triangle method

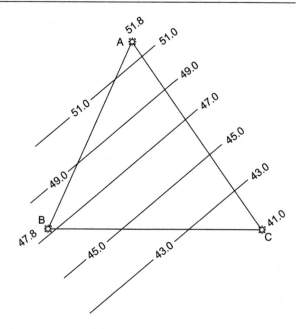

tion is generally known as the method of hydrologic triangle. To explain it, the following example may be considered:

The positions of three observation wells A, B and C (Fig. 12.7) are plotted on the base map. The heights of the water table above standard datum are given as A = 51.8, B = 41 and C = 47.3 as shown in Fig. 12.7. It is required to roughly depict the direction of groundwater flow by geometrical extrapolation on the base map.

Figure 12.7 shows the method of extrapolation using data from the three observation wells. A triangle connecting the positions of the three wells is constructed. The distance between the point of highest potential and the one of lowest potential is divided to scale and the calculated potentials are plotted on the line AB. The same is done with the line BC connecting the lowest and middle potential points. Eventually, straight lines are drawn connecting equal potentials on both lines. These represent the equipotential lines, perpendicular to which the general direction of flow (NW to SE) is likely to lie in the case of isotropy. These conditions, however, are restricted, as mentioned before, to aquifers of isotropic nature. In the case of anisotropy, resulting from structural conditions of the aquifer, complications may make corrections essential. Modification methods can be designed using instructions from a standard manual of hydrogeology.

12.1.2
Mathematical Modelling of Fluid Flows in Soil

Mathematical models, as noted before, are either analytical, i.e. made of equations having unique solutions, or stochastic, in which for every input there is a range of possible outputs, reflecting randomness or uncertainty. Problems related to soil pollution can be seen as potential objects for both modelling approaches. However, only deterministic models will be dealt with here.

Models dealing with natural phenomena mostly describe the behaviour of a segment or a subsystem of nature. In dealing with fluid movements in soil, it is clear that we are dealing here with the subsystem comprising the aquifer and the fluids percolating within. Any change or perturbation in the parameters, characterising one of the two elements of the system, will have a response shown by a change in the behaviour of the system as a whole. For example, an increase or decrease of the pore space or a change in its geometry will result in a change of the flow speed or the quantity of fluid passed or retained. Also a change of the properties of the fluid, such as density or viscosity, will produce a change in the behaviour of the system. Not only the parameters characterising the system elements in this narrow sense are important, also universal parameters belonging to the system Earth as a whole will play an important role in determining the behaviour of this subsystem. The most prominent of these are the forces resulting from Earth's gravity and attraction between material bodies. In the following, a short account of the forces acting on the system aquifer/groundwater will be given.

Forces acting on this system can be viewed as belonging to two main classes:

- flow-retarding forces, and
- flow-accelerating forces.

Flow-retarding forces are those forces resulting from the attraction of water to the surfaces of solids in the substrate. In this group, forces resulting from attraction by adsorption or capillary action (*matric suction*), and forces resulting from osmotic attraction (*osmotic suction*) are most relevant.

Flow accelerating forces are mainly those ones resulting from gravitational attraction towards the centre of the Earth; they are summed up in the so-called *gravitational potential*. Some other universal forces may be affecting the system, yet their effects are not relevant for practical purposes. The above-mentioned forces, which may be called *the system forces*, are integral parts of the system producing potentials that may be summarised as follows:

- P_m = attraction potentials (known as *matric potential*)
- P_o = potential resulting from attraction of solute ions to water molecules (known as *osmotic potential*)
- P_g = potential resulting from gravity attraction (known as *gravitational potential*)

The total potential of the system is calculated by algebraically summing up all these potentials. By assigning negative signs to the retarding forces (P_m and P_o) we get the following equation:

$$P_{total} = P_g - P_o - P_m$$

This means that P_{total} will have a positive value if, and only if,

$$P_g > P_o + P_m$$

Out of this result it is clear that flow can only occur if the gravitational potential is higher than all retardative ones.

Darcy's Law – The Principal Mathematical Model of Groundwater Flow

Henry Darcy (1803–1858) – a French engineer – was the first to construct a mathematical model for fluid flows in porous media, connecting all parameters of the system in an analytical equation. Through a series of experiments, he found the following relations:

In a porous medium, Q (total flow rate, standing for the total discharge of fluid per unit time; $cm^3 s^{-1}$) is directly proportional to:

- the cross-sectional area in cm^2
- density of fluid, $g\ cm^{-3}$
- h_2-h_1, the head loss (difference between hydraulic heads in two points, up and down stream)

It was also found that Q is inversely proportional to:

- length of the flow path (cm)
- dynamic fluid viscosity (mPa s)

Writing this in mathematical shorthand, gives the following relation:

$$Q = \frac{\rho A(h_1 - h_2) K}{\eta l}$$

where K is a proportionality constant (the permeability constant/hydraulic conductivity), which was found to be dependent on the hydraulic properties of the medium and its lithology. This shows that all parameters describing the two main elements of the system (fluid/porous medium – or fluid/soil) are connected in an analytical relation that allows calculation of different values, which might be required in solving problems arising from soil pollution.

For water (η and $\rho = 1$), Darcy's law takes the following form:

$$Q = \frac{AHK}{l}$$

which can be written:

$Q = KAH / l$

Where $H = h_2 - h_1$ (*head loss*) = difference between hydraulic heads in two wells, up and down stream (see Fig. 12.8). H is defined as the hydraulic gradient and is dimensionless since h_2, h_1 and l have all length dimensions (m).

Putting $J = H / l$, and rearranging in the last equation, we get:

$$K = \frac{Q}{AJ}$$

or $Q = KAJ$

Now Q has the dimensions $m^3 s^{-1}$ and A is in m^2, so we get for K the dimensions:

$m^3 s^{-1} m^{-2} = m s^{-1}$ (distance/unit time)

Physically, these are velocity dimensions and give the hydraulic conductivity of the medium in distance per unit time. Hydraulic conductivity values are specific for different lithologic units. They can be determined in the laboratory through samples taken from the aquifer or found by reference to standard manuals where K-values for different geologic materials are tabulated. Table 12.1 is an example showing hydraulic conductivity of some selected sediments.

Computations Based on Aquifer Model

In Fig. 12.8 a sand stone aquifer is represented by the inclined rectangle. Two wells (*Well 1* and *Well 2* in the figure) are drilled 60 m apart. The top casings lie at 87 m above sea level for *Well 1* and 90 m above sea level for *Well 2*. Systematic measurement of water table levels showed that the depth to the water table in *Well 1* is an average of 6 m, while in *Well 2*, the average depth lies by 4.5 m.

To apply Darcy's mathematical model (Darcy's law) to this aquifer, we may find K from the tables (see Table 12.1), yet the most difficult problem in this case is to find A, the cross-sectional area perpendicular to the direction of flow (see Fig. 12.9).

This problem can be solved by applying one of the following methods:

- Drilling of inspection wells to determine the dimensions of the aquifer, which is very expensive and time consuming.
- Determining the thickness of the aquifer by computation from the geologic map, which is a good method for achieving an approximate result. On a local level, as the one discussed here, this method is safe enough to provide satisfactory results. It needs however some knowledge in geologic computations.

Table 12.1 Permeability constants for some selected sediments

Sediment	K (m s^{-1})
Gravel	$10^{-1} - 10^{-2}$
Coarse sand	$\sim 10^{-3}$
Medium sand	$10^{-3} - 10^{-4}$
Fine sand	$10^{-4} - 10^{-5}$
Sandy silt	$10^{-5} - 10^{-7}$
Silty clay	$10^{-6} - 10^{-9}$
Clay	$< 10^{-9}$

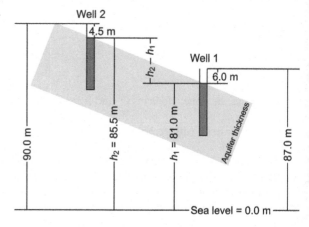

Fig. 12.8 Schematic representation of the sandstone aquifer discussed in the example

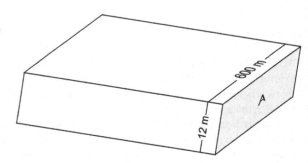

Fig. 12.9 Aquifer dimensions – flow front is represented by the area A, perpendicular to the axis of flow

For the example discussed, if the thickness of the sandstone bed (aquifer) was 12 m and if it was extending for 600 m in the breadth, the cross-sectional area A, perpendicular to the direction of flow, would be $12 \times 600 = 7\,200$ m² (see Fig. 12.9).

So the computation would continue as follows:

$Q = KJA$

- $K = 10^{-4}$ m s^{-1} = $10^{-4} \times 60 \times 60 \times 24 = 8.64$ m day^{-1}.
- $J = (85.5$ m $- 81$ m$) / 60$ m $= 0.075$ (dimensionless)

Therefore:

$Q = 8.64$ m day$^{-1} \times 0.075 \times 7\,200$ m² $= 4\,665.6$ m³ day^{-1}.

This means that the total flow rate (volumetric flow rate) in this sandstone aquifer amounts to about 4 666 cubic metre per day.

Another way of dealing with this model, in the case of investigating pollution problems, is by using the results of the chemical analysis of the water samples taken during monitoring works as follows.

Geochemical Dimension of the Model

If at the beginning of computations in the example mentioned above, a polluting source was found to lie uphill somewhere in the vicinity of *Well 2*, and if chemical analysis has shown an average concentration of 32 ppm of the pollutant in the groundwater, the concentration of the contaminant found by chemical analysis can be mathematically expressed as:

$$C = R_c / Q$$

where C = concentration, R_c = rate of release of the contaminant to groundwater, and Q = total flow rate

If it is known that the rate of release of the contaminant to the groundwater was 150 kg day^{-1}, then the concentration 32 ppm (0.032 kg m^{-3}) = 150 / Q, then

$$Q = 150 / 0.032 = 4673 \text{ m}^3 \text{ day}^{-1}$$

Putting this in Darcy's equation ($Q = KJA$), we get

$$4673 = 8.64 \times 0.075 \times A$$

Then A = 4673 / (8.64 × 0.075) = 7211 m^2, which is almost the same value for the cross-sectional area as calculated from the geologic map.

On the other hand, calculating the cross-sectional area from the geologic map helps in calculating the rate of contaminant release, by using the concentration known from the results of chemical analysis.

As shown here, the flexibility of this model allows the calculation of various factors that might be required for solving problems arising from soil pollution, and might supply vital information for designing remediation measures. Such information includes the area of contaminated surface, the breadth and dimensions of plumes, as well as the rate of the volumetric water flow through the aquifer per unit time.

The Effective Velocity (Interstitial Velocity) of Flow

The flux of water q is defined as the flow through a unit cross-sectional area, perpendicular to a line representing the axis of pathway. It may be calculated from Darcy's equation by taking A = 1; accordingly, the flux in the foregoing example may be calculated in the following way:

If $Q = KJA$, then $q = KJ$ (since A = 1; lower case q is used here to differentiate the flux from the total volumetric discharge Q).

This gives a value for $q = 8.64 \times 0.075 = 0.648$

In case the whole pore space was full, the flux q would be related to the effective velocity of flow v (*interstitial velocity*) by the relation: $q = v\varphi$, where φ is the porosity of the aquifer.

Assuming the porosity in the sandstone aquifer was 0.35, the effective velocity of flow in the foregoing example would be calculated in the following manner:

The effective velocity of flow $v = q / \varphi = 0.648 / 0.35 = 1.85$ m day^{-1}.

This means that, despite the huge volume of water flowing through the total volume of the aquifer, the front is moving very slowly with a velocity of 1.85 m per day. To reach a point 500 m away from the source of pollution, it needs 270.27 days (500/1.85) – about 9 months.

The above-mentioned relations can be used in different ways to calculate aquifer parameters according to the available information. For example, if Q and A are known, then dividing Q/A gives q, which again may be used to calculate other unknowns if required.

The foregoing discussion, however, shows that the model, as demonstrated here, is subject to several constraints, which may be summarised in the following:

- Full homogeneity of lithologic, hydraulic and tectonic properties throughout the aquifer (isotropy) is assumed.
- The model is one-dimensional, depending on computations that deal with discrete points lying on a one-dimensional axis.

As for homogeneity and isotropic character of the aquifer, the problem cannot be easily solved, and unless careful tectonic and petrographic analyses are done, the results of computations remain approximate, if not erroneous, in complicated cases. However, the problem of space dimensions may be mathematically treated, especially as several computer programs are now offered that make it possible to deal with complicated computations in relatively short times.

To begin with, let us take the case of the hydraulic gradient discussed in the foregoing example. We have here two discrete points representing two wells. These two points have only reference positions along the axis of flow l. In the Cartesian plane they may have the coordinates x_1 and x_2; h_1 and h_2 – the corresponding heads at the two points may as well be represented by y_1 and y_2 (see Fig. 12.10).

Accordingly, the head loss $h_2 - h_1$ can be given by $y_2 - y_1$ and the hydraulic gradient J would be given by $(y_2 - y_1)/(x_2 - x_1)$.

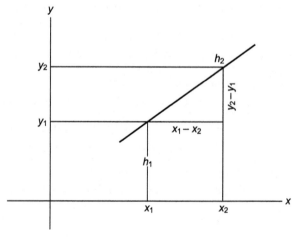

Fig. 12.10 The relation between the hydraulic h and the coordinates on the axis of flow

12.1 · Types of Models

If, for any point between x_1 and x_2, there exists a relation that allows the calculation of y, when x is known, one says that y is a function of x or $y = f(x)$, meaning that y depends on the independent variable x. The function $y = f(x)$ is also said to be one-dimensional, because y depends on one single variable. If the function $y = f(x)$ holds, i.e. is represented by points all over the range $x_1y_1-x_2y_2$, we say that the function is continuous. It is also said to be differentiable, i.e. one can find the slope of the tangent to the curve, known as the *derivative*, or the rate of change of the dependent variable with respect to any change of the independent one, everywhere on the curve representing the function. In our case, this is dy/dx = rate of change of h by changing l. This is written dy/dx or y' or $f'(x)$ and is simply known as the *first derivative* of the function. Higher derivatives may be obtained by further differentiation. Notations such as d^2y/dx^2, $f''(x)$, y'' or y''' are usually used to characterise derivatives of orders higher than one.

Returning to our original problem of the relation between the hydraulic head h and the distance between two points or their coordinates on a given axis (here x), we state (without proof) that the relation that allows the calculation of $h(y)$ on any point of the x-axis was found to be any relation satisfying the equation d^2x/dy^2 = 0.

This is a differential equation known as *Laplace's equation*, the solution of which gives another equation – the one regulating the relation between y and x. An example may be given by the equation: $y = 5x - 17$, dh/dl = 5 and d^2h/dl^2 = 0. So, this equation may be a possible expression of the relation between h and l. We observe here that h is calculated with respect to one variable only, l, i.e. the relation is one-dimensional, and is represented by a straight line equation.

The Continuous One-Dimensional Darcy's Model

It has been shown that the hydraulic head is connected to the coordinate on the horizontal axis with a continuous function, the first derivative of which equals $J =$ dy/dx or specifically dh/dx; since we are dealing with the head loss rather than with the total change in h, dh/dx will take a negative sign, becoming –dh/dx. Using this new relation in Darcy's law gives:

$$Q = -K \frac{dh}{dx} A$$

The interstitial velocity will be

$$v = -\frac{K \, dh}{\varphi \, dx}$$

where j is the porosity; since $v = KJ/j$.

The Two-Dimensional Darcy's Model

Suppose that hydraulic heads in two wells A and B, lying at a distance X from each other, were connected by a mathematical relation, which allows the calculation of the

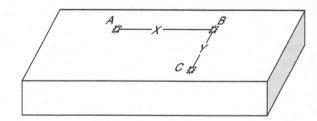

Fig. 12.11 Two-dimensional relation of h with respect to x and y

hydraulic heads at any point on the line AB; and suppose that a third well C, lying at a distance Y from the line AB (see Fig. 12.11), was found not to obey this relation and that it was found that the hydraulic head here was depending upon both distances X and Y, meaning that h is not only a function of X but a function of X and Y – something that we may write in mathematical form as $h = f(X, Y)$.

A function of two variables is said to be a two-dimensional function. It can have a derivative, with respect to any of the variables calculated under the assumption that the other variable remains constant. Such a derivative is known as a *partial derivative*. In the present case, the partial derivative of h, with respect to Y, may be denoted by $\partial h/\partial Y$ (i.e. the rate of change of h with respect to Y, so long X remains constant), while the derivative with respect to X is written as $\partial h/\partial X$, also meaning the rate of change of h with respect to X, so long as Y remains constant. These two derivatives represent the coordinates of a two-dimensional vector (H_x, h_y), commonly known as the gradient of the function $y = f(x, y)$ and is denoted by the symbol $\nabla(h_X, h_Y)$, or simply ∇h (read del h).

Accordingly, the equation $q = -KJ$ may be written as $q = K(-\nabla h)$ or $q = -K\nabla h$.

Similar to the case of the one-dimensional Darcy's model, we state here without proof that in any two-dimensional region of an isotropic aquifer of homogeneous conductivity, if the head function h is continuous and depends on two variables x and y, the equation describing the function will be one that fulfils the relation:

$$\partial^2 h/\partial x^2 + \partial^2 h/\partial y^2 = 0$$

This is a second-degree differential equation (including second derivatives), known as the two-dimensional Laplace equation. The solution of this equation is, unfortunately, not so easy like the first-degree equation of the last section. For the solution we need additional information about at least two points within the domain of the function. These additional pieces of information are known as *the boundary conditions*. In practical work, solutions at different boundary conditions are found using special computer programs (simulation programs) and the results are tabulated for use under different conditions. This technique of modelling is known as *the simulation technique*.

The Three-Dimensional Case

Analogous to the derivation of the two-dimensional case, if h was dependant on three variables x, y and z, we would be confronted with a three-dimensional function, denoted by $h = f(x, y, z)$. As in the case of the two-dimensional function, this one (if continuous) would deliver partial derivatives with respect to x, y and z. The vector, having these coordinates ($\partial h/\partial x, \partial h/\partial y, \partial h/\partial z$), is known as the gradient of the function and is denoted by ∇h (read del h). Sometimes it is denoted by grad h. The equation giving h at any point of the region where the function is continuous is also subject to the condition:

$$\partial^2 h/\partial x^2 + \partial^2 h/\partial y^2 + \partial^2 h/\partial z^2 = 0,$$

which is the three-dimensional Laplace equation.

Simulation solutions are also carried out using special computer programs.

Flows in the Vadose Zone

We have so far discussed water flows in the saturated zone using Darcy's law. Application of the derived relations to the unsaturated zone (vadose zone) was found to be dependent upon the degree of saturation according to the following relations:

$$v = \left[\left(\frac{S-S_0}{1-S_0}\right)^a \frac{d_e^2}{c} \frac{\varphi^3}{(1-\varphi)^2} \frac{g}{\eta}\right] \mathrm{grad}\, h$$

$$K(\Theta) = k\left(\frac{S-S_0}{1-S_0}\right) = k\left(\frac{\Theta-\Theta_0}{n-\Theta_0}\right)^a$$

whereby:

- v = interstitial velocity
- S_0 and Θ_0 represent the rest saturation and rest water content at which the hydraulic conductivity = 0
- a = an exponent which is commonly taken as 3
- d_e = effective grain size
- c = a conditional factor depending on the geometry of grains
- φ = the porosity
- g = gravitational acceleration
- η = dynamic viscosity of the water
- grad = del operator of the vector
- $K(\Theta)$ = unsaturated hydraulic conductivity
- Θ = total potential

Part V
Soil Remediation

Remediation is the logical consequence, drawn from the results of sampling, monitoring and model construction, when, in the area of investigation, pollution is found to be at a risk level that requires intervention. Its goal is, in the first place, to bring a polluted soil to a sustainable environmental condition, at which the risk arising from toxic pollutants is reduced to a minimum. Remediation techniques may be applied to the soil in place (in situ) or to the soil after being transported to special facilities. It may be mainly done by applying physical and chemical methods, or may be carried out using bioremediation methods. Yet in all cases and under all conditions, remediation is the last step in a soil conservation plan, based on systematic diagnostic efforts including monitoring and modelling of the actual soil condition. A successful remediation plan should make use of all observations and results of the three diagnostic steps: sampling, monitoring and model construction, as it was repeatedly mentioned before.

Chapter 13

Planning and Realisation of Soil Remediation

A successful remediation plan has to be based on the information gained during the preliminary diagnostic works, which are done before taking the decision to start a remediation project. The following checklist should be worked out and carefully studied before starting with the actual design of the technical work.

1. What are the types and chemical nature of the pollutants determined at the site?
   - Organics
   - Inorganic substances
2. What are the dimensions and scale of pollution?
   - Is the pollution localised?
   - Is it of the dispersed type?
   - How urgent is the remediation plan?
3. What is the risk level?
   - Low risk level
   - Medium level
   - High risk level
4. Which technical measures are thought to be most suitable for carrying out this project?
   - In place (in situ)?
   - Ex situ? And if yes, should the soil be transported to a special facility? Should the remediation be carried out in a prepared bed system, or in a tank?
5. Are there any financial restrictions on choosing the technical method of remediation?
6. Which method is technically suitable and financially fitting in the economic framework of the project?

13.1
Categories of Pollutants

A pollutant can in fact be any environmentally harmful substance that is accidentally or on purpose transported to the soil. Yet, for the sake of planning, a common, rather simplified classification, enables the selection of the suitable remediation method. The following checklist helps limit into a few categories the uncountable substances that may pollute the soil, having similar chemical and physical properties, and in most cases, having comparable degrees of response to a given remediation process.

- Are the pollutants in the present case chemically characterised and identified? If yes, are they solids, NAPLs or leachates?
- Are they organics or non-organics? If organic, are they aliphatics or aromatics? If aromatics are they halogenated?
- How high/low are their molecular weights? (from the tables)
- Are they volatile or of low volatility?
- Are they (for organics) polar or non-polar? What is their degree of solubility (high, low)?
- Is there any information about their biodegradability?
- If inorganic, of which category are they: metals, metal cations, waste–acids/alkalis? Are they easily oxidisable compounds? Are there any inorganic cyanides?

The careful utilisation of this checklist, as we will see later, is a very important step on the right way to select the appropriate method of remediation. Other factors, such as the scale of pollution and financial restrictions, may impose a revision of the decisions taken at this stage.

13.2
Scale of Pollution

Results of sampling and chemical investigation supply enough information about the spatial dimensions of the pollution case which is supposed to be treated. Pollution cases may be, according to their spatial dimensions, classified into the following two main types:

1. *Localised pollution cases.* These are cases resulting from spill accidents, where materials spilled are known and the risk is at a minimum when quick measures are taken. In these cases, remediation is mostly carried out in situ. Material safety data sheets supply information on the pollutant or the hazardous material forming the spill, so that immediate actions can be taken. One speaks also of localised pollution when the source of pollution is known, such as leaking tanks, landfills or old industrial facilities. In such cases, pollution would be spreading from the source in a flow pattern, which is more or less localised and showing concentrations that decrease with increasing distance from the source of pollution. The flow pattern and rate of decrease in contaminant's concentration with increasing distance from the source, can be characterised by careful sampling, investigation and mapping of the results.
2. *Diffused pollution cases.* Pollutants entering the soil will try to spread in both horizontal and vertical directions, whereby the dimensions of transport and diffusion will depend upon the saturation of the soil and upon its hydraulic and lithological character. When pollutants reach the groundwater, its further transport will, as it was said before, depend upon the lithological character of the aquifer. Some aquifers, due to this property, may be selective in transporting material reaching the saturated zone. As an example, one may consider the case of nitrates reaching the saturated zone in a fine-grained lime stone aquifer (micrite, chalk). In this environment, the nitrates can persist for very long times and can attain very high concentrations through accumulation. They are safe from being denitrified by dentrification bacteria, simply because the fine-grained chalk has very small pore sizes; these impede penetration

of such microorganisms into the aquifer. The same phenomenon is also responsible for the accumulation of many substances in the groundwater. Such contaminants, which have been seeping and accumulating in soil over long periods, form immense amounts of pollution that may extend over huge spatial dimensions. Accumulation of same scales may also occur when materials that were bound by complex formation on humic substances are released, due to a change of the chemical environment, and then dispersed within the soil. Such cases of wide spatial dimensions, which are not localised or characterised by a source and flow patterns, are normally described as diffuse cases of pollution. Their characterisation, mapping and remediation, needs more detailed planning and technical installations than localised cases.

13.3
Risk Level

Risk levels of contaminants should be determined according to the information collected on the chemical and physical properties of the potentially toxic material and its degree of dispersion in the area of investigation. Information on bioavailability and mobility of the material may indicate a low risk level, if the toxic material is in an immobile form or a non bio-available form with no impact on the environmental conditions. However, continuous monitoring is important in such cases, where no immediate risk exists, yet possible problems are expected on change of the chemical or physical conditions. In cases where immediate risk exists, such as after spill accidents or the discovery of old toxic deposits resulting from old landfills, or military or industrial sites, measures for remediation should be started or carried out within a short time.

13.4
Remediation Technologies

According to the scale of pollution, the risk level and the financial and time constraints on the remediation project, treatment of the soil may take place immediately in place (in situ), or the soil may be transported to special facilities where remediation may be carried out in special reactors or vessels, which are specially designed for this purpose *(in tank method)*. An example of this process is the washing of heavily polluted soils in special tanks. The polluted soil may also be transported and spread on a surface prepared to prevent the spread of contamination in lateral and vertical directions. Beds arranged in this way form the so-called *prepared beds*, upon which the remediation process will take place. This method is especially suitable for soils contaminated by oil products. Generally speaking, however, four classes of remediation technologies are known. These are:

- Chemical and physical methods
- Biological methods
- Fixation methods (also storing and immobilisation)
- Thermal destruction methods

Figure 13.1 shows the main types of remediation technologies in a schematic way.

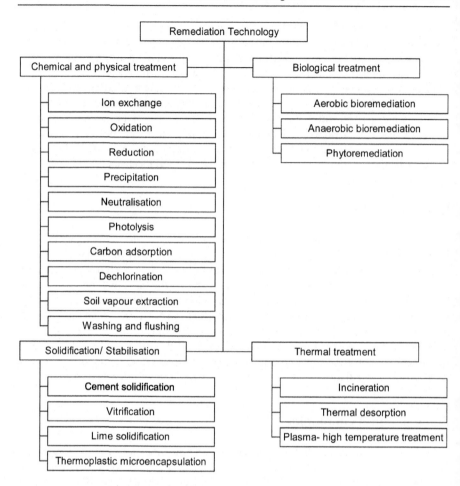

Fig. 13.1 Common remediation technologies

Some of the above-mentioned methods may require specific technical installations (tanks and prepared beds), others may be suitable for use in place (in situ), while others may be suitable for all three operational modes of remediation. Table 13.1 roughly shows the specific mode, or modes, suitable for each of the above-mentioned techniques.

From the table, one can clearly see that technologies like vacuum extraction and soil flushing are mainly done in situ, while all bioremediation methods are suitable for all operational modes. This fact plays a role in the financial planning of the remediation project and should be taken into account. The main decisive role, however, is played by the effectiveness of the method to the type of pollution encountered. In the following, each of the above-mentioned technologies will be shortly described.

Table 13.1 The different operational modes with their corresponding remediation techniques based on Boulding (1995)

Operational mode(s)	Suitable remediation technique
In situ	Soil vacuum extraction (SVE), soil flushing
In situ or in prepared beds	Carbon adsorption, ion exchange
In situ or in tank	Thermal stripping, dechlorination, cement solidification, vitrification, lime solidification, thermoplastic microencapsulation
All (in situ, in tank or in bed)	Neutralisation, oxidation, bioremediation (all methods)
In prepared bed	Photolysis
In prepared bed or in tank	Precipitation, reduction, carbon adsorption, ion exchange
In tank	Pyrolysis, infrared, rotary kiln, fluidised bed, soil washing

13.4.1
Chemical and Physical Remedial Techniques

The aim of all chemical and physical methods of remediation is to change the chemical environment in a way that prevents the transport of toxic substances to other elements of the soil system, examples here can be given by transport to plants, to groundwater, or to soil organisms. Such preventive measures may include decreasing mobility, change of chemical constitution or any of the factors on which it has been elaborated in Chap. 8. Chemical and physical methods of remediation include the following.

Oxidation

Oxidation is a common, highly effective remediation technology for soils contaminated by toxic organic chemicals and cyanides. Oxidising agents used in this technology include a wide range of substances, among which the most common are hydrogen peroxide, ozone and potassium permanganate. All three methods, according to EPA (2000), are of high rates of efficiency (over 90%) after only a short time in many cases. For example, efficiency reaches >90% for unsaturated aliphatic compounds, such as trichlorethylene (TCE), as well as for aromatic compounds such as benzene. The reaction, if hydrogen peroxide is the oxidising agent used, takes place according to the following equation:

$$3H_2O_2 + C_2HCl_3(TCE) \rightarrow 2CO_2 + 2H_2O + 3HCl$$

Ozone destruction of toxic contaminants takes place in the following manner:

$$O_3 + H_2O + C_2HCl_3 \rightarrow 2CO_2 + 3HCl$$

Still using the same example, i.e. oxidation of trichlorethylene, the reaction, if potassium permanganate (KMnO$_4$) is used, takes the following path:

$$2KMnO_4 + C_2HCl_3 \rightarrow 2CO_2 + 2MnO_2 + 2KCl + HCl$$

Oxidation technology has been successfully used for in situ remedy at source areas as well as for flume treatment. It is mostly used for benzene, ethylbenzene, toluene and xylene (BTEX) as well as for PAHs, phenols and alkenes.

Ion Exchange, Chelation and Precipitation

Soil components with high CEC values are capable of binding positively charged organic chemicals and metals in a way that makes them chemically immobile and thus reduces the risk imposed by them on the soil environment. The addition of soil conditioners, such as synthetic resins, zeolites or clays, may help increase the CEC characteristics of the soil and thus enhance the binding of positively charged contaminants on the negative functional groups of the soil matter.

Newly developed resins are now used for sites contaminated with complex materials, such as toxic metals combined with organic matter or even radioactive contaminants. Such multipurpose resins are capable of carrying out the process in one step. An example may be given by the resin developed with the help of the American Department of Defence (DoD), which depends on polymers derived from naturally occurring humic acids (Sanjay et al. 2006).

According to the authors, this resin is effective in cleaning sites contaminated with various types of pollutants, especially when the aquifers are contaminated with NAPLs. The treatment material is put within permeable reactive barriers (PRB) constructed to intercept the path of the contaminated groundwater plume.

Photolysis

Photolytic degradation technology depends upon degrading the organic contaminants with ultraviolet radiation. This may be carried out using artificial UV light or just by exposing the soil to sunlight, which may be sufficient for degrading shallow soil contaminants. The process can be carried out in situ or in prepared beds. However, deeply contaminated soils must be excavated and transported to special facilities, where the process is carried out in special tanks. A combination of photolytic degradation and bioremediation may be achieved by adding microorganisms and nutrients to the soil after the photolytic treatment.

Photolytic treatment using UV technologies is a very effective method, due to the fact that UV rays actually destroy the contaminants leaving no residue.

UV photons are used in this technology to break the chemical bonds in volatile organic substances (VOCs), such as trichlorethylene (TCE), toluene, benzene, etc.

The selection of UV sources should be done according to the absorption band of the target organic contaminant, as every toxic contaminant has an optimal wavelength for photodissociation. Wekhof (1991) gives the following as examples for effective UV treatment:

- benzene: 184 nm
- acetone: 220 nm and 318 nm
- TCE: 280 nm

In the case of complex organic substances, which may dissociate into other toxic compounds having different absorption bands, the process should be repeated with suitable lamps until only non-toxic substances remain at the end of the process.

Adsorption on Granulated Active Carbon (GAC)

This technology depends upon the tendency of most organic compounds to adsorb on the surface of activated carbon. Adsorption tendency increases with the molecular weight boiling point of the organic material. Thus we find that the technology of adsorption on granular activated carbon (GAC) is best suited for volatile organic compounds, hydrocarbons of high molecular weights, halogenated volatile organic compounds (VOC) and their halogenated forms, as well as some explosives and pesticides.

Remediation through adsorption on activated carbon is a method that can be carried out in the liquid phase (as in treatment of groundwater), or in the gas phase (as in treating off-gases from soil vapour extraction remediation methods). As a matter of fact, one of the earliest applications of this method was the use of GAC in adsorbing military gases by gas masks in the First World War.

Adsorption on activated carbon is a process carried out ex situ in special tanks or in prepared beds. It is principally used to treat toxic gases, solvents and organically based odours. However, impregnation of activated carbon with additional chemicals may be helpful in controlling some inorganic contaminants such as hydrogen sulphide, mercury or radon.

Reductive Dechlorination

Reductive dechlorination is a quite effective technology, with the help of which chlorine in polychlorinated organic compounds can be removed or substituted. It is mostly used to treat volatile chlorinated compounds by passing the heated gases containing the contaminants through layers of noble metal catalysts, triggering off a reductive reaction that destroys the halogen bond. An example of this may be given by the change of trichlorethylene into ethane. Reductive dechlorination of organic compounds may also be accomplished by redox active soil components such as iron oxides or Fe(II) bearing clays. This is a low cost technology with good effectiveness.

Reductive microbial dechlorination (Klasson et al. 1996) is a method combining the benefits of biotechnology with the known abiotic methods of dechlorination, by adding microorganisms to the prepared beds where the remediation takes place. These microorganisms enhance the process of dechlorination.

Soil Vapour Extraction (SVE)

Soil vapour extraction is a popular technology for the remediation of soils. It is a relatively simple process to remove volatile and easily evaporated organic contaminants within the vadose zone, i.e. contaminants persisting or accumulated above the groundwater table. The technical process of this technology involves injecting clean air into the unsaturated zone to effect a separation of organic vapours from the soil solution, by partitioning these vapours between the soil solution and the soil air. The vapours joining the soil air are then removed via vacuum extraction wells. Figure 13.2 shows a schematic view of an SVE arrangement.

Needless to say, the effectiveness of soil vapour extraction will depend principally on the degree of water saturation in the treated soil, as well as on the physical and chemical properties of the extracted contaminant, such as vapour pressure and volatility.

Vapour extracted by this method may be further treated by carbon adsorption or any other suitable method that may help to dispose of the toxic gases collected.

To enhance the extraction in this technology, heated air or steam may be injected into the soil. Reports on the use of steam at sites of defunct gas stations show high efficiency performance at a reasonably low cost. Adding an air sparging system to the technical installations of SVE makes this technology also suitable for removing contaminants from the saturated zone.

Soil Washing

In this technique, polluted soil is scrubbed by water through mechanical agitation to remove the hazardous contaminants or reduce their volume. It makes use of the selective binding of contaminants to fine material (silt and clay) rather than to coarse soil material (sand and gravel). Adding chemical additives or surfactants to the water may enhance this process. After separating the two soil fractions, fine material carrying the major part of contaminants is further treated by other methods of remediation to get rid of the separated contaminants (see Fig. 13.3), while the coarse material, if cleaned up, may be returned to the plot.

Soil washing belongs to the category of *volume reduction techniques* in which the contaminants are concentrated in a relatively small mass of material. It is used to treat soils contaminated by a wide range of pollutants, ranging from metals to oil products and pesticides.

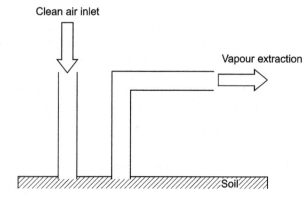

Fig. 13.2 A schematic diagram to explain the technical arrangements required for soil vapour extraction (SVE)

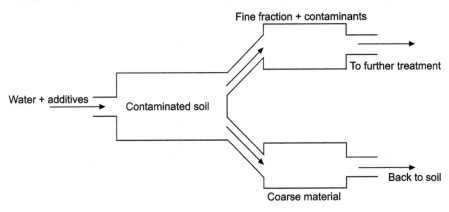

Fig. 13.3 Schematic diagram showing the different steps in soil washing

Soil Flushing

Soil flushing is a remediation method used for in situ treatment of inorganic and organic contaminants. Known sometimes as *the cosolvent flushing method*, this technique depends upon injecting a solvent mixture, such as water and alcohol or surfactants, into the vadose or saturated zone. The leachate, i.e. the solvent with leached contaminants, is drawn from recovery wells to be treated above ground or disposed of. The flushing technique is mainly used to treat soils contaminated by inorganics, including radioactive contaminants. It may also be used to treat VOCs, SVOCs, pesticides and fuel remnants. It must, however, be mentioned that flushing may not be effective for soils with low permeability. Also, the costs for above-ground treatment of the leachates may raise the financial burden of the remediation project. Figure 13.4 is a diagrammatic illustration of the process.

13.4.2
Biological Treatment (Bioremediation)

Biological treatment of contaminated soils is a remedial technique making use of naturally occurring microorganisms in the soil, which are capable of degrading toxic materials while carrying out their daily biological activities. Examples of such organisms include bacteria and yeast. As explained before, some bacteria are capable of digesting a wide range of organic contaminants that are otherwise very difficult to separate or degrade by any of the known technical methods.
Bioremediation is an easy and effective method resulting in the changing of organic contaminants, such as fuels or other oil products, into carbon dioxide and water. However, the time required for complete remediation will depend upon whether the process is carried out in situ, or in special facilities where excavated soil material is transported. Ex situ technologies are normally faster and more effective than in situ processes.

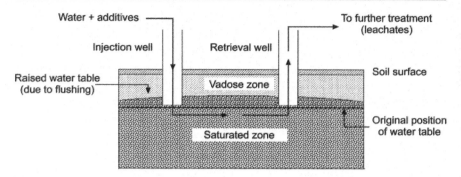

Fig. 13.4 Diagrammatic illustration of soil flushing techniques

13.4.2.1
In Situ Bioremediation Techniques

In situ bioremediation techniques are mainly used to treat non-halogenated semi-volatile organics, such as diesel fuel and heavy oils, beside other materials that are susceptible to metabolism by microorganisms. This technique, sometimes known as *aerobic bioremediation*, is accomplished by introducing oxygen and nutrients to the soil in order to enhance biodegradation of the contaminants. Two technical methods are used to create the suitable life conditions for the microorganisms. These are:

1. *Bioventing.* In this method, atmospheric air is injected through special wells into the soil above the water table, i.e. in the vadose zone, to supply the oxygen required for the microorganisms.
2. *Peroxide injection.* Here, oxygen is introduced in a liquid form through injection of hydrogen peroxide into the soil. However, this method is only applied to sites where the groundwater is already contaminated, so as to avoid unknown consequences resulting from the contamination of the groundwater by this chemical in areas of limited pollution.

13.4.2.2
Ex Situ Bio-Remedial Methods

Ex situ bio-remedial methods, i.e. those methods carried out away from the pollution site, are normally faster than the in situ methods. They are applicable for a wider range of contaminants, yet they are more expensive and may, in some cases, need pre-treatment as well as post treatment measures, in order to achieve the optimum effectiveness. According to whether the treatment takes place in special tanks or in prepared beds, ex situ bioremediation consists of two main technologies: *slurry phase treatment* and *solid phase remediation.*

1. *Slurry phase treatment.* In this technology, the polluted soil is excavated and transported to special facilities where it is mixed with water in special tanks (bio reactors). Oxygen and nutrients are later added, and the so formed mixture is thoroughly

mixed to form a thin slur. Temperature, nutrients and oxygen concentrations are controlled so that the organisms may have the best conditions in which to sustain their bioactivities, leading to the degradation of the pollutants.

2. *Solid phase treatment.* Here the polluted soil is treated above the ground in prepared beds. Despite the benefit of being less expensive than the slurry bed treatment, it is not so effective and needs more time and space to prepare the beds. Three main techniques are commonly used to carry out this remediation method:
   - *Land farming.* The soil is excavated and spread on a pad with a built-in system for collecting any possible leachates. The so formed bed is regularly mixed and turned over in order to facilitate aeration and enhance biological activity. Nutrients are added if required, since a lack of nutrients and oxygen may lead to retardation of the biodegradation processes.
   - *Soil biopiles.* The excavated soil is heaped in piles of several metres in height. Air is blown through the pile to enhance degradation activities by the microorganisms. If required, nutrients are also added. Due to emissions from the piles, the whole process is sometimes carried out in inclusions that control any volatile contaminants.
   - *Composting.* Composting is an aerobic process during which organic matter is decomposed by microorganisms producing heat, carbon dioxide, water vapour and humus.

In the composting technique, biodegradable waste or contaminated soil is mixed with bulking materials, such as straw, to facilitate circulation of air and water required for the biological activities of the microorganisms. Nutrients are also added if required. Biodegradation of the waste or contaminants takes place some times in *static piles composting heaps*, where the soil is heaped in piles, which are periodically aerated with blowers or vacuum pumps. It may also be carried out in *mechanically agitated special tanks* where aeration takes place through agitation. Another technique through which composting may be carried out is the one known as *window composting*. In this technique soil is spread in long piles so that it is exposed to atmospheric air and the photolytic effects of sunlight. Organic matter degradation by microorganisms takes place in these heaps assisted by atmospheric oxygen and humidity.

Figure 13.5 shows in a schematic way the most common processes in biological treatment of polluted soils and the relations that connect them. Metabolic processes and enzymatic reactions were already explained in Chap. 9.

13.4.2.3
Phytoremediation

Phytoremediation is the term applied for the utilisation of plants as collectors or agents for remediation of lands contaminated by organic or inorganic pollutants. The same may also be used to remove contaminants from groundwater. As a matter of fact, the major part of these processes takes place in the rhizosphere, i.e. in the root parts of the plant, rather than in any other parts.

In general, phytoremediation is a passive technology, which has been gaining increasing popularity in the last few decades. It can be used in sites contaminated by

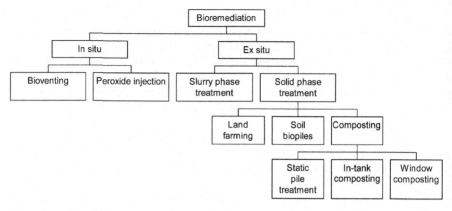

Fig. 13.5 A schematic diagram showing the different technologies of bioremediation

inorganic as well as organic pollutants. In the latter case, moderately hydrophobic material may respond best. Examples of these are toluene, benzene, PAHs, xylenes, ethylbenzene and many chlorinated solvents. Hydrophobicity, which is normally measured as the log of the octanol-water partitioning coefficient (known as K_{ow}) is, accordingly, set by some authors as a measure of the root capacity to absorb organic matter.

Phytoremediation consists of two main categories depending upon whether the removal of contaminants or stabilisation of geochemical conditions in the soil are contemplated. These will be discussed here under the headings:

- Phytoextraction
- Phytostabilisation

Phytoextraction
Many technologies are grouped under the general heading of *phytoextraction*, of which the following will be briefly introduced:

Phytoextraction by Direct Uptake and Accumulation
Plants capable of absorbing heavy metals are normally in a position to accumulate them in some of their shoot tissues – a property that lends them the name *hyperaccumulators*.

Various hyper accumulators are mentioned in the literature. Some of these are known to be in a position to accumulate large amounts of heavy metals. Of these we may mention the Indian Mustard *Brassica juncea* (Fig. 13.6b), which is capable of collecting two tons of lead per hectare in one planting (Bishop 1995); or the alpine pennycress *Thelaspi caerulescens* (Fig. 13.6d), formerly known as *Thelaspi alpestre*, which may accumulate zinc in amounts reaching 2 000–4 000 mg kg^{-1} (Brown et al. 1995). A further example may be given by the so-called Bahia grass *Paspallum notatum* (Fig. 13.6c), which is a good hyperaccumulator for caesium. The cactus variety *Aeollanthus subacaulis* (Fig. 13.6a) is also known to be capable of extracting and accumulating copper from soils.

13.4 · Remediation Technologies

Fig. 13.6 a *Aeollanthus subacaulis* (from Malaisse et al. 1999, Biotechnol. Agron. Soc. Environ 3(2):104–114); **b** *Brassica juncea* (Indian mustard) (from USDA-NRCS Database – Briton NL, Brown A *"Illustrated flora of the northern States and Canada"*, vol 12, p. 193); **c** *Paspallum notatum* (Bahia grass) (from USDA-NRCS-Database – Hitchcock AS (Rev. A. Chase) 1950 *"Manual of the Grasses of the United States"*); **d** *Thlaspi caerulescens* (alpine pennycress) (from Mary Ellen (Mel) Harte, www.forstryimages.org)

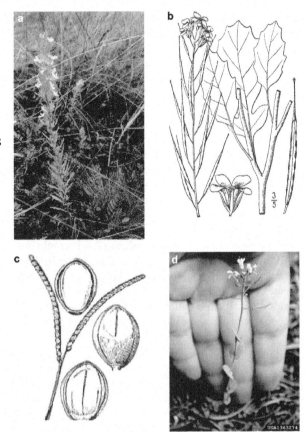

One drawback of hyperaccumulators is the fact that they are capable of extracting pollutants up to about 60 cm depth; in case the contamination is deeper than that, deep rooted poplar trees are preferred. Such trees, despite the probability of recycling heavy metals back to the soil through leaf litter, are capable of extracting heavy metals by sequestration and are often used to extract deep lying heavy metal pollutants.

Rhizofiltration and Phytosorption
Some plants and algae are capable of extracting heavy metals from solutions by a process similar to filtration. Such processes take place mainly in the root system and are known as *rhizofiltration*. It takes place more readily in water than in soil and that is why it is mainly used to extract metals or radioactive matter from water.

Other than in phytoextraction processes, here only the roots, where the metal accumulation has taken place, are harvested and disposed of. Two main steps are characteristic for this method: first the plants are grown in greenhouses until the root system is well developed, then in the second step, contaminated water replaces the water in the tanks; the plants then take up the water and the contaminants along with it. In the case of aquatic plants, a similar process takes place, whereby the plants collect and accumulate toxic elements from the ambient aquatic medium. Such processes, which depend in their rates and efficiency for remediation upon various biological and chemical factors, are collectively known as *phytosorption*. They may be carried out by aquatic plants as well as algae, and can be effective in adsorbing many heavy metals such as cadmium, zinc, copper and nickel.

Special methods are used for the removal of organic contaminants from soil, the most important of which can be summarised as follows:

- *Phytodegradation*. In this process, plants are used to degrade or break down organic contaminants through biochemical reactions taking place within the plant. These are mainly enzymatic reactions, as those discussed in some detail in Chap. 9.

  The degradation process may take place inside the plant or may be carried out within the rhizosphere, as explained before. It is effective in removing (degrading) a wide number of organic compounds, including organic solvents, pesticides, explosives, petroleum, as well as other aromatic compounds. An example may be given by the degradation of organic solvents, such as ethylene dibromide (EDB) or trichlorethylene (TCE), which were found to be metabolised by the tropical tree *Leuceana leucocephala* (Doty et al. 2003). Further work showed that a wide variety of plants are capable of taking up and metabolising a number of organic solvents. Castro et al. (2003) found that sunflowers (*Helianthus annuus*) can take up and incorporate benzotriazoles out of hydroponic solutions in which they were grown.

  Some pesticides were also found to be susceptible to phytodegradation, as was shown by Li et al. (2002), Knuteson et al. (2002) and many others.

  Some explosives can also be degraded in soils by the action of plants. Several authors have reported on the degradation of 2,4,6-trinitrotoluene (TNT) by plants like Johnson grass, wild ryegrass, or hybrid poplar. As examples of these works, we may mention Wang et al. (2003) and van Aken et al. (2004).

  Remediation of petroleum-based contaminants by plants. Phytodegradation of organic compounds may be enhanced if their metabolic action is accompanied by bacterial activities. Practical work by Godsy et al. (2003) has confirmed the mechanisms lying behind this technology. The efficiency of this method in degrading crude oil was substantiated by the work of another group of scholars (Banks et al. 2003).

  For an excellent review of the literature dealing with phytodegradation of organic compounds in general, the reader is referred to Newman and Reynolds (2004).

- *Enhanced rhizosphere biodegradation*. In this process, plants work hand in hand with soil organisms in order to break down organic pollutants. Microorganisms carrying out biodegradation of the polluting material are supported by the root system (rhizosphere) that produces nutrients (e.g. alcohols, sugars) needed for energy requirements of the organisms. This cooperative process keeps the microorganisms at a level sufficient for them to carry out their life activities and secure a continuous degradation of the toxic contaminants. As a matter of fact, the cooperative character of the microbe-plant community goes beyond one being dependent upon the other for support or symbiotic cooperation on a nutrition basis, to the extent that the consequences of the change in environmental conditions are a result of the respiratory activities of the microbe-plant complex community. One of these consequences is the decrease of oxygen concentration in the ambient space, leading to the creation of a reducing environment. Such an environment is favourable for the different pollutant's degrading reactions, as for example the transformation of highly chlorinated compounds.

The above-mentioned technologies for remediation are nowadays recognized and used for the remediation of a wide spectrum of organic contaminants, such as polycyclic aromatic hydrocarbons, petroleum compounds, explosives, as well as various chlorinated solvents.

- *Organic pumps.* As some trees (e.g. poplar, cotton woods) are capable of pulling out large volumes of groundwater, they may be compared with pumps continuously pumping water out of the soil. This action decreases the tendency of contaminants to penetrate the saturated zone and reach the groundwater.
- *Phytovolatilisation.* With the help of some naturally occurring, or genetically modified plants, some elemental forms of metals (e.g. arsencic, mercury and selenium) may be absorbed out of the soil and biologically converted into gaseous species, eventually being released into the atmosphere. Despite the apparent ease of this technology, some authors believe it to be rather harmful to the environment, since the gaseous forms of the metals released into the atmosphere may themselves be poisonous, such is the case with mercury or selenium (Watanabe 1997). Other authors, however, argue that the addition of these species into the atmosphere would not contribute significantly to the atmospheric pool (Heaton et al. 1998).

Phytostabilisation
The contamination of soil with heavy metals may in many cases be closely related to other problems dealing with soil transport and erosion. These could represent the real factors behind the concentration of heavy metals in the soil, as it has been shown during many discussions in the present book. That is why these problems have to be taken into account when planning a project of phytoremediation for soils contaminated with heavy metals. A practical approach in this case may simply be the addition of soil amendments. In other cases, the planting of vegetation may stabilise the soil, hold it in place, and provide a vegetation cover, which will stop transport and reduce the soil loss in fine and organic matter which lead to concentration of heavy metals in the site. Many processes used to master stabilisation problems were discussed in Chap.5.

13.4.3
Solidification/Stabilisation Methods

This is a group of technologies aimed at immobilising or stabilising contaminants in the soil and to prevent them from entering the environment, either by enclosing them into a solid mass or converting them to the least soluble, mobile or toxic form. Various technologies are known that secure safe performance of these processes. The following are the most successful among them:

- *Bitumen-based solidification.* In this technology, the contaminated material is embedded in molten bitumen and left to cool and solidify. The contaminants thus encapsulated in the molten bituminous mass are changed to an immobile form that cannot enter the environment.
- *Encapsulation in thermoplastic materials.* Thermoplastic materials (e.g. modified sulphur cement) are molten and mixed with the contaminated material in special

tanks and vigorously mixed to form a homogenous slurry fluid. After cooling, the resulting solid may safely be disposed of.

- *Polyethylene extrusion.* The contaminated soil is mixed with polyethylene binders, heated and then left to cool. The resulting solid may be disposed of, or used in other ways.
- *Pozzolan/Portland cement.* Pozzolanic-based materials (e.g. fly ash, kiln dust, pumice) are mixed with the contaminated matter in presence of water and alkali additives. In this environment, heavy metals may precipitate out of the slurry. The restmass solidifies enclosing the remaining organic contaminants.
- *Vitrification.* In this process the contaminated soil is encapsulated into a monolithic mass of glass. Vitrification may be carried out in situ or ex situ. In situ vitrification is performed by introducing graphite electrodes into the soil and heating them electrically with powerful generators to temperatures between 1 600–1 800 °C. At these temperatures the soil melts and forms a glass block. Upon cooling, organic contaminants are pyrolysed and reduced to gases during the melting process, while heavy metals remain enclosed in the stabilised glass mass. This method has also been successfully used in treating soils contaminated by radioactive materials.

Vitrification may also be done in special appliances where contaminated soil would be molten in presence of borosilicate and soda lime to form a solid glass block.

13.4.4
Thermal Treatment

Volatilisation and destruction of contaminants by thermal treatment is a very effective technique. It is achieved by heating the contaminated soil in kilns to temperatures between 400 and 700 °C, followed by further treatment of the kiln off-gas at higher temperatures (800–1 200 °C) to secure total oxidation of the organic volatile matter. Thermal treatment includes various technologies, the most important of which are:

- *Incineration.* In this technology, contaminants are combusted at high temperatures (970–1 200 °C). It is particularly effective for halogenated and other refractory organic pollutants. Properly operated incinerators may be of very high destruction and removal efficiency (DRE), reaching to as much as 99.9999%, which is normally required for PCBs and dioxins.

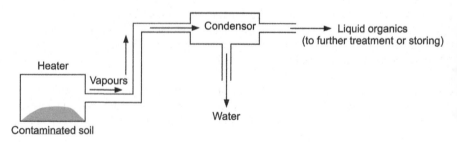

Fig. 13.7 Schematic diagram of a thermal desorption system

13.4 · Remediation Technologies

- *Thermal desorption.* This is the process by which organic contaminants are volatilised under controlled conditions by heating the contaminated soil to temperatures up to 600 °C. Under these conditions, contaminants of low boiling points vaporise to be afterwards collected and further treated. Other than incineration, this technology aims to physically separate the contaminants from the soil (Fig. 13.7).
- *Plasma high-temperature metals recovery.* At high temperatures (plasma activated) metal fumes are purged, and then later recovered and recycled. This is suitable for soil as well as for groundwater.

References

Abriola LM (1988) Multiphase flow and transport models for organic chemicals: a review and assessment. Electric Power Research Institute, Pal Alt, CA (EPRI EA-5976)

Addiscott TM, Wagenet RJ (1985) Concepts of solute leaching in soils: a review, modeling approaches. J Soil Sci 36:411–424

Agassi M, Steinberg I, Morin J (1981) Effect of electrolyte concentration on and soil sodicity on infiltration rate and crust formation. Soil Sci Soc Am J 51:309–314

Aiken GR, McKnight DM, Wershaw RL, MacCarthy P (eds) (1985) Humic substances in soil, sediment, and water. Wiley-Interscience, New York

Aken B van, Yoon JM, Schnoor JL (2004) Biodegradation of nitro substituted explosives 2,4,6-trinitrotoluene, hexahydro-1,3,5-trinitro-1,3,5-triazine and octahydro-1,3,5,7-tetranitro-tetrazocine by a phytosymbiotic *Methylobacterium* sp., associated with poplar tissues (Populus deltoids x nigra DN34). Appl Environ Microbio 70:508–517

Alloway BJ, Ayres DE (1997) Chemical principles of environmental pollution, 2nd edn. Blackie Academic and Professional, London

American Association of State Highways and Transportation (AASHTO) (1978) Standard specifications for transportation materials and methods of sampling and testing AASHTO Designation: M, 12th edn. AASHTO, Washington, DC, 145–173

American Petroleum Institute (API) (1982) The migration of petroleum products in soil and groundwater – principles and countermeasures. API, Washington, DC (Publication 4149)

Amundson R, Harden J, Singer M (eds) (1994) Factors of soil formation: a fiftieth anniversary retrospective. Soil Science Society of America, Madison, WI (Special Publication 33,160 pp)

Anger WK, Setzer JV (1979) J Toxicol Env Health 5:793

Antonov N (1985) Chemical weapons at the turn of the century. LN 72-96, pp 30–33

Armson KA (1979) Forest soils, properties and processes. University of Toronto Press, Toronto

Arnold F (1992) A performance comparison of different analytical and numerical saturated zone contaminant transport models. Ground Water Management 9:21–29 (Proceedings 5th International Conference on Solving Ground Water Problems with Models)

Arozarena MM et al. (1989) Stabilization/solidification of CERCLA and RCRA wastes: physical tests, chemical testing procedures, technology screening, and field activities. EPA/625/6-89/022

Arshad MA, Lowery BA, Grossman B (1996) Physical tests for monitoring soil quality. In: Doran JW, Jones AJ (eds) Methods for assessing soil quality. Soil Science Society of America, Madison, WI (Special Publication 49)

Asphalt Institute (1969) Soil manual. College Park MD (Manual Series 10)

Atkins PV (1978) Physical chemistry. Oxford University Press

ATSDR (2001) Toxicological profile for mustard gas. Draft for public comment. Update. Agency for Toxic Substances and Disease Registry, Atlanta, GA

Attewell PB, Farmer IW (1976) Principles of engineering geology. John Wiley and Sons, New York

Austin ME (1972) Land resource regions and major land resource areas of the United States (exclusive of Alaska and Hawaii). Soil Conservation Service, U.S. Department of Agriculture, Washington, DC (Agricultural Handbook 296)

Averett RC, Leenheer JA, McKnight DM, Thorn KA (1989) Humic substances in the Suwannee River, Georgia: interactions, properties, and proposed structures. U.S. Geological Survey, Denver, Colorado (Open-File Report 87-577)

Avery BW (1980) Soil classification for England and Wales. Soil Survey Technical Monograph No. 14, Harpenden

Baas-Becking LGM (1925) Studies on the sulphur bacteria. Ann Bot 39:613–650

Bachmann A (1993) Soil remediation at Schweizerhalle: a case study. Soil Environment 1:695–712

Bailey GD (compiler) (1987) Bibliography of soil taxonomy, 1960–1979. CAD International, Tucson, AZ (World literature on USDA soil taxonomy)

Baker EW (ed) (1985) Proceedings of the First Meeting of the International Humic Substances Society, Estes Park, Colorado, August 1983. Organic Geochem 8:1–146

Banks MK, Kulakow P, Schwab AP, Chen Z, Rathbone K (2003) Degradation of crude oil in the rhizosphere of sorghum bicolor. Int J Phytoremed 5:225–234

Barles RW, Daughton CG, Hsieh D (1979) Accelerated parathion degradation in soil inoculated with acclimated bacteria under field conditions. Arch Environ Con Tox 8:647–660

Barrow CJ (1991) Land degradation. Cambridge University Press, Cambridge

Battelle (1993) Technical resource document: Solidification/stabilization and its application to waste materials. EPA/530/R-93/012

Beck P (1996) Soil and groundwater remediation techniques. Geosci Can 23:22–40

Beck AJ, Jones KC, Hayes MHB, Mingelgrin U (1993) Organic substances in soil and water: natural constituents and their influences on contaminant behaviour. The Royal Society of Chemistry, Cambridge

Beck AJ, Wilson SC, Alcock RE (1995) Kinetic constraints on the loss of organic chemicals from contaminated soils: implications for soil quality limits. Crit Rev Env Sci Tech 25:1–43

Becker G (1987) Special Issue of Humic Substances Research from the Third International Meeting of IHSS, Oslo, Norway, August 4–8, 1986. Sci Total Environ 62:1–505

Becker G, Gjessing E (1986) Abstracts of Oral and Poster Papers of the Third International Meeting of IHSS, August 4–8, 1986, Oslo, Norway

Bell FG (1992) Engineering properties of soils and rocks, 3rd edn. Butterworth-Heinemann, Boston, MA

Benazon N (1995) Soil remediation: a practical overview of Canadian cleanup strategies and commercially available technologies. Hazardous Materials Management 7:10

Bertoldi M de, Sequi P, Lemmes B, Papi T (eds) (1996) The science of composting. Parts 1 and 2. Blackie (a division of Chapman and Hall), London

Bewley RJF, Campbell R (1978) Scanning electron microscopy of oak leaves contaminated with heavy metals. Trans Brit Mycol Soc 71:508–511

Bewley RJF, Campbell R (1980) Influence of zinc, lead and cadmium pollutants on the microflora of hawthorn leaves. Microb Ecol 6:227–240

Biaski M, Clarkowska J, Profus D, Rostkowska J, Skowerska M, Uryga G, Witalinski W (1977) Influence of industrial emissions on the quantity appearance of the soil fauna. Prezgl Zool 11:310–328

Bicki T, Felson AS (1994) Remediation of pesticide contaminated soil at agrochemical facilities. Mech Pestic Mov Ground Water, pp 81–99

Biddulph O (1960) Foliar entry and distribution of radioisotopes in plants. In: Caldecott RS, Snyder LA (eds) Radioisotopes in the biosphere. University of Minnesota, pp 73–85

Biggins PDE, Harrison RM (1978) Identification of lead compounds in urban air. Nature (Lond) 272: 531–532

Bingham FT (1979) Bioavailability of cadmium to food crops in relation to heavy metal content of sludge amended soil. Environ Health Perspect 28:39–43

Bingham FT, Page AL, Mahler RJ, Ganje TJ (1976) Yield and cadmium accumulation of forage species in relation to cadmium content of sludge-amended soil. J Environ Qual 5:57–60

Birkeland PW (1984) Soils and geomorphology. Oxford University Press, New York, NY

Bishop MS (1960) Subsurface mapping. Wiley, New York, NY

Bishop JE (1995) Pollution fighters hope a humble weed will help reclaim contaminated soil. Wall Street J, Aug, 7

Bitton G (1975) Adsorption of viruses onto surfaces in soil and water. Water Res 9:473–484

Blackmer GL, Geary GL, Vickrey TM, Reynolds RH (1976) Cadmium contamination in the Texas high plains. J Environ Sci Health 11:357–366

Blatt H, Middleton G, Murray R (1980) Origin of sedimentary rocks, 2nd edn. Prentice-Hall, Englewood dins, NJ

Block C, Dams R (1976) Study of fly ash emission during combustion of coal. Environ Sci Technol 10: 1011–1017

Blokker PC (1972) A literature survey on some health aspects of lead emissions from gasoline engines. Amos Environ 6:1–18

Bloom PR (1981) Metal-organic matter interactions in soil. Soil Science Society of America, Madison, WI (Chemistry of the soil environment, Special Publication 40, pp 129–150)

Boggess WR (ed) (1977) Lead in the environment. National Science Foundation, Washington

Boggess SF, Willavize S, Koeppe DE (1978) Differential response of soybean varieties to soil cadmium. Agron J 70:756–760

Bolter E, Hemphill D, Wixson B, Butherus D, Chen R (1972) Geochemical and vegetation studies of trace substances from lead smelting. Trace Substances in Environmental Health, vol 6. University of Missouri, Missouri, pp 79–86

References

Booz, Alien & Hamilton Inc (1989) Determining soil response action levels based on potential contaminant migration to ground water: a compendium of examples. EPA/540/2-89/057

Boulding JR (1994) Description and sampling of contaminated soils: a field guide. Lewis Publishers, Chelsea

Boulding JR (1995) Practical handbook of soil, vadose zone and ground water contamination: assessment, prevention and remediation. CRC Press, Inc

Bouma AH (1969) Methods for the study of sedimentary structures. Wiley-Interscience, NewYork, NY

Bowen HJM (1979) Environmental chemistry of the elements. Academic Press, London

Bowles JE (1978) Engineering properties of soils and their measurement. McGraw-Hill, New York, NY

Bowles JE (1984) Physical and geotechnical properties of soils, 2nd edn. McGraw-Hill, New York, NY

Brady NC (1974) The nature and properties of soils, 8th edn. Macmillan, New York, NY

Brandt CS (1972) Plants as indicators of air quality. In: Thomas WA (ed) Indicators of environmental quality. Plenum Press, New York, pp 101–107

Brewer R (1976) Fabric and mineral analysis of soils, 2nd edn. Krieger Publ Co, Melbourne, FL

Brewer R, Sleeman JR (1988) Soil structure and fabric. CSIRO, Division of Soils, Adelaide, Australia

Bridges EM (1991) Waste materials in urban soils. In: Bullock P, Gregory PJ (eds) Soils in the urban environment. Blackwell, Oxford, pp 28–46

Broker G, Gliwa H (1978) Emission factors for the emission of lead, zinc, cadmium and mercury from factories. Schriftenreihe Landesanstalt Immissionsschutz. 43:7–11

Brooks RR (1971) Bryophytes as a guide to mineralisation. New Zealand J Bot 9:674–677

Brooks RR (1972) Geobotany and biogeochemistry in mineral exploration. Harper and Row, London

Brooks RR (1979) Indicator plants for mineral prospecting – a critique. J Geochem Explor 12:67–78

Brooks RR (1980) Accumulation of nickel by terrestrial plants. In: Nriagu JO (ed) Nickel in the environment. Wiley and Sons, New York, pp 407–430

Brooks RR, Lyon GL (1966) Biogeochemical prospecting for molybdenum in New Zealand. New Zealand J Sci 9:706–718

Brooks RR, Radford CC (1978) Evaluation of background and anomalous copper and zinc concentrations in the 'copper plant' Polycarpaea spirostvlis and other Australian species of the genus. Proc Australias Inst Min Metall 268:33–37

Brooks RR, Wither ED (1977) Nickel accumulation by *Rinorea bengalensis* (Wall.) J Geochem Explor 7:295–300

Brooks RR, Lee J, Jaffre T (1974) Some New Zealand and New Caledonian plant accumulators of nickel. J Ecol 62:493–499

Brooks RR, Lee J, Reeves RD, Jaffre T (1977) Detection of nickeliferous rocks by analysis of herbarium specimens of indicator plants. J Geochem Explor 7:49–57

Brooks RR, Holzbecher J, Ryan DE (1981) Horsetails (*Equisetum*) as indirect indicators of gold mineralisation. J Geochem Explor 16:21–26

Brown SL, Chaney RL, Angle JS, Baker AJM (1995) Zinc and cadmium uptake by hyperaccumulator *Thelaspi careulescens* and metal tolerant *silene vulgaris* grown on sludge amended soils. Environ Sci Technol 29:1581

Bruce EL (1999) Environmental transport processes. John Wiley and Sons, London

Bucheli TD, Blum F, Desaules A, Gustafsson JP (2004) Polycyclic aromatic hydrocarbons, black carbon, and molecular markers in soils of Switzerland. Chemosphere 56:1061–1076

Buffle J (1988) Complexation reactions in aquatic systems: an analytical approach. Ellis Horwood, Chichester (Ellis Horwood Series in Analytical Chemistry)

Bullock P, Murphy CP (eds) (1983) Soil micromorphology. AB Academic Publishers, Berkhamstead, UK (vol 1: Techniques and applications, vol 2: Soil genesis)

Bullock P, Fedoroff N, Jongerius A, Stoops G, Tursina T (1985) Handbook for soil thin section description. Waine Research Publications, Wolverhampton, UK

Buol SW, Hole FD, McCracken RJ (1989) Soil genesis and classification, 3rd edn. Iowa State University Press, Ames, IA

Bureau of Reclamation (1969) Soil as an engineering material. Research Report No. 17, USA

Burken JG, Schnoor JL (1996) Phytoremediation: plant uptake of atrazine and role of root exudates. J Environ Eng 122:958–963

Burnett JW (1991) Anthrax. Cutis 48:113–114

Burton KW, John E (1977) A study of heavy metal contamination in the Rhonda Fawr, South Wales. Wat Air Soil Pollut 7:45–68

California Department of Toxic Substances Control (1994) Preliminary endangerment assessment guidance manual. California Environmental Protection Agency, Sacramento, CA

Calvet R (1989) Adsorption of organic chemicals in soils. Environ Health Perspect 83:145–177

Camarero S, Sarkar S, Ruiz-Dueñas FJ, Martinez MJ, Martinez AT (1999a) Description of a versatile peroxidase involved in the natural degradation of lignin that has both manganese peroxidase and lignin peroxidase substrate interaction sites. J Biol Chem 274:10324–10330

Camarero S, Sarkat S, Ruiz-Duenas FJ, Jesus M, Swift RS, Spark KM (eds) (1999b) Understanding and managing organic matter in soils, sediments, and waters. International Humic Substances Society, St. Paul, MN, USA

Caroll D (1959) Ion exchange in clays and other minerals. Bull Geol Soc Am 70:754

Carry B, Harris P (2001) Erosion and control in cropping lands. The State of Queensland, Department of Natural Resources and Mines (NRM facts, Land series L13, available at *www.nrm.qld.gov.au/factsheets/pdf/land/LM13,w.pdf*)

Casenave A, Valentin C(1989) Les état de surface de la zone Sahélienne. Influence sur l'infiltration. ORSTOM, Collection Didactiques

Casenave A, Valentin C (1992) A runoff capability classification system based on surface features criteria in semi-arid areas of west Africa. J Hydrol 130:231-249

Castro S, Davies LC, Erickson LE (2003) Phytotransformation of benzotriazoles. Int J Phytoremed 5:245-266

Chambers LD et al. (1990) Handbook of in situ treatment of hazardous waste contaminated soils. EPA/540/2-90/002

Chen Y, Tarchintzki JT, Brouwer J, Morin J, Banin A (1980) Scanning electron microscope observations of soil crusts and their information. Soil Sci 130:49-55

Choppin G (1988) Chemistry of actinides in the environment. Radiochim Acta 43:82-83

Christmann RF, Gjessing ET (eds) (1983) Aquatic and terrestrial humic materials. Ann Arbor Science Publishers, Michigan

Christopherson RW (1992) Geosystems: an introduction to physical geography, Maxwell Macmillan Canada Inc.

Clapp CE, Hayes MHB, Senesi BN, Griffith SM (eds) (1996) Humic substances and organic matter in soil and water environments. International Humic Substances Society, St. Paul, MN

Clapp CE, Hayes MHB, Senesi BN, Bloom PR, Jardine PM (eds) (2001) Humic substances and chemical contaminants. Soil Science Society of America, Madison, WI, USA

Cockerham LG (ed) (1994) Basic environmental toxicology. Lewis Publishers, Boca Raton, FL

Cole MA, Zhang L, Liu X (1995) Remediation of pesticide-contaminated soil by planting and compost addition. Compost Sci Util 3:20

Committee on Risk Assessment Methodology (CRAM) (1993) Issues in risk assessment. National Academy Press, Washington, DC

Compton JAF (1988) Military chemical and biological agents: chemical and toxicological properties. Telford Press, Caldwell, NJ, pp 10-12

Conway RE (ed) (1982) Environmental risk analysis for chemicals. Van Nostrand Reinhold, New York, NY

Cremeens DL, Brown RB, Huddleston JH (eds) (1994) Whole regolith pedology. Soil Science Society of America, Madison, WI (Special Publication 34, 136 pp)

Cruikshank JG (1972) Soil geography. Halstead Press Division, John Wiley and Sons, New York, NY

Cullinane MJ, Jones LW, Malone PG (1986) Handbook for stabilization/solidification of hazardous wastes. EPA/540/2-86/001

Curry DS (1990) Assessment of empirical methodologies for predicting ground water pollution from agricultural chemicals. In: Fairchild DM (ed) Ground water quality and agricultural practices. Lewis Publishers, Chelsea, MI, pp 227-245

Cutright TJ, Lee S (1995) In situ soil remediation: bacteria or fungi? Energ Source 14:413-419

Czupyma G et al. (1989) In situ immobilization of heavy-metal-contaminated soils. Noyes Data Corporation, Park Ridge, NJ

Daniels RB, Hammer RD (1992) Soil geomorphology. John Wiley and Sons, New York

Davies G, Ghabbour EA (eds) (1998) Humic substances: structures, properties and uses. Royal Society of Chemistry, Cambridge

DeLuca T, Johnson P (1990) RAVE: relative aquifer vulnerability evaluation. Montana Department of Agriculture, Helena, MT (MDA Technical Bulletin 90-01, 4 pp)

Dennen WH, Moore BR (1986) Geology and engineering. W.C. Brown, Dubuque, IA

Dennis RM, Dworkin D, Lowe WL, Zupko AJ (1992) Evaluation of commercially available soil-washing processes for site remediation. J Hazard Indust Waste 24:515-525

Department of the Army (1975) Military chemistry and chemical compounds. FM 3-9/AFM 355-7, pp 3-4

Derome J et al. (2002) Submanual on soil solution collection and analysis. Elaborated by expert manual on soil (available from *http://www.icp.forests.org*)

Desaules A, Sprengart J, Wagner G, Muntau H, Theocharopoulos S (2001) Description of the test area and reference sampling at Dornach. Sci Total Environ 264:17-26

DiGuilio DC (1992) Evaluation of soil venting application. Ground Water Issue Paper, EPA/540/S-92/004

Dobrin MB, Milton B, Savit CH (1988) Introduction to geophysical prospecting, McGraw-Hill Inc., NY
Doty SL, Shang TQ, Wilson AM, Moore AL, Newman LA, Strand SE, Gordon MP (2003) Metabolism of the soil and groundwater contaminants, ethylene dibromide, and trichlorethylene, by the tropical leguminous tree, *Leuceana leucocephala*. Water Res 37:441–449
Douglas LA (ed) (1990) Soil micromorphology: a basic and applied science. Elsevier, New York (Papers presented at the 8th International Working Meeting of Soil Micromorphology in San Antonio, TX)
Douglas LA, Thompson ML (eds) (1985) Soil micromorphology and soil classification. Soil Science Society of America, Madison, WI (Special Publication 15, 216 pp)
Drozd J, Gonet SS, Senesi N, Weber J (eds) (1997) The role of humic substances in the ecosystems and in environmental protection. PTSH – Polish Society of Humic Substances, Wroclaw, Poland
Duchaufour P (1988) Pédologie. Masson, Paris
Duda AM, Johnson RJ (1987) Targeting to protect groundwater quality. J Soil Water Conserv 42:325–330
Dunne T, Dietrich WE Brunengo MJ (1978) Recent and past erosion rates in semi-arid Kenya. Z Geomorphol Suppl 29:130–140
Dury GH (1960) Map interpretation, 2nd edn. Pitman, London, UK
Eastern Research Group, Inc. (1991) Summary report on issues in ecological risk assessment. Proceedings of a Colloquium Series March–July, 1990. Prepared for Risk Assessment Forum, U.S. Environmental Protection Agency, Washington, DC
Electric Power Research Institute (EPRI) (1985) Sampling design for aquatic ecologic monitoring. EPRI, Palo Alto, CA (EPRI EA-4302, five volumes)
Ellis S, Mellor A (1995) Soils and environment. Routledge Physical Environment Series
Elwell HA (1978) Modelling soil losses in Southern Africa. J Agr Eng Res 23:117–127
Environmental Protection Agency (1992) Proceedings of the 4th Forum on Innovative Hazardous Waste Treatment Technologies. San Francisco, CA
EPA (2000) 542-N-00-006, September 2000 (Issue 37)
Erbas-White I, San Juan C (1993) Development of a matrix approach to estimate soil clean-up levels for BTEX Compounds. Ground Water Management 17:74–84 (Proceedings API/NWGA Hydrocarbon Conference)
Evans R (1980) Mechanics of water erosion and their spatial and temporal controls, an empirical viewpoint. In: Kirkby MJ, Morgan RPC (eds) Soil erosion. Chichester, Wiley, pp 109–128
Evans R (1981) Potential soil and crop losses by erosion. In: Proceedings SAWMA Conference on Soil and Crop Loss: Developments in Erosion Control. National Agricultural Centre, Stoneleigh, UK
Evanylo GK, McGuinn R (2000) Agricultural management practices and soil quality measuring, and comparing laboratory and field test kit indicators of soil quality attributes (*http//www.Ext.vt.edu/pubs/compost/452-400*, 24 May 2001)
Fairbridge RW, Finkl CW Jr (eds) (1979) The encyclopaedia of soil science. Part 1: Physics, chemistry, biology, fertility, and technology. Dowden, Hutchinson and Ross, Stroudsburg, PA
Fanning DS, Fanning MCB (1989) Soil: morphology, genesis and classification. John Wiley and Sons, London
FAO-AGL(2000) Report available from *www.fao.org/AGL*
FAO-UNESCO (1974) Legend of the soil map of the world. UNESCO, Paris
FAO-UNESCO (1989) Soil map of the world: revised legend. International Soil Reference and Information Centre, Wageningen
Farrar WE (1995) Anthrax: from Mesopotamia to molecular biology. Pharos 58:35–38
Federal Aviation Administration (FAA) (1967) Airport paving. FAA, Washington, DC
Federal Energy Management Agency, U.S. Department of Transportation and U.S. Environmental Protection Agency (FEMA/DOT/EPA) (1989) Handbook of chemical hazard analysis procedures. (available from Federal Emergency Management Agency, Publications Department, 500 C St., SW, Washington, DC 20472)
Felsot AS (1991) Testing the use of landfarming to remediate pesticide-contaminated soil excavated from agrochemical facilities. Fourth Chemical Congress of North America, New York
Felsot A, Dzantor EK, Case L, Liebl R (1990) Assessment of problems associated with landfilling or land application of pesticide waste and feasibility of cleanup by microbiological degradation. Report from Illinois Department of Energy and Natural Resources
Finkl CW (1982) Soil classification. Hutchinson Ross Publishing Company, Stroudsburg, Penn
Fish JR, Revesz R (1996) Microwave solvent extraction protocol for chlorinated pesticides from soil. Science 14:230–234
Fishman MJ, Hem JD (1976) Lead content of water. In: Levering TG (ed) Lead in the environment. U.S. Geological Survey Professional Paper 957:35–41
Fitchko J (1989) Criteria for contaminated soil/sediment cleanup. Pudvan Publishing, Northbrook, IL
Fortescue JAC (1980) Environmental geochemistry. A holistic approach. Springer-Verlag, Berlin (Ecological Studies 35)

Fotyma M, Mercik S (1992) Chemia rolna. PWN, Warszawa
Fox RD (1996) Critical review: physical/chemical treatment of organically contaminated soils and sediments. J Air Waste Manage 46:391
Francis JD, Brower BL, Graham WF (1981) National statistical assessment of rural water conditions. U.S. Environmental Protection Agency, Washington, DC
Franke S (1967) Manual of military chemistry. Volume 1: Chemistry of chemical warfare agents. Deutscher Militärverlag, Berlin (East). (Translated from German by U.S. Department of Commerce, National Bureau of Standards, Institute for Applied Technology, NTIS no. AD-849 866)
Franz DR, Jahrling PB, Friedlander AM, McClain DJ, Hoover DL, Byrne WR et al. (1997) Clinical recognition and management of patients exposed to biological warfare agents. J Am Med Assoc 278:399–411
Frimmel FH, Christman RF (eds) (1988) Humic substances and their role in the environment. Wiley-Interscience, Chichester
Frimmel FH, Abbt-Braun G, Heumann KG, Hock B, Luedemann HD, Spiteller M (eds) (2002) Refractory organic substances in the environment. Wiley-VCH
Frischknecht FC, Muth L, Grette R, Buckley T, Kornegay B (1983) Geophysical methods for locating abandoned wells. U.S. Geological Survey (Open-File Report 83-702)
Fuhriman DK, Barton JR (1971) Ground water pollution in Arizona, California, Nevada and Utah. EPA 16060 ERU 12/71 (NTIS PB211 145)
Gaffney JS, Marley NA, Clark SB (eds) (1996) Humic and fulvic acids. American Chemical Society, Washington, DC (ACS Symposium Series 651)
Galer C (1988) Technology screening guide for treatment of CERCLA soils and sludges. EPA/540/2-88/004
Gan J, Koskinen WC, Becker RL (1995) Effect of concentration on persistence of alachlor in soil. J Environ Qual 24:1162–1169
Garland GA, Grist TA, Rosalie E (1995) The compost story: from soil enrichment to pollution remediation. Bio-Cycle 36:53–56
Gass TE, Lehr JH, Heiss HR Jr (1977) Impact of abandoned wells on ground water. EPA/600/3-77-095 (NTIS PB272-665)
Gelher LW (1993) Stochastic subsurface hydrology. Prentice-Hall, Englewood Cliffs, NJ
Ghabbour AE, Davies G (eds) (1999) Understanding humic substances: advanced methods, properties and uses. Royal Society of Chemistry, Cambridge
Ghabbour AE, Davies G (eds) (2000) Humic substances versatile components of plants, soil and water. Royal Society of Chemistry, Cambridge
Ghabbour AE, Davies G (eds) (2001) Humic substances: structures, models and functions. Royal Society of Chemistry, Cambridge
Ghabbour AE, Davies G (eds) (2003) Humic substances: nature's most versatile materials. Taylor and Francis, New York
Ghabbour AE, Davies G (eds) (2005) Humic substances: molecular details and applications in land and water conservation, Taylor and Francis, New York
Ghadiri H, Rose CW (eds) (1992) Modeling chemical transport in soils: natural and applied contaminants. Lewis Publishers, Chelsea, MI
Ghiorse WC, Balkwill DL (1985) Microbiological characterization of subsurface environments. In: Ward CH, Giger W, McCarty PL (eds) Ground water quality. Wiley Interscience, New York, pp 386-401
Godsy EM, Warren E, Pganelly W (2003) The role of microbial reductive dechlorination of TCE at a phytoremediation site. Int J Phytoremed 5:73–88
Goerlitz DF, Brown E (1972) Methods for analysis of organic substances in water. U.S. Geological Survey (Techniques of Water-Resources Investigations TWRI 5-A3)
Gonzalez JM, Laird DA (2004) Role of smectite and Al-substituted goethites in the catalytic condensation of arginine and glucose. Clay Clay Mineral 52:443–450
Gorelick S (1983) A review of distributed parameter groundwater management modeling methods. Water Resour Res 19:305–319
Green A van, Chase Z (1998) Recent mine spill adds to contamination of southern Spain. EOS Transactions of the American Geophysical Union 79:449–455
Greeson PE, Ehlke TA, Irwin GA, Lium BW, Slack KV (eds) (1977) Methods for collection and analysis of aquatic biological and microbiological samples. U.S. Geological Survey (Techniques of Water-Resources Investigations TWRI 5-A4)
Griffith JB et al. (1976) Attenuation of pollutants in municipal landfill leachate by clay minerals. Illinois State Geological Survey (Environmental Geology Notes 78)
Griffiths RA (1995) Soil washing technology and practice. J Hazard Material 40:175–189
Guamaccia JF et al. (1992) Multiphase chemical transport in porous media. Environmental Research Brief, EPA/600/S-92/002

Gustafson D (1993) Pesticides in drinking water. Van Nostrand Reinhold, New York, NY
Guswa JH, Lyman WJ, Donigian AS Jr, Lo TYR, Shanahan EW (1984) Groundwater contamination and emergency response guide. Noyes Publications, Park Ridge, NY
Guy HP (1969) Laboratory theory and methods for sediment analysis. U.S. Geological Survey (Techniques of Water-Resources Investigations TWRI 5-C1)
Hach Company (1991) Handbook for waste analysis, 2nd edn. Hach Company, Loveland, CO
Haimes YY, Bear J (eds) (1987) Groundwater contamination: use of models in decisionmaking. Reidel Publishing Co., Dordrecht, The Netherlands
Haimes YY, Snyder JH (eds) (1986) Groundwater contamination. Engineering Foundation, New York
Hall DGM, Reeve MJ, Thomson AJ, Wright VF (1977) Water retention, porosity and density of field soils. Soil Survey of England and Wales, Harpenden (Technical Monograph 9)
Hallberg GR (1986) Overview of agricultural chemicals in ground water. In: National Water Well Association (ed) Agricultural impacts on ground water – a conference. National Water Well Association, Dublin, OH, pp 1–66
Hamaker JW, Thompson JM (1972) Adsorption. In: Göring CAI, Hamaker JW (eds) Organic chemicals in the soil environment, vol 1. Marcel Dekker Inc., New York
Hammann M, Desaules A (2003) Manual: sampling and sample pre-treatment for soil pollutant surveys. Swiss Agency for Environment, Forests and Landscape (SAFEL), CH-3003 Bern (Environment in practice VU-4814-E, 100 pp)
Handa R, Bhatia S, Wali JP (1996) Melioidosis: a rare but not forgotten cause of fever of unknown origin. Br J Clin Pract 50:116–117
Hardy EP, Krey PW, Volchock HL (1973) Global inventory and distribution of fallout plutonium. Nature 241:444–445
Harris T, Johns J, Mirriman J, Merricks J (1997) Groundwater pollution primer. (available at *http://ewr.cee.vet.edu*)
Hassall KA (1982) The chemistry of pesticides: their metabolism, mode of action and uses in crop protection. Verlag Chemie
Hassett JJ, Banwart WL(1989)The sorption of non polar organics by soils and sediments .In: Sawhney BL, Brown K (eds) Reactions and movements of organic chemicals in soils. Soil Science Society of America, Madison, WI (Special Publication 22)
Hatayama HIC, Chen JJ, de Vera ER, Stephens RD, Storm DL (1980) A method for determining the compatibility of hazardous wastes. EPA/600/2-80/076
Hayes MHB, Swift RS (eds) (1984) Volunteered papers – 2nd International Conference. International Humic Substances Society, University of Birmingham, Birmingham, UK
Hayes MHB, Wilson WR (eds) (1997) Humic substances, peats and sludges. Royal Society of Chemistry, Cambridge
Hayes MHB, MacCarthy P, Malcolm RL, Swift RS (eds) (1989) Humic substances II: In search of structure. Wiley-Interscience, New York
Heaton ACP, Rugh CL, Wang N, Meagher RB (1998) Phytoremediation of mercury and methyl mercury – polluted soils using genetically engineered plants. J Soil Contam 7:497–510
Hem JD (1972) Chemistry and occurrence of cadmium and zinc in surface water and ground water. Water Resour Res 8:661–679
Hem SC, Melancon SM (1986a) Vadose zone modeling of organic pollutants. Lewis Publishers, Chelsea, MI
Hem SC, Melancon SM (1986b) Guidelines for field testing, soil fate and transport models: final report. EPA/600/7-86/020 (NTIS PB86-209400) [PRZM, SESOIL, PESTAN]
Hessen DO, Tranvik LJ (eds) (1998) Aquatic humic substances – ecology and biogeochemistry. Springer-Verlag, Berlin
Hillel D (1982) Introduction to soil physics. Academic Press, San Diego, CA
Hinkel ME, Denton EH, Bigelow RC et al. (1978) Helium in soil gases of the Roosevelt hot springs, known geothermal resource area, Beaver County, Utah. U.S. Geological Survey J Res 6:563–569
Hoddinott KB, O'Shay TA (eds) (1993) Application of agricultural analysis in environmental studies. American Society for Testing and Materials, Philadelphia, PA (ASTM STP 1162)
Holdgate MW (1979) A perspective of environmental pollution. Cambridge University Press, Cambridge
Holman C (1993) Bioremediation attacks on-site contaminants. World Wastes 36:6
Hölting B (1980) Hydrogeologie: Einführung in die Allgemeine und Angewandte Hydrogeologie. Ferdinand Enke Verlag, Stuttgart, Germany
Holton GJ (1988) Thematic origins of scientific thought: Kepler to Einstein. Harvard University Press, Cambridge, Massachusetts
Howard PH, Saxena J, Durkin PR, Ou LT (1975) Review and evaluation of available techniques for determining persistence and routes of degradation of chemical substances in the environment. EPA 560/5-75/006

http://www.uh.edu/engines/espi1190.htm: Leinhard JH Engines of our ingenuity No. 1190
http://www.alenafix.com/old-fbg/articles/huff-cw.html (Understanding the Chemical War Convention)
http://www.r-biopharm.de/Food/Seaweed/PSPdata.html
Huang C, Bradford JM (1990) Portable laser scanner for measuring soil surface roughness. Soil Sci Soc Am J 54:1402–1406
Hudson NW (1981) Soil conservation. Batsford, London
Hudson NW (1993) Field measurement of soil erosion and run-off. Food and Agriculture Organisation of the United Nations (FAO-UN), Rome
Hudson NW, Jackson DC (1959) Results achieved in the measurement of erosion and run-off in southern Rhodesia. Proceedings of the Third Inter-African Soils Conference, Dalaba, pp 575–583
Huff C (1997) Understanding the chemical war convention. (available at *http://www.alnafix.com*)
Hunt B (1983) Mathematical analysis of groundwater resources. Butterworth, Stoneham, MA
Hunt JR, Sitar N, Udell KS (1988) Nonaqueous phase liquid transport and cleanup. II. Experimental studies. Water Resour Res 24:1259–1269
Hunter GB (1992) Extraction of pesticides from contaminated soil using su-percritical carbon dioxide. U.S. EPA, pp 100–110
Hutton JH, Trentini AJ (1994) Thermal desorption of polynuclear aromatic hydrocarbons and pesticides contaminated soils at an Ohio superfund site: a case study. Proceedings of Annual Meeting of Air and Waste Management Association 87(14B), 29 pp
Hwang ST, Falco JW, Nauman CH (1987) Development of advisory levels for polychlorinated biphenyls (PCBs) cleanup. EPA/600/6-86/002
Imhoff PT, Frizzel A, Miller CT (1995a) CMR News, School of Public Health, University of North Carolina at Chape Hill, 2:1–4
Imhoff PT, Gleyzer SN, McBride JF, Vancho LA, Okuda I, Miller CT (1995b) Cosolvent-enhanced remediation of residual dense nonaqueous phase liquids: experimental investigation. Environ Sci Technol 29:1966–1976
Information Management Staff, Office of Solid Waste and Emergency Response (IMS/OSWER) (1990) Report of the usage of computer models in hazardous waste/superfund programs, phase, final report. U.S. Environmental Protection Agency, Washington, DC
Irgolic KJ, Martell AE (eds) (1985) Environmental inorganic chemistry. VCH Publishers, Deerfield Beach, FL
Irvine DEG, Knights B (1974) Pollution and the use of chemicals in agriculture. Butterworth, London
Isbell RF (1996) The Australian soil classification. Australian soil and land survey handbook. CSIRO Publications, Melbourne
Ishimure M (1990) Paradise in the sea of sorrow – English translation by Livet Monnet, Yamaguchi Publishing House (c/o Japan Publications Trading Co. Ltd., Tokyo)
Ishiwatari R (1990) Abstracts of oral and poster papers at the Fifth International Meeting of IHSS, August 6–10, 1990, Nagoya, Japan
Ishiwatari R, Lowe L, McKnight D, Shinozuka N, Yonebayashi K (1992) Special Issue: Advances in Humic Substances Research, a Collection of Papers from the Fifth International Meeting of IHSS, Nagoya, Japan, 6-11 August, 1990. Sci Total Environ 117/118:1–591
Jackson M (1958) Soil chemical analysis. Prentice-Hall, Englewood Cliffs, NJ
Jackson M (1979) Soil chemical analysis – advanced course, 2nd edn. Madison, WI (published by author)
James SC, Kovalick WW, Bassin J (1995) Technologies for treating contaminated land and groundwater. Chem Ind 13
Jaycock MJ, Parfitt GD (1981) Chemistry of interfaces. Ellis Horwood, Chichester, UK
Jenkins SH (ed) (1979) The agricultural industry and its effects on water quality. Pergamon Press, New York, NY
Jenne EA (ed) (1979) Chemical modeling in aqueous systems: speciation, sorption, solubility, and kinetics. American Chemical Society, Washington, DC (ACS Symposium Series 93)
Jockel W, Hartje J (1995) Die Entwicklung der Schwermetallemissionen in der Bundesrepublik Deutschland von 1985 bis 1995. TÜV Rheinland e.V. Köln (Forschungsbericht 94-104 03524)
Jóhannesson B (1960) The soils of Iceland. University Research Institute, Department of Agriculture, Reykjavik (Reports series B, No 13)
John TJ (1996) Emerging and re-emerging bacterial pathogens in India. Indian J Med Res 103:4–18
JuryWA, Fluhler H (1992) Transport of chemicals through soil: mechanism, models and field applications. Adv Agron 47:141–201
Kaufman MI, Mckenzie DJ (1975) Upward migration of deep well waste injection fluids in Floridan aquifer, South Florida. U.S. Geological Survey J Res 3:261–271

References

Keller A, Desaules A (2003) The Swiss Soil Monitoring Network: regular measurements of heavy metals in soil and field balances. Workshop proceedings: Assessment and Reduction of Heavy Metal Inputs into Agro-Ecosystems (AROMIS), November 24–25, Kloster Banz, Germany, pp 73–77

Keller E, Rickabaugh J (1992) Effects of surfactant structure on pesticide removal from a contaminated soil. J Hazard Indust Waste 24:652–661

Keller A, Desaules A, Schwab P, Weisskopf P, Scheid S, Oberholzer H (2005) Monitoring soil quality in the long term: examples from the Swiss Soil Monitoring Network. Annual Proceedings of the Austrian Soil Science Society, Agroscope FAL Reckenholz, Eidg. Forschungsanstalt für Agrarökologie und Landbau, 8046 Zürich

Kempton H, Davis A (1992) Remediation of solvent-contaminated soils by aeration. J Environ Qual 21:121–128

Kennedy VC, Brown TC (1965) Experiments with a sodium ion electrode as a means of studying cation exchange rates. Clay Clay Mineral 13:351–352

Killham K (1994) Soil ecology. Cambridge University Press, New York, NY

Kimball D, Siegel L, Tyier P (1993) Covering the map: a survey of military pollution sites in the United States. Physicians for Social Responsibility, Washington, DC

Kirk TK, Farrell RL (1987) Enzymatic "combustion": the microbial degradation of lignin. Annu Rev Microbiol 41:465–505

Kirkby MJ (1980) The problem. In: Kirkby MJ, Morgan RPC (eds) Soil erosion. Wiley, Chichester, pp 1–16

Kittrick JA (ed) (1986) Soil mineral weathering. Van Nostrand Reinhold, New York, NY

Klasson T, Burton JW, Evans BS, Reves ME (1996) Anaerobic dechlorination of PCBs. Biotechnol Prog 12:310–315

Klötzli FA (1993) Ökosysteme. Aufbau, Funktionen, Störungen. G. Fisher, Stuttgart, Germany

Knight-Ridder (1996) Knight-Ridder Information Database Catalog – Spring, 1996

Knuteson SL, Whitwell T, Klaine SJ (2002) Influence of plant age and size on simazine toxicity and uptake. J Environ Qual 31:2096–2103

Kohl S, Rice JA (1996) The binding of organic contaminants to humin. Proceedings of HSRC/WERC Joint Conference on the Environment. Kansas State University Press, Albuquerque, NM, pp 364–368

Kononova MM (1966) Soil organic matter. Pergamon Press, Elmsford, NY

Kopp JF, Kroner RC (1968) Trace metals in water in the United States, October 1, 1962–September 30, 1967. U.S. Department of the Interior, Federal Water Pollution Control Administration

Kotterman MJJ, Vis EH, Field JA (1998) Successive mineralization and detoxification of benzo[a]pyrene by the white rot fungus *Bjerkandera* sp. Strain BOS 55 and indigenous microflora. Appl Environ Microbio 64:2853–2858

Kozlovskiy FI (1972) Structural functions and migrational landscape geochemical processes (translation). Pochvovedeniye 4:122–138 (citation after Fortescue 1980)

Kubat J (ed) (1992) Humus, its structure and role in agriculture and environment. Elsevier, New York, NY (Proceedings 10th International Symposium Humus et Planta, Prague, August 1991)

Kubiena WL (1938) Micropedology. Collegiate Press, Ames, Iowa

Kubiena WL (1948) Entwicklungslehre des Bodens. Springer, Wien

Kubiena WL (1953) Bestimmungsbuch und Systematik der Böden Europas. Ferdinand Enke Verlag, Stuttgart

Kubiena WL (Hrsg) (1967) Die mikromorphologische Bodenanalyse. Ferdinand Enke Verlag, Stuttgart

Kubiena WL (1970) Micromorphological features of soil geography. Rutgers University Press, New Brunswick, New Jersey

Kumada K (1987) Chemistry of soil organic matter. Japan Scientific Societies Press, Tokyo and Elsevier, Amsterdam

Kuman S, Mukerji KG, Lal R (1996) Molecular aspects of pesticide degradation by microorganisms. Crit Rev Microbiol 22:1–26

Kurfürst U, Desaules A, Rehnert A, Muntau H (2004) Estimation of measurement uncertainty by the budget approach of heavy metal content in soils under different land use. Accredit Qual Assur 9:64–75

Kuritz T, Wolk CP (1995) Use of filamentous cyanobacteria for biodegradation of organic pollutants. Appl Environ Microbiol 61:234–238

Kuznetsov SI (ed) (1962) Geologic activity of microorganisms. Transactions of the Institute of Microbiology No. DC (translated from Russian), Consultants Bureau, New York

Kuznetsov SI, Ivanov MV, Lyalikova NN (1963) Introduction to geological microbiology. McGraw-Hill, New York, NY

Laine MM, Jorgensen KS (1996) Straw compost and bioremediated soil as inocula for the bioremediation of chlorophenol-contaminated soil. Finish Environmental Agency 62:1507–1513

Laitinen A, Michaux A, Aaltonen O (1994) Soil cleaning by carbon dioxide: a review. Environ Technol 15:715–727
Lamar RT, Dietrich DM (1990) In situ depletion of pentachlorophenol from contaminated soil by *Phanerochaete* spp. Appl Environ Microbiol 56:3093–3100
Lance JC, Gerbe CP, Melnick JL (1976) Virus movement in soil columns flooded with secondary sewage effluent. Appl Environ Microbiol 32:520–526
La Spina J, Palmquist R (1992) Catalog of contaminant databases: a listing of databases of actual or potential contaminant sources. Washington State Department of Ecology, Olympia, WA
Latimer WM (1952) Oxidation potentials, 2nd edn. Prentice-Hall, Englewood Cliffs, NJ
Lee MH, Lee CW (2000) Association of fallout-derived ^{137}Cs, ^{90}Sr, and ^{139,140}Pu with natural organic substances in soils. J Environ Radioactivity 47:253–262
Lee HS, Nagy S (1988) Quality changes and non-enzymatic browning intermediates in grapefruit juice during storage. J Food Sci 53:168–176
Li H, Sheng G, Sheng W, Xu O (2002) Uptake of trifluralin and lindane from water by ryegrass. Chemosphere 48:335–341
Lichtfouse É (1999) A novel model of humin – analysis 27, N° 5. EDP Sciences, Wiley-VCH
Lindsay WL (1979) Chemical equilibria in soils. John Wiley and Sons, New York, NY
Lischer P, Dahinden R, Desaules A (2001) Quantifying uncertainty of the reference sampling procedure used at Dornach under different soil conditions. Sci Total Environ 264:119–126
Liu D, Chawla VK (1976) Polychlorinated biphenols (PCB) in sewage sludges. In: Hemphill DD (ed) Trace substances in environmental health. University of Missouri Press, Columbia, MO, pp 247–250
Lloyd JW, Heathcote JA (1985) Natural inorganic hydrochemistry in relation to groundwater. Oxford University Press, New York, NY
Loehr R (1989) Treatability potential for EPA listed hazardous chemicals in soil. EPA/600/2-89/011
Lopez-Avila V, Young R, Benedicto J, Ho P, Kim R, Beckert WF (1995) Extraction of organic pollutants from solid samples using microwave energy. J Anal Chem 67:2096–2102
Lowery B, Arshad RA, Lal R, Hickey WJ (1996) Soil water parameters and soil quality. In: Doran Jw, Jones AJ (eds) Methods of assessing soil quality. Soil Science Society of America, Madison, WI (Special Publication 49)
MacCarthy P, Clapp CE, Malcolm RL, Bloom PR (1990) Humic substances in soil and crop sciences: selected readings. Soil Science Society of America and American Society of Agronomy, Madison, WI
Malaisse F, Baker AJM, Ruelle S (1999) Diversity of plant communities and leaf heavy metal content at Luiswishi copper/cobalt mineralization, Upper Katanga, Dem. Rep. Congo. Biotechnol Agron Soc Environ 3:104–114
Malcolm RL, Kennedy VC (1970) Variation of cation exchange capacity and rate with particle size in stream sediments. J Wat Pollut Control Fed 42
Manahan SE (1993) Fundamentals of environmental chemistry. Lewis Publishers, Boca Raton, FL
Mann MJ, Groenendijk KE (1996) The first full scale soil washing project in the USA. Environ Prog 15:108–111
Martijn A, Bakker H, Schreuder RH (1993) Soil persistence of DDT, dieldrin, and lindane over a long period. B Environ Contam Tox 51:178–184
Martin MH, Coughtrey PJ (1982) Biological monitoring of heavy metal pollution – land and air. Maxwell Macmillan Canada Inc., Applied Science Publishers, London, New York
Marshall MR, Jeongmok K, Cheng-I W (2000) Enzymatic browning in fruits, vegetables and sea foods. FAO
Mauron J (1981) The Maillard reaction in food; a critical review from the nutritional standpoint. Progr Food Nutr Sci 5:5–35
McAdams CL (1994) New technologies in soil remediation. Waste Age 24:36
McBride MB (1994) Environmental chemistry of soils. Oxford University Press, New York
McFarland MJ, Salladay D (1996) Composting treatment of Alachlor impacted soil amended with the white rot fungus *Phanerochaete chrysosporium*. Hazard Waste Hazard 13:363–373
McGinnis GD, Borazjani H, McFariand LK, Pope DF, Strobel DA (1988) Characterization and laboratory soil treatability studies for creosote and pentachlorophenol sludges and contaminated soil. EPA/600/2-88/055
McGovern GW, Thomas W, Christopher GW (1995–2001) Biological warfare and its cutaneous manifestations – the electronic textbook of dermatology. (available at *textbook@telemedicine.org*)
Means JL et al. (1994) The application of solidification/stabilization to waste materials. Lewis Publishers, Boca Raton, FL
Meckes MC, Engle SW, Kosco B (1996) Site demonstration of Terra-Kleen response group's mobile solvent extraction process. J Air Waste Manage 46:971–977
Miller D (1994) Remediation of a central Arizona pesticide applicators airstrip with thermal desorption techniques. Proceedings of the 87th Annual Meeting of the Air and Waste Management Association 14B

Miller C, Valentine RL, Roehl RL (1996) Chemical and microbiological assessment of pendimethalin-contaminated soil after treatment with Fenton's agent. Water Res 30:2579–2586
Mirsal IA (1995) Carbonate rocks in time and space. J Geol Soc Philippines L(2):61–75
Mitreteksystems – http://www.mitretek.org/mission/envene/chemical/agents/chemagent.html
Mohr DH, Merz PH (1995) Application of a 2D air flow model to soil vapor extraction and bioventing case studies. Ground Water 33:433–444
Moore W (1995) Cleaning up America's soil: an overview of the processes, equipment, and legalities involved with contaminated earth cleanup. Constr Equipment 91:48 (Special report: Soil remediation)
Morgan RPC (1980) Implications. In: Kirkby MJ, Morgan RPC (eds) Soil erosion. Wiley, Chichester, pp 253–301
Morgan RPC (1995) Soil erosion and conservation, 2nd edn. Longman Group Ltd., Harlow, UK
Morin J, Benyamin Y, Michaelis A (1981) The dynamics of soil crusting by rainfall impact and the water movement in the soil profile. J Hydrol 52:321–335
Mückenhausen E (1965) The soil classification system of the Federal Republic of Germany. Pedologie Spec Ser 3:57–74
Muntau H, Rehnert A, Desaules A, Wagner G, Theocharopoulos S, Quevauviller P (2001) Analytical aspects of the CEEM soil project. Sci Total Environ 264:27–49
Munz C, Bachmann A (1993) Documentation of an environmentally acceptable soil restoration. Altlastensanierung 93:1135–1142
Mute A (ed) (1986) Methods of soil analysis. Part 1: Physical and mineralogical methods, 2nd edn. American Society of Agronomy, Madison, WI (Agronomy Monograph 9, 1188 pp)
Nappipieri P, Bollag JM (1991) Use of enzymes to detoxify pesticide contaminated soils and waters. J Environ Qual 20:510–517
Nash JH (1987) Field studies of in situ soil washing. EPA/600/2-87/110
Nash JH, Rosenthal S, Wolf G, Avery M (1992) Potential reuse of petroleum-contaminated soil: a directory of permitted recycling facilities. EPA/600/R-92-096
NEA (2002) Chernobyl: assessment of radiological and health impacts. Nuclear Energy Agency, Paris
Neider R (1986) Die radiologischen Auswirkungen des Reaktorunglücks von Tschernobyl in der Bundesrepublik Deutschland. Forschung Aktuell 19:45–49 (Sonderheft Tschemobyl, Zeitschrift der TU Berlin, Pocahontas Press, Blacksburg, Virginia)
Newman LA, Reynolds CM (2004) Phytodegradation of organic compounds. Curr Opin Biotech 15:225–230 (available at www.sciencedirect.com)
Nichols AB (1993) Pesticides remediation opportunities. Environmental Protection 4
Nuy EGJ (1993) Halocarbon-containing soil successfully thermally remediated removal of dioxins and pesticides possible without excessive emissions. Poly-tech Tijdschr 48:50–51
Oberlander PL (1989) Fluid density and gravitational variations in deep bore-holes and their effect on fluid potential. Ground Water 27:341–350
OECD (1995) Environmental data. Compendium
Olhoeft GR (1986) Direct detection of hydrocarbons and organic chemicals with ground penetrating radar and complex resistivity. Proceedings of the NWWA-API Conference Petroleum Hydrocarbons and Organic Chemicals in Ground Water, Houston
Orlov DS (1985) Humus acids of soils. A. A. Balkema, Rotterdam, The Netherlands
O'Shay TA, Hoddinott KB (eds) (1994) Analysis of soils contaminated with petroleum constituents. American Society for Testing and Materials, Philadelphia, PA (ASTM STP 1221, 120 pp)
Ostroff AG (1965) Introduction of oilfied water technology. Prentice-Hall Engelwood, Cliffis, NJ
Oughton DH, Salbu B, Riise G, Lien H, Østby G, Noren A (1992) Radionuclide mobility and bioavailability in Norwegian and Soviet soils. Analyst 117:481–486
Page AL, Miller RH, Keeney DR (eds) (1982) Methods of soil analysis. Part 2: Chemical and microbiological properties, 2nd edn. American Society of Agronomy, Madison, WI (ASA Monograph 9)
Palmer M (1985) Methods manual for bottom sediment sample collection. EPA/905/4-85/004
Palmer CD, Johnson RL (1989) Physical processes controlling the transport of non-aqueous phase liquids in the subsurface. In: EPA (ed) Transport and fate of contaminants in the subsurface. (EPA/625/4-91/1026, Chapter 10)
Parizek RR (1971) Impact of highways on the hydrogeologic environment. In: Coats R (ed) Environmental geomorphology. State University of New York, Bighamton
Parr JF, Papendick RI, Hornick SB, Meyer RE (1992) Soil quality: attributes and relationship to alternative and sustainable agriculture. Am J Alternative Agr 7:5–11
Paxman J, Harris R (1982) A higher form of killing: the secret story of chemical and biological warfare. Hill and Wang, New York
Pendergrass S, Prince J (1991) Chemical dechlorination of pesticides at a Superfund site in Region II. Proceedings of Annual Meeting of Air and Waste Management Association 84:14
Perdue EM, Gjessing ET (eds) (1990) Organic acids in aquatic ecosystems. Wiley-Interscience, Chichester

Perelman AI (1967) Geochemistry of epigenesis. Plenum, New York
Perry RD, Featherston JD (1997) *Yersinia pestis* – etiologic agent of plague. Clin Microbiol Rev 10:35–66
Pettyjohn WA, Hounslow AW (1983) Organic compounds and ground-water pollution. Ground Water Monit Rev 3:41–47
Pettyjohn WA, Studlick JRJ, Bain RC, Lehr JH (1979) A groundwater quality atlas of the United States. National Demonstration Water Project
Pfaff JD (1981) Methods for the determination of chemical contaminants in drinking water – instructors handbook. EPA/430/1-81/023
Piccolo A (ed) (1996) Humic substances in terrestrial ecosystems. Elsevier, Amsterdam
Pierson DH (1975) The passage of nuclear weapons debris through the atmosphere. In: Chadwick MJ, Goodman JT (eds) The ecology of resource degradation and renewal. John Wiley and Sons, New York (Proceedings Symposium of the British Ecological Society, 10–12 July 1973, pp 266)
Plumb RH Jr (1981) Procedures for handling and chemical analysis of sediment and water samples. U.S. Army Engineer Waterways Experiment Station, Vicksburg, MS (Technical Report EPA/CE-81/1)
Portier R, Roy M (1989) Design of in situ biodegradation systems for persistent pesticide remediation. 197th American Chemical Society Meeting, Dallas, Texas
Povoledo D, Golterman HL (eds) (1975) Humic substances, their structure and function in the biosphere. Centre for Agricultural Publishing and Documentation, Wageningen (Proceedings of an international meeting held at Nieuwersluis, the Netherlands, May 29–31, 1972)
Prentiss AM (1937) Chemicals in war. A treatise on chemical warfare. McGraw Hill, New York
Pritchard PH, Bourquin AW (1984) The use of microcosms for evaluation of interactions between pollutants and microorganisms. Adv Microbial Ecol 7:133–215
Provost LP, Elder RS (1985) Choosing cost-effective QA/QC programs for chemical analysis. EPA/600/4-85/056
Pye VL, Kelley J (1984) The extent of groundwater contamination in the United States. In: Geophysics Study Committee et al. (eds) Groundwater contamination. National Academy Press, Washington, DC, pp 23–33
Quinlan M, Peery D (1996) The guidance of remediation projects using gas chromatographic field screening for pesticides or polychlorinated biphenyls. AT Onsite 2:171–177
Rack KD, Leslie AR (eds) (1993) Pesticides in urban environments: fate and significance. American Chemical Society, Washington, DC (ACS Symposium Series 522:385)
Radian Corporation (1988) FGD chemistry and analytical methods handbook. 2: Chemical and physical test methods, revision 1. Electric Power Research Institute, Palo Alto, CA (EPRI CS-3612)
Raghavan R, Coles E, Dietz D (1990) Cleaning excavated soil using extraction agents: a state-of-the-art review. EPA/600/2-89/034
Raghavan R, Coles E, Dietz D (1991) Cleaning excavated soil using extraction agents: a state of the art review. J Hazard Mat 26:81–87
Rainwater FH, Thatcher LL (1960) Methods for collection and analysis of water samples. U.S. Geological Survey (Water-Supply Paper 1454)
Rashid MA (1985) Geochemistry of marine humic compounds. Springer-Verlag, New York
Rees TJ (1993) Glutathione-S-transferase as a biological marker of aquatic contamination. Research Thesis in Applied Toxicology, Portsmouth University, UK
Rhode Island Department of Environmental Management (RIDEM) (1992) Inventory of potential sources of groundwater contamination in wellhead protection areas. RIDEM, Providence, RI (RIDEM Guidance Document, 38 pp)
Rice JA, MacCarthy P (1990) A model of humin. Environ Sci Technol 24:1875–1877
Richard RP (1973) Impact of highways on the hydrogeologic environment. In: Coates DR (ed) Environmental geomorphology and landscape conservation, volume III. Benchmark papers in Geology
Rima DR, Chase B, Myers BM (1971) Subsurface waste disposal by means of wells – a selected annotated bibliography. U.S. Geological Survey (Water-Supply Paper 2020)
Rivers DB, Frazer FR (1988) Enzyme stabilization for pesticide degradation. Nucl Chem Waste Man 8:157–163
Rose C (1985) Acid rain falls on British woodlands. New Sci 1482:52–57
Rose A, Hawkes HE, Webb JS (1979) Geochemistry in mineral exploration, 2nd edn. Academic Press, London
Rose J, Boardman J, Kemp RA, Whiteman CA (1985) Palaeosols and the interpretation of the British Quaternary stratigraphy. In: Richards KS, Arnett RR, Ellis S (eds) Geomorphology and soils. George Allen and Unwin, London, pp 348–375
Ross GJ (1980) The mineralogy of spodosols. In: Theng BKG (ed) Soils with variable charge. New Zealand Society of Soil Science, Lower Hutt, pp 127–143
Ross S (1989) Soil processes: a systematic approach. Routledge, London

Rosswall T (ed) (1973) Modern methods in the study of microbial ecology. Swedish Natural Science Research Council, Stockholm (Bulletins from the Ecological Research Committee 17)
Roth TM (1996) Innovative in-situ remediation techniques. Part 1: Soil remediation. Remediation Management 2:24
Rowell DL (1994) Soil science: methods and application. Longmans, Harlow
Rowell DL, Wild A (1985) Causes of soil acidification: a summary. Soil Use Manage 1:32–33
Rozov NN, Ivanova EN (1967) Classification of the soils of the USSR. Sov Soil Sci 2:147–156
Sahle-Demessie E, Meckes MC, Richardson TL (1996) Remediating pesticide contaminated soils using solvent extraction. Environ Prog 15:293–300
Saiz-Jimenez C (1988) Abstracts of oral and poster papers. Fourth International Meeting of IHSS, October 3–7, 1988, Matalascanas Beach, Huelva, Spain
Saiz-Jimenez C, Rosell RA, Albaiges J (1989) Special issue of humic substances research from the Fourth International Meeting of IHSS, Matalascanas Beach, Huelva, Spain, October 3–7, 1988. Sci Total Environ 81/82:721–723
Sanjay HG, Daman W, Fataftah A (2006) A new multipurpose remediation media – environmental protection. (available at *www.stevenspublishing.com*)
Salomons W, Förstner U (1984) Metals in the hydrocycle. Springer-Verlag, Berlin
Salonen K, Kairesalo T, Jone RI (eds) (1992) Dissolved organic matter in lacustrine ecosystems: energy source and system regulator. Kluwer Academic Publishers, Dordrecht
Sanchez PA, Benites JR (1987) Low-input cropping for acid soils of the humid tropics. Science 238: 1521–1527
Sanders PF (1995) Calculation of soil cleanup criteria for volatile organic compounds as controlled by the soil-to-groundwater pathway: comparison of four unsaturated soil zone leaching models. J Soil Contam 4:1–24
Santamaría L, Amézaga J (1999) On the origin and remediation of the Aznallcóllar toxic spill – an independent assessment. Summary of the discussions of the Doñana Forum until May 1999 (available at *http://www.wise-uranium.org*)
Santos MCD, St Arnaud RJ, Anderson DW (1986) Quantitative evaluation of pedogenic changes in Boralfs (Gray Luvisols) of east central Saskatchewan. Soil Sci Soc Am J 50:1013–1019
Saull M (1990) Nitrates in soil and water. New Sci 127(1734):1–4
Scalf MR, Keeley JW, LaFevers CJ (1973) Ground water pollution in the south central states. EPA R2-73/268 (NTIS PB222 178)
Scalf MR, Dunlap WJ, Kreissl JF (1977) Environmental effects of septic tank systems. EPA/600/3-77-096 (NTIS PB272-702)
Schaetzl RJ (1991a) Distribution of spodosol soils in southern Michigan: a climatic interpretation. Ann Assoc Am Geogr 81:425–442
Schaetzl RJ (1991b) A lithosequence of soils in extremely gravelly, dolomitic parent materials, Bois Blanc Island, L. Huron. Geoderma 48:305–320
Schaetzl RJ, Barrett LR, Winkler JA (1994) Choosing models for soil chronofunctions and fitting them to data. Eur J Soil Sci 45:219–232
Schäfer et al. (1996) Pollution risk assessment for abandoned military sites in Berlin. Stadt Berlin
Schlegel HG (1974) Production, modification and consumption of atmospheric-trace gases by microorganisms. Tellus 26:11–20
Schmid P, Gujer E, Zennegg M, Bucheli TD, Desaules A (2005) Correlation of PCDD/F and PCB concentrations in soil samples from the Swiss soil monitoring network (NABO) to specific parameters of the observation sites. Chemosphere 58:227–234
Schnitzer M (1991) Soil organic matter – the next 75 years. Soil Sci 151:41–58
Schnitzer M, Khan SU (1972) Humic substances in the environment. Marcel Dekker Inc., New York
Schnitzer M, Kodama H (1977) Reactions of minerals with soil humic substances. In: Dixon JB, Weed SB (eds) Minerals in soil environments. Soil Science Society of America, Madison, WI, pp 741–770
Schofield NJ, Bari MA (1991) Valley reforestation to lower saline groundwater tables: results from Stene's farm, western Australia. Austr J Soil Res 29:635–650
Schunke E, Zoltai SC (1988) Earth hummocks (thufur). In: Clark MJ (ed) Advances in periglacial geomorphology. Wiley, Chichester, pp 231–245
Schwartz J, Liu Y (1997) Process for decreasing chlorine content of chlorinated hydrocarbons. Assignee: The Trustees of Princeton University (Patent No. 5,606,135)
Schwertmann U, Murad E, Schuize DG (1982) Is there Holocene red-dening (hematite formation) in soils of axeric temperate areas? Geoderma 27:209–223
Seelig BD (2000) Salinity and sodicity in North Dakota soils. NDSU Extension Service, North Dakota State University, EB 57, May 2000 (available at *http://www.ext.nodak.edu*)
Selby MJ (1993) Hillslope materials and processes. Oxford University Press

Semer R, Reddy KR (1996) Evaluation of soil washing process to remove fixed contaminants from a sandy loam. J Hazard Mat 45:45–47
Senesi N, Miano TM (1992) Abstracts of oral and poster papers. Sixth International Meeting of IHSS, September 20–25, 1992, Monopoli (Bari), Italy
Senesi N, Miano TM (eds) (1994) Humic substances in the global environment and implications for human health. Elsevier, Amsterdam
Shacklette HT et al. (1971) Elemental composition of surficial materials in the conterminous United States. U.S. Geological Survey (Professional Paper 574-D; includes: Al, Ba, Be, Bo. Ca, Ce, Cr. Co, Cu, Ga, Fe, La, Pb, Mg, Mo. Ne, Ni, Nb, P. K, Sc. Na, Sr, Ti. V, Y, Yb, Zn, Zr)
Shacklette HT et al. (1973) Lithium in surficial materials of the conterminous United States and partial data on cadmium. U.S. Geological Survey (Circular 673)
Shacklette HT et al. (1974) Selenium, fluorine, and arsenic in surficial materials of the conterminous United States. U.S. Geological Survey (Circular 692)
Sharma J (1979) Manual of analytical quality control for pesticides and related compounds in humans and environmental samples: a compendium. EPA/600/1-79/008
Sharom MS, Miles JRW, Harris CR, McEwen FI (1980) Behavior of 12 insecticides in soil and aqueous suspensions of soil and sediment. Water Res 14:1095–1100
Shimp JF, Tracy JC, Davis LC et al. (1993) Beneficial effects of plants in the remediation of soil and groundwater contaminated with organic materials. Crit Rev Env Sci Tec 23:41–77
Shineldecker CL (1992) Handbook of environmental contaminants. Lewis Publishers, Chelsea, MI
Shishido T, Ulsui K, Fukami J (1972) Oxidative metabolism of diazinon by microsomes from rat liver and cockroach fat body. Pestic Biochem Phys 2:27–38
Shoji S, Ping CL (1992) Wet andisols. Proceedings 8th International Soil Correlation Meeting, Louisiana and Texas, October 6–20, 1990, pp 230–234
Shoji S, Nanzyo M, Dalgren RA (1993) Volcanic ash soils: genesis, properties, and utilization. Elsevier, Amsterdam
Shotyk W (1992) Organic soils. In: Martini IP, Chesworth W (eds) Weathering, soils and paleosols. Elsevier, Amsterdam, pp 203–224
Siag A, Fournier DJ, Waterland LR (1993) Pilot-scale incineration of contaminated soil from the Chemical Insecticide Corporation Superfund site. EPA Report, pp 192
Sigua GC, Isensec AR, Sadeghi AM (1993) Influence of rainfall intensity and crop residue on leaching of atrazine through intact no-till soil cores. Soil Sci 156:225–232
Silka LR, Swearingen TL (1978) Manual for evaluating contamination potential of surface impoundments. EPA-570/9-78-003 (NTIS PB85-211433)
Silvestri A, Razalis M, Goodman A, Vasquez P, Jones AR Jr (1981) Development of an identification kit for spilled hazardous materials. EPA/600/2-8/194
Sims RC, Doucette WJ, McLean JE, Grenney WJ, Dupont RR (1988) Treatment potential for 56 EPA listed hazardous chemicals in soil. EPA/600/6-88/001
Sims JL, Sims RC, Matthews JE (1989) Bioremediation of contaminated soils. EPA/600/9-89/073
Sims JL, Sims RC, Dupont RR, Matthews JE, Russell HH (1993) In situ bioremediation of contaminated unsaturated subsurface soils. EPA/540/S-93/501 (Engineering Issue Paper, 16 pp)
Singh UP, Orban JE (1989) Innovative isolation of oily hazardous wastes. International Conference on Physiochemical Biology 2:992–1010
Singh G, Kathpal TS, Spencer WF, Dhankar JS (1991) Dissipation of some organochlorine insecticides in cropped and uncropped soil. Environ Pollut 70:219–239
Skidmore EL, Williams JR(1991) Modified EPIC wind erosion model. In: ASA-CSSA-ASSA (eds) Modelling plant and soil systems. ASA-CSSA-ASSA, Madison, WI (Agronomy Monograph 31, pp 457–469)
Skougstad MW et al. (eds) (1979) Methods for determination of inorganic substances in water and fluvial sediments, 2nd edn. U.S. Geological Survey (Techniques of Water-Resources Investigations TWRI 5-A1)
Smith KA (ed) (1991) Soil analysis: modern instrumental methods, 2nd edn. Marcell Dekker, New York, NY
Smith RK (1994) Handbook of environmental analysis. Genium Publishing, Corn Schenectady, NY
Smith KA, Mullins CE (1991) Soil analysis: physical methods. Marcel Dekker, New York, NY
Smith WE, Smith AM (1972) Life (June 2), pp 74–79
Smith WE, Smith AM (1975) Minamata. Holt, Reinhart and Winston, New York
Snyder JL, Grob RL, McNally ME, Oostdyk TS (1994) Supercritical fluid extraction of selected pesticides from fortified soils and determination by gas chromatography with electron capture detection. J Environ Sci Heal A 29
Soil Conservation Service (SCS) (1984) Procedures for collecting soil samples and methods of analysis for soil survey. U.S. Government Printing Office (Soil Survey Investigations Report 1)

References

Soil Survey Division Staff (1993) Soil survey manual. Soil Conservation Service, U.S. Department of Agriculture (Handbook 18)
Soil Survey Staff (1992) Keys to soil taxonomy. Soil Management Support Services Technical Monograph, Soil Taxonomy, Agricultural Handbook No. 436, U.S. Department of Agriculture, F975
Soil Survey Staff (1999) Soil taxonomy, a basic system of soil classification for making and interpreting soil surveys, 2nd edn. U.S. Department of Agriculture
Sparks DL (1989) Kinetics of soil chemical processes. Academic Press, San Diego
Sparks DL (1995) Environmental soil chemistry. Academic Press, San Diego
Stabnikova EV, Selezneva MV (1996) Use of the biological preparation Lestan for cleaning soils of oil hydrocarbons. Prikl Biokhim Mikrobiol 32:219–223
Steinberg CEW (2003) Ecology of humic substances in freshwaters. Springer-Verlag, Heidelberg
Stevenson FJ (1994) Humus chemistry, 2nd edn. Wiley, New York
Stigliani WM, Anderberg S (1993) In: Ayres RU, Sitaottis UE (eds) Industrial metabolism – restructuring for sustainable development. United Nations University Press, Tokyo
Stockham J, Fochtman EG (1979) Particle size analysis. Ann Arbor Science Publishers, Ann Arbor, MI
Stockholm International Peace Research Institute (1971) The problem of chemical and biological warfare: a study of the historical technical, military, legal, and political aspects of CBW and possible disarmament measures. Vol 1: The rise of CB wappons. Humanities Press, New York
Sud RK, Kuman P, Narula N (1986) Role of bacterization in environmental decontamination from pesticides. Environ Ecol 4:156–157
Suffet IH, MacCarthy P (eds) (1989) Aquatic humic substances: influence of fate and treatment of pollutants. American Chemical Sociey, Washington, DC (Advances in Chemistry Series 219)
Sundstrom GC (1994) Risk-based remediation of pesticide contamination. Proceedings of Annual Meeting of Air and Waste Management Association 87, 15 pp
Takei A, Kobski S, Fukushima Y (1981) Erosion and sediment transport measurement in a weathered granite mountain area. In: IHAS (ed) Erosion and measurement transport. Proceedings of the Florence Symposium (IAHS Publication 133:493–502)
Tan KH (2003) Humic matter in soil and the environment. Principles and controversies. Marcel Dekker Inc., New York
Tang Z, Lei TL, Yu J, Sheinberg I, Mamedov AI, Ben-Hur M, Levy GJ (2006) Runoff and interril erosion in sodic soils treated with dry PAM and phosphogypsum. Soil Sci Soc Am J 70:679–690
Tarradellas J, Brecon G (1997) Chemical pollutants in soils: In: Tarradellas J, Bitton G, Rossel D (eds) Soil ecotoxicology. Lewis Publishers, Chelsea, MI
Tarradellas J, Bitton G, Rossel D (1997) Soil ecotoxicology. Lewis Publishers, Chelsea, MI
Terce M, Calvet R (1977) Some observations on the role of Al and Fe and their hydroxides in the adsorption of herbicides by montmorillonite. Z Pflanzenkund Pflanzenschutz (Sonderheft VIII)
Thatcher LL, Janzer VJ, Edwards KW (1977) Methods for determination of radioactive substances in water and fluvial sediments. U.S. Geological Survey (Techniques of Water-Resources Investigations TWRI 5-A5)
Theng BKG (1979) Formation and properties of clay-polymer complexes. Elsevier, Amsterdam
Theng BKG, Churchman GJ, Newman RH (1986) The occurrence of interlayer clay organic complexes in two New Zealand soils. Soil Sci 142:262–266
Theocharopoulos SP, Wagner G, Sprengart J, Mohr ME, Desaules A, Thompson ML, Scharf RL (1994) An improved zero-tension lysimeter to monitor colloid transport in soils. J Environ Qual 23:378–383
Theocharopoulos SP, Wagner G, Sprengart J, Mohr ME, Desaules A, Christou M, Quevauviller P (2001a) Comparative soil sampling in the Dornach site (Switzerland) for soil three-dimensional description. Sci Total Environ 264:63–72
Theocharopoulos SP, Wagner G, Sprengart J, Mohr ME, Desaules A, Christou M, Quevauviller P (2001b) European soil sampling guidelines for soil pollution studies. Sci Total Environ 264:51–62
Thorn KA, Folan DW, MacCarthy P (1989) Characterization of the IHSS standard and reference fulvic and humic acids by solution state ^{13}C and HNMR spectrometry. U.S. Geological Survey (Water-Resources Investigations Report 89-4196)
Thurman EM (1985) Organic geochemistry of natural waters. Martinus Njhoff/Dr. W. Junk Publishers, Dortrecht
Timmons DM (1990) In situ verification of mercury, arsenic, pesticides and PCB-bearing waste. June 24–29, Pittsburgh, PA, p 11
Tipping E (2002) Cation binding by humic substances. Cambridge University Press
Titball RW, Turnbull PCB, Hutson RA (1991) The monitoring and detection of *Bacillus anthracis* in the environment. J Appl Bacteriol 70:9–18
Tomas JR (1993) Glutathione-S-transferase as a biological marker of aquatic contamination. Research Thesis in Applied Toxicology, Portsmouth University, UK

Topp GC, Reynolds WD, Green RE (eds) (1992) Advances in measurement of soil physical properties: bringing theory into practice. Soil Science Society of America, Madison, WI (Special Publication 30)

Troxler WL, Goh SK, Dicks LWR (1993) Treatment of pesticide-contaminated soils with thermal desorption. J Air Waste Manage 43:1610–1619

Truett JB, Holberger RL, Barrett KW (1983) Feasibility of in situ solidification/stabilization of landfilled hazardous wastes. EPA/600/2-83/088

Trüper HG (1982) Microbiological processes in the sulphur cycle through Chernobyl in the Bundesrepublik Deutschland. Forschung Aktuell 19:45–49 (Zeitschrift der TU Berlin, Sonderheft Tschernobyl, Pocahontas Press)

Tüfekçioglu A, Küçük M (2004) Soil respiration in young and old oriental spruce stands and in adjacent grasslands in Artvin, Turkey. Türk J Agric For 28:429–434

UNESCO (1974) Soil map of the world. Food and Agricultural Organization, United Nations, Rome and Paris

UN/ECE ICP-Forest. PCC Hamburg and Prague (1994) UN/ECE International Co-operative Programme on Assessment and Monitoring of Air Pollution Effects on Forests. Manual of methods and criteria for harmonising sample assessment, monitoring and analysis of the effects of air pollution on forests, 3rd edn

United States Army (1967) The unified soil classification system. U.S. Army Waterways Experiment Station, Office of the Chief of Engineers (Technical Memorandum 3-357)

UNSCEAR (2000a) United Nations Scientific Committee on the Effects of Atomic Radiation Reports to the General Assembly of the United Nations with Annexes, vol I. United Nations, New York

UNSCEAR (2000b) United Nations Scientific Committee on the Effects of Atomic Radiation Reports to the General Assembly of the United Nations with Annexes, vol II. United Nations, New York

Unterberg W, Williams RS, Balinsky AM, Reible DD, Wetzel DM, Harrison DP (1987) Analysis of modified wet-air oxidation for soil detoxification. EPA/600/2-87/079

Upsky D, Tuva W, Dorrier R, Johnson B, Gardner M (1989) Methods for evaluating the attainment of cleanup standards. Vol 1: Soils and solid media. EPA/230/2-89/042

USCWC (2003) Mustard gas. (available at: *http://ntp niehs.nih.gov*)

U.S. Department of Agriculture(1975) Agricultural Handbook 436. Washington, DC

U.S. Department of Energy (DOE) Various dates. The Environmental Survey Manual DOE-EH-0053: Vol 1 (August 1987, Chapter 8, 2nd edn., January 1989 – Sampling and analysis phase); Vol 2 (August 1987 – Appendices A, B and C); Vol 3 (2nd edn., January 1989 – Appendix D, Parts 1, 2 and 3 – Organic and inorganic analysis methods and non-target list parameters); Vol 4 (2nd edn., January 1989 – Appendix D, Part 4 – Radiochemical analysis procedures); Vol 5 (2nd edn., January 1989 – Appendices E: Field sampling, F: Quality assurance, G: Decontamination, H: Sample management, sample handling, transport and documentation, J: Health and safety, K: Sampling and analysis plan)

U.S. Environmental Protection Agency (EPA) (1979) Survey of solidification/stabilization technology for hazardous industrial wastes. EPA/600/2-79-056

U.S. Environmental Protection Agency (EPA) (1980) Guide to the disposal of chemically stabilized and solidified wastes. EPA/SW-872

U.S. Environmental Protection Agency (EPA) (1986) Systems to accelerate in situ stabilization of waste deposits. EPA/540/2-86/002

U.S. Environmental Protection Agency (EPA) (1988) Technological approaches to the cleanup of radiologically contaminated superfund sites. EPA/540/2-88/002

U.S. Salinity Laboratory Staff (1954) Diagnosis and improvement of sline and alkali soils. USDA, U.S. Government Printing Office, Washington, DC (Agricultural Handbook 60)

Utchfield JH, Clark LC (1973) Bacterial activities in ground waters containing petroleum products. American Petroleum Institute, Washington, DC (API Publication 4211)

Valentin C (1991) Soil crusting in two alluvial soils of the northern Niger. Geoderma 48:201–222

Vyas BN, Mistry KB (1981) Influence of clay mineral type and organic matter content on the uptake of ^{239}Pu and ^{241}Am by plants. Plant Soil 59:75–82

Wagner G, Sprengart J, Desaules A, Muntau H, Theocharopoulos S, Quevauvillier P (1999) Comparative evaluation of European methods for sampling and sample preparation of soils – design and state of the CEEM soil project. In: Kiene A (ed) Nachbereitung und Auswertung des Europäischen Workshops "Soils in Europe – Elaboration and Application of Investigation Meth-ods". Deutsches Institut für Normung (DIN) e.V., D-10787 Berlin (Abschlussbericht des UBA-Forschungsvorhabens 304 03 002, pp 48–58)

Wagner G, Mohr M-E, Sprengart J, Desaules A, Theocharopoulos S, Muntau H, Rehnert A, Lischer P, Quevauviller P (2000) Comparative evaluation of European methods for sampling and sample preparation of soils. European Commission, Office for Official Publications of the European Communities, Luxembourg (BCR information series, EUR 1971, 206 pp)

Wagner G, Desaules A, Muntau H, Theocharopoulos S, Quevauviller P (2001a) Harmonisation and quality assurance in pre-analytical steps of soil contamination studies – conclusions and recommendations of the CEEM soil project. Sci Total Environ 264:103–117

Wagner G, Lischer P, Theocharopoulos S, Muntau H, Desaules A, Quevauviller P (2001b) Quantitative evaluation of the CEEM soil sampling intercomparison. Sci Total Environ 264:73–101

Wagner G, Mohr M-E, Sprengart J, Desaules A, Muntau H, Theocharopoulos S, Quevauviller P (2001c) Objectives, concept and design of the CEEM soil project. Sci Total Environ 264:3–15

Wagner G, Quevauviller P, Desaules A, Muntau H, Theocharopoulos S (eds) (2001d) Comparative evaluation of European methods for sampling and sample preparation of soils. Sci Total Environ 264

Waksman SA (1932) Principles of soil microbiology. The Williams and Wilkins Company, Baltimore, MD

Wang C, Lyon DY, Hughes JB, Bennet GN (2003) Role of hydroxylamine intermediates in the phytotransformation of 2,4,6-trinitrotoluene by *Myrophyllum aquaticum*. Environ Sci Technol 37:3595–3600

Wannemacher RW Jr, Wiener SL (1997) Trichothecene mycotoxins. In: Sidell FR, Takafuji ET, Franz DR (eds) Medical aspects of chemical and biological warfare. Office of The Surgeon General, United States Army, Falls Church, VA, pp 655–676

Watanabe ME (1997) Phytoremediation on the brink of commercialization. Environ Sci Technol 31:182–186

Watters RL, Hakonson TE, Lane LJ (1983) The behavior of actinides in the environment. Radiochim Acta 32:89–103

Weber J (2005) Definition of soil organic matter. http://www.Humin-tech.com

Wekhof A (1991) Treatment of contaminated water, air and soil with UV flashlamps. Environ Progr 10

Wentz JA, Taylor ML (1990) APEG treatment process and results for chemically degrading PCBs, PCDDs, and pesticides in contaminated soil. Proceedings of 7th National Conference on Hazardous Waste Material, pp 392–396

White RE (1997) Introduction to the principles and practice of soil science. Blackwell, Oxford

WHO (1993) Guidelines for drinking water quality. Vol 1: Recommendations. WHO, Geneva

Wilkins C (1978) The distribution of lead in soils and herbage of Prembrokeshire. Environ Pollut A5:23–30

Wilson WR (ed) (1991) Advances in soil organic matter research: the impact on agriculture and the environment. Royal Society of Chemistry, Cambridge

Wischmeier WH (1976) Use and misuse of the universal soil loss equation. J Soil Water Conserv 31:5–9

Woodruff NP, Siddoway FH (1965) A wind erosion equation. Soil Science Society of America Proceedings 29:602–608

Woods End Research (1997) Guide to Solvita testing and managing your soil. Woods End Research Laboratory Inc., Mount Vernon, ME

Yaron B (1978) Some Aspects of Surface interactions of clays with organophosphorus pesticides. Soil Science Society of America Proceedings 36:583–586

Yaron B, Calvet R, Prost R (1996) Soil pollution – processes and dynamics. Springer-Verlag, Berlin

Yih RY, McRae DH, Wilson HF (1968) Mechanism of selective action of 3',4'-dichloropropionanilide. Plant Physiol 43:1291–1296

Young AL, Reggiani GM (1988) Agent orange and its associated dioxin. Elsevier, Amsterdam

Zacher D (1982) Soil erosion. Elsevier, Amsterdam

Zanchi C (1983) Influenze dell'azione battente dellapioggia e del ruscellamento nel processo erosivo e variazioni dell'erodibilità del suolonei diversi periodi stagionali. Annali Isituto Sperimentale per lo Studio e la Difesa del suolo 14:347–358

Zepp RG, Sonntag CH (eds) (1995) The role of non-living organic matter in the Earth's carbon cycle. Wiley, Chichester

Ziechmann W (1993) Humic substances. BI Wissenschaftsverlag, Mannheim

Zingg AW (1940) Degree and length of land slope as it affects soil loss in run-off. Agr Eng 21:59–64

Index

A

accessory soil minerals 23
acetaldehyde 124
acid
 –, naturally occuring 4
 –, rain 150
  –, formation 151
 –, waters 7
acid-base equilibrium 200
acidification, anthropogenic factors 111
acidity
 –, exchangeable 112
 –, relation to crust formation 109
acrisols 89
ACTET 42
actinolite, $Ca_2(Mg,Fe)_5(OH)_2Si_8O_{22}$ 18
Actinomycetes 28
active cations, in soil water 44
addition 9–10
 –, organic 10
adsorbents, in soil medium 177
adsorption 121
 –, factors 185
 –, on active carbon 271
 –, on clay minerals 54
adsorptive retention 177
aeration 49
Afghanistan 163, 167
Africa 163
Agent Orange 144, 154
A-horizon 13, 58
AIDS 168
Al/Si-ratios 22
alachlor 144
albeluvisols 89
albolls 79
aldose 30
aldrin 140
alfisols 64, 67, 84
 –, formation 66
algae 27
aliphatic compounds 29
 –, nitrogen free 29
alisols 89
alkalinity 51
allophane 69

alloys 119
Alnazcollar mine 126
alpha decay 128
alteration 175
aluminium 69, 127
 –, hydrolysis 51
 –, hydroxy 52
 –, in drinking water 127
 –, mobilisation 111
 –, oxide, surface charge 52
 –, oxyacids, reactions 51
 –, silicates, conversion to clay minerals 4
alumino-silicates 51
Alzheimer (senile dementia) 127
Amadori rearrangement 37
amalgams 118
America 110–111
American Indians 164
Americans 154
amines 33
amino
 –, acids 33
  –, polymerisation 33
 –, sugars 33
amiton 161
ammonia (NH_3) 221
amphibole 4, 18
 –, oxidation 4
 –, examples 18
anatase, TiO_2 23
andisols 68–69
 –, formation 68
andosols 88
anhydrite ($CaSO_4$), conversion to gypsum 4
aniline 144
Annelida 27
annelids, unsegmented 25
anthraquinone 32
anthrepts 78
anthrosols 88
anti-lewisite 159
ants 27
aqualfs 66
aquands 68
aquents 72
aquepts 77
aquerts 86

aquic 61
-, regime 43
aquifer model 255
aquods 83
aquolls 79
aquox 81
aquults 86
arenosols 89
arents 73
aridic 61
-, regime 43
aridisols 70-71
-, formation 70
ARI-Russian peat borer 231
aromatic compounds, nitrogen free 30
arsenic 120, 126, 148, 280
arsine gas (AsH_3) 127
arthropods 25
-, categories 25
As^{3+} 123
As^{5+} 123
ascorbic acid 37
Asia Minor 118
atrazine 144
Australia 86, 167
Austria 134

B

Bacillus anthracis 166
background radiation 129
bacteria 27, 164, 166
Bacteriophyta 28
badgers 25
Bahia grass 276
Balkans 171
barbituric acid 147
basalt 17
base
  -, leaching 110
  -, map 226
  -, saturation
    -, andisols 70
    -, percentage 53
    -, ultisols 85
Basel 125
Becquerel 128
beetles 25
Beggiatoa sp. 221
Belgium 120
benzene 271
benzimidazole 146
  -, structure 147
benzoquinone 32
Berlin 36, 163
beryl, $Be_3Al_2Si_6O_{18}$ 17
beryllium 127
beta decay 128
BHC *see* Lindane
B-horizon 13, 58
Bible 153

bioaccumulation 122
bioavailability, of heavy metals 122
biodegradation 206
biological
  -, monitoring, planning 242
  -, warfare (BW) 164
bioremediation 273
  -, ex situ 274
  -, in situ 274
biota, role in soil formation 10
bioturbation 81
bioventing 274
bipyridyl 144
bird sanctuaries 126
bismuth 120
blue-green algae 28
Bolimow 153
bone disorders 117
Bordeaux mixture 146
botulinum toxins 172
Brassica juncea 277
British 164
Brucella melitensis 168
brucite $[Mg_3(OH)_6]_n$ 18, 22
buffering capacity 112, 201
Burkholderia pseudomallei (Pseudomonas) 167

C

Ca^{2+} 51
cadmium 117, 123, 126, 148, 151
caking 187
calabar 142
calcids 70
calcisols 89
calcite 193
  -, encrustations 24
calcium 111
  -, in acidic soils 51
cambids 70
cambisols 89
Camelford District 127
capillary water 43
cappings 108
Captafol 146
Captan 146
carbamate 142
  -, poisoning 142
  -, naturally occuring 142
carbamic acid, NH_2-COOH 142
carbaryl 217
carbohydrates 29
  -, classification 29
carbonate, horizons 7, 49, 51, 70, 89, 111-113, 201, 203
carbonic acid 13
carcasses of animals 166
cartilage 34
cassiterite (SnO_2) 118
catechol
  -, colouration of plants 39

Index

–, oxidation 39
Catholic University 159
cation
  –, bonding by 54
  –, bridging 55
  –, exchange capacity (CEC) 52, 148
  –, organic soil 53
  –, values in clay rich soils 62
cellulose 40
centipedes 25
centrifuge drainage 233
ceramics 118
CH_3Hg 124
chain silicates 16–18
  –, structure 18
chelation 205, 270
chemical
  –, adsorption 178
  –, equilibrium 10, 183, 247
  –, mobility 199
  –, potential 122
  –, properties 50
  –, warfare 153
  –, weapons 155
Chemical Weapons Convention (CWC) 155
chemisorption see chemical adsorption
chemolithotrophic bacteria 220
Chernobyl 129, 135
chernozems 89
China 171
Chisso Corporation 124
chitin 34
  –, structure 35
Chlamydia 171
Chlamydobacteria 28
chlorine gas 153
chlorite 22
  –, structure 22
chloroform 35
chloropicrin 162
cholinesterase 142
C-horizon 11
chromium 127
cinnabar (HgS) 119
classification
  –, chemical criteria 62
  –, environmental criteria 60
  –, FAO-UNESCO sytem 88
  –, national systems 93
  –, other systems 92
clay 11, 107, 148
  –, flakes, booklets 9
  –, formation and alteration 22
  –, minerals 177
    –, basic structural elements 21
    –, classification 19
  –, organic complexes 54–55
climatic factors 10
CO_2
  –, in protein decomposition series 44
  –, in soil waters 44

coal petrographers 38
coatings 6
  –, hydrous oxides 55
cobalt, simple chloride 203
cocci see spherical bacteria
Cold War 154, 158
  –, soil pollution 162
colour
  –, characteristic 4
  –, chart, Munsel 47
  –, red, yellow or brown 13
  –, soil profile 47
compaction
  –, alleviation 108
  –, causes 107
  –, physical changes 107
  –, testers 107
complex
  –, formation, and bioavailability 205
  –, substances 35
complexing agents 13
concentration gradients 6
condensation 30
  –, reactions, reducing sugars 36
Congo 171
Congo-crimean haemorrhagic fever 171
conservation strategies 105
consistence, of soil grains 48
consumers 11, 25
contaminants, transport 189
copper 119, 123, 148, 151
  –, aquocomplex 203
  –, -zinc alloys 119
Coppersmith 119
Cornwall 127
cox 148
Coxiella burnetii 171
Cr^{3+} 123
Crete 119
Crimea 171
crop rotation 107
crustaceans 119
crusting 108
crust 107
  –, biological 109
  –, depositional 109
  –, formation, avoidance and alleviation 110
  –, mechanisms of formation 109
  –, structural 108
cryalfs 66
cryands 68
cryepts 79
cryerts 86
cryic 61
cryids 70
cryods 83
cryolls 79
cryopedogenesis 75
cryoturbation 73, 75, 81
^{137}Cs 133, 136
cultivation pans 107

Curie 128
cyanides 148
cyanogen chloride 162
cycle, biogeochemical 6
cyclodiene 142
cyclosilicate 17
Cyprus 119

D

Dalapon 146
dancing cats, disease 124
Darcy's
 –, equation, geochemical dimension 257
 –, Law 254
Darwin 27
data acquisition 227
DDT 140
deaminases 209
decay, organic matter 10
decomposers 11, 25
DEFIC 43
defoliation 154
degassing 10
degradation, chemical 110
dehydrases 218
dehydrogenases 216
density 96, 190
deoxyribonucleic acid 34, 168
desert soils 51
desertification 98
detergents 117
diagenesis 44, 53
diagenetic changes, early 53
diagnostic horizon 57–58
 –, subsurface 58
diaspore ($Al_2O_3 \cdot H_2O$) 23
dieldrin 140
diffuse double layer (DDL)
 –, model 177
 –, theory 177
diffusion 121
 –, coefficient 190
dihydric phenols 32
dimethylphosphoramidocyanidic acid 156
dioptase 17
dioxin 152, 281
diquat 144
disaccharides 30
dissolution-precipitation 202
diuron 144
DNA see deoxyribonucleic acid
dolomite 194
Doñana 126
double layer 52
doublet structure 18
drainage 49
DRE (distroy removal efficiency) 281
Dukuchaev, V.V. 57
durids 70
durisols 89
dynamic equilibrium 122

E

Earth
 –, crust 15, 117
 –, pillars 27
Ebola 168, 171
 –, Haemorrhagic Fever 171
Egypt 31, 98, 119
 –, north-western 51
Egyptians 153
Ehrlichia 171
electronic industries 118
eluviation 13
encapsulation 280
enchytraeid 27
enstatite ($MgSiO_3$) 18
entisols 71
 –, formation 73
environment, upper surface 5
enzymatic transformations 207
enzymes
 –, classification 208
 –, lignin degrading 217
epipedon 58
 –, subsurface units 58
erodibility 100
erosion 10, 100
 –, background 100
 –, human activities 101
erosivity 100
ether 35
ethylene dibromide (EDB) 278
EU 151
Eubacteria 27
Europe 111
exchangeable
 –, base content 53
 –, cations 51
Exodus 7:20–21 153
explosives 163
 –, degradation in soil 278
extra landscape flow (ELF) 8, 10, 188

F

FAO 57
Fe^{2+}, oxidation of 28
feldspars 23
 –, as rock constituents 23
 –, chemical weathering 5
 –, main types 23
ferralsols 89
ferrihydrite 69
fibrists 75
field
 –, capacity 43, 49
 –, water holding capacity 98
fish 119, 122, 124, 126, 153
fleas 164
flow
 –, negative 10
 –, patterns, Fortescue 8

Index

–, subsurface 10
fluid
  –, flows, mathematical modelling 252
  –, transport, modelling 250
fluometuron 144
fluorine 127
fluvents 73
fluvisols 88
fly ash 148
foliage, chemical investigation 243
folists 75
Fortescue, flow patterns 8
foxes 25
framework silicates 16, 23
France 125, 151
Francisella tularensis 167
Fritz Haber 153
fructose 30
fuel spills 190
fulvates 54
fulvic acid 35, 40, 53
  –, modes of formation 41
  –, structure 40
fungi 30, 172
fungicides 139, 146

G

gallium 120
  –, arsenades 127
garnet 17
gaseous phase 44
gelepts 78
gelisols 73–75
  –, formation 75
Genoese 164
geochemical flow 8
geologic cycle 4
geological survey 226
geophysical methods 235
Germany 125–126, 148, 163
gibbsite (Al_2O_3) 23
gibbsite ($Al_2OH_6)_n$ 18, 22–23
glass industry 127
gleysols 88
glucosamine 33
glucose 30
glutathione 213
glyceraldehyde 29
glycine 144
glycoside 30
  –, formation 36
glycosylmine, N-substituted 36
glyphosate 144
goethite (Fe_2O_3) 23
gradient, partial pressure 10
grain coats 9
gravitational water 43
Great Britain 120
great group 63
groundwater, different zones 237
Guadalquivir 126

gypsids 70
gypsisols 89
gypsum 110, 112, 114

H

haemoglobin 117
halloysite 20
heavy metal 117, 121, 277, 280
  –, biochemical
    –, classification 125
    –, effects 123
  –, biologically
    –, essential 123
    –, non-essential 123
  –, poisoning 124
  –, uptake 122
Helmholtz 178
hematite (Fe_2O_3) 23
hemimorphite 17
hemists 76
Henna 32
heptachlor 140
herbicides 139, 144, 154
hexagonal symmetry, pseudo 18
highways 149
Hilliers peat borer 229
Hiroshima 131
histamine 146
histels 73
histosols 75–77, 88
  –, formation 76
  –, landscape 76
Holland 126
humates 54
humic
  –, acid 35–36, 53
    –, model (Stevenson) 38
    –, structural model 39
    –, theories of origin 40
  –, substances 169, 186
    –, fractionation 36
humification 11, 81
humin 35, 41
  –, Lichtfouse model 41
humods 83
humults 86
humus 11–13, 23–24, 32, 47, 51, 79, 217, 275
  –, definition by Brady 23
Hungary 134, 151
hydrargyrum *see* mercury
hydrated ferric oxide 23
hydration 4
hydraulic conductivity, apparent 190
hydrogen
  –, bacteria 222
  –, bonding 54–55
  –, cyanide 162
hydrolases 209
hydrologic
  –, maps 226
  –, triangle, method 252

hydrolysis 4, 29, 206
hydromica 20
hydrophobic substances 122
hydroquinone 32
hydroxybenzene *see* phenols
hydroxyhydroquinone 32
hygroscopic water 43
hyperaccumulators 276–277
hyperthermic 61
hyphae 28

I

^{131}I 133, 136
I.G. Farbenindustrie 154
igneous rocks 17, 23
illite 20–21
illuviation 6, 9, 13
imidazole 146
immonium ions 37
imoglite 69
inceptisols 76, 78
 –, formation 79
incineration 280
India 86
indium 120
industrial
 –, revolution 148
 –, waste 124
infertility 127, 172
infiltration 175, 188
inosilicates 17
insect larvae 25
insecticides 139
interlayer 20
 –, bonding, clays 22
ion
 –, exchange 270
 –, exchangers 20
 –, selectivity 182
ionic pollutants, adsorption 182
iron 11, 22, 69, 148
 –, amorphous oxihydride 203
 –, oxidation 6
 –, redox chemistry 204
 –, silicates, conversion to clay minerals 4
isomerases 219
isomorphic substitution 23
isomorphous substitution 51
isopods 25
isothermic 61
isotherms, Langmuir 179

J

Japan 124
Japanese 167

K

K$^+$ 51
Kampfstoff Lost 157

kandic horizons 81
kaolinite 20–22
 –, in humid climates 22
kastanozems 89
Kenya 142
ketosamine, formation 37
Kozlovskiy, flows in landscape 6
land use 227
landscape
 –, geochemical flow (LGF) 188
 –, prism, flows in 6
laterite 82
Latin America 163
lattice
 –, expandable 20
 –, tetrahedral 15
layers 19
leaching processes 11
lead 117, 148, 151
lewisites 158
lice 171
lichens 3
ligases 218
lignin theory 38
lignolytic enzymes 38
liming 111
 –, materials 112
Lindane 140
linings, in interstitial space 6
lipases 210
lipids 35
liquid phase 42
lithotrophic bacteria, classification 221
litter 24, 110, 150, 277
lixisols 89
loss 9–10
 –, of motor control 124
lowsone 32
 –, structure 33
Lumbricid worms 27
luvisols 89
lyases 218
lysimeter
 –, funnel type 232
 –, plate type 232
 –, suction 230
 –, zero-tension 230
lysogenic phase 168

M

macrofauna 25
macroscopic dispersion 192
magnesium 13, 111
 –, in acidic soil 51
Maillard reaction 36
 –, initial stage 37
main migrational cycle (MMC) 188
major environmental accidents 124
manganese 11, 22, 120
 –, amorphous oxihydrides 203
 –, mobilisation 111

–, redox chemistry 204
Manila 171
Marismas 126
MAST *see* mean annual soil temperature
material
   –, loss, particular 10
   –, transfer 6
mean annual soil temperature (MAST) 60
Megascolecides australis 27
melanization 81
melioidosis 167
mercury 117–120, 123–125, 280
   –, interaction with soil 119
   –, uptake by humans 119
   –, volatile compounds 119
mesic 61
mesofauna 25
metabolism 122
metallurgical industries 117, 119, 127
metals, reactions with organic matter 53
metasilicate 17
methane 33
   –, oxidising bacteria 222
methoxychlor 217
methyl alcohol 30
methylated tin 118
Mg^{2+} 51
mica plates, detrital 24
micelles 52
micrite 266
microfauna 25
microorganisms 27
microscopic dispersion 190
Middle East 171
migration
   –, air (gas) 8
   –, cycle, main 6
millipedes 25
Minamata 124
mineral matter 15
mineralisation 8
miscarriage 127
mixed layer-clay 20
models 247–248
   –, analogue 248
   –, mathematical 248
   –, types 248
moisture regimes 43, 61
moles 25
mollisols 79–80
   –, formation 80
molluscicides 142
molluscs 25
   –, large 25
monitoring 225
   –, biological 241
   –, flow directions 238
   –, groundwater 237
   –, hydraulic heads 239
monosaccharide 30
   –, structure 31
montmorillonite 20–21

–, in water logged terrain 23
Moses 153
mucoprotein 34
mustard gas 156
mustards 156
mycelium 28
mycotoxins 164, 172

N

Na^+ 51
Nagasaki 131
nahcolite ($NaHCO_3$) 51
naphthaquinone 32
NAPLs *see* non-aqueous phase liquids
NATO 163
natrification processes, in soil 44
natron ($Na_2CO_3 \cdot 10H_2O$) 51
nematocides 142
nematodes 25, 142
nerve agents 154
nervous system 117
   –, disruption 119
nesosilicates 16
neurotoxin 124, 153
Neuve-Chappelle 153
nickel 117
nicotine, structure 143
Ni-Schrapnell 153
nitisols 89
nitrifying bacteria 110, 221
Nitrobacter 221
nitrogen
   –, fixation 6
   –, fertilisers 111
   –, mustards 158
Nitrosomonas europaea 221
non-aqueous phase liquids (NAPLs)
   –, behaviour in soils 194
   –, denser than water (DNAPLs) 196
   –, lighter than water (LNAPLs) 195
   –, spills, water-saturated soils 195
nontronite 20
NO_x 148
N-substituted glycosylamine, formation during pedogenesis 37
nuclear
   –, accidents 133
   –, debris 133
nucleic acids 33
nucleotides 34

O

Oder 154
O-horizon 11
olivine, $(Mg,Fe,Mn)_2SiO_4$ 16
o-quinone 32
orders 63
organic
   –, compounds, free from nitrogen 28
   –, conditioners 97

–, debris, decomposed 11
–, material
   –, flow 8
   –, soil 6
–, matter 23
   –, classification of 29
   –, content in gelisols 73
   –, dead 28
   –, of low C/N-ratio 96
–, mercury 124
–, pumps 280
–, soil matter, interaction with mineral components 53
–, substances, in soil waters 44
organochlorine compounds 144
organochlorines 140
organophosphorus compounds 139, 155, 187
   –, fluoronited 155
   –, sulphonated 156
organotins, structure 147
Orientia 171
orthels 74
orthents 73
orthods 83
orthosilicates 16
   –, classification 17
osteomalacia 117
osteoporosis 117
outer sphere complexes 203
overburden
   –, map 226
   –, wind blown 11
oxidases 216
oxidation
   –, below the water table 4
   –, in the vadose zone 4
   –, photochemical reactions 150
   –, -reduction 55
   –, state 123
   –, type 202
oxidoreductases 216
oxisols 81–82
   –, characterisation 81
   –, formation 81

P

paints 118
PAHs *see* polycyclic aromatic hydrocarbons
Pancreas disease 120
Paracelsus 120
parent
   –, material 10, 117
   –, rock 4
Paspallum notatum *see* Bahia grass
PCBs 281
PCDDs *see* dioxins
peat bogs 36
ped 47
pedogenesis 3, 8
pedogenic
   –, cycle, active 8

–, processes 7
pedolisation 10
pedological maps 226
pedoturbation 81
PEG (polyethylene glycol) 105
pergelic 61
permeable reactive barriers (PRB) 270
perox 81
peroxidases 216
perudic regime 43
pest 164
pesticide 96, 125
   –, classification 139
   –, definition 137
   –, degradation
      –, hydolases 210
      –, transferases 213
pH 51, 111, 113, 119, 121, 148, 150, 187
   (*see* soil acidity)
phaeozems 89
phase, mineral-solid 15
phenol 30
   –, mono-, di- and tri- 31
phenolic aldehydes 40
phenoloxidase 40
phenoxyacetic acid 144
phosgene ($COCl_3$) 161
phosphorus, deficiency 111
photocopier paper 120
photographs 227
photolithotrophic bacteria 220
photolysis 270
phthalic acid 146
phthalimide 147
phyllosilicates 18–19
   –, structure 19
physotigmine 142
phytodegradation 278
   –, petroleum-based contaminants 278
phytoextraction 276
phytoremediation 275
phytosorption 278
phytostabilisation 276
phytovolatilisation 280
pigmentation, of animals 32
plague 164, 167
planosols 89
plant
   –, indicators 113
   –, nutrients 8
plastics 118, 124
plinthite 81
plinthosols 89
plumbum nigrum 117
plutonium 130
   –, in organic matter 131
podzols 89
Poland 120, 151
pollutant 199
   –, agrochemical 137
   –, alteration 199
   –, categories 265

Index

-, chemical change 199
-, major types 117
-, urban sources 147
pollution 175
 -, chemical warfare 153
 -, diffused 266
 -, highway activities 150
 -, localised 266
 -, mechanisms 175
 -, physical processes 175
 -, scale of 266
 -, sewage sludge 151
 -, sources 137
 -, transport activities 148
 -, waste 151
polyacrylamide (PAM) 110
polychlorodibenzofuran (PCDF) 151
polycyclic aromatic hydrocarbons (PAHs) 149, 151
polyethylene extrusion 280
polyhydroxyaldehydes 29–30
polyhydroxyketones 29
polypeptides see proteins
polyphenol
 -, theory 39
  -, after Stevenson 41
polysaccharides 30
 -, hydrolysis 31
pore
 -, geometry 192–193
 -, size, distribution 49
porosity 49
Portland cement 280
positron emission 128
potassium 13
potentiometric maps 251
POTET 42
potworms see enchytraeid
power
 -, generation 148
 -, stations
  -, coal-fired 148
  -, nuclear 129
Poxviridae 170
precipitation 187, 270
 -, carbonates 8
pressure solution 6
primary amines, condensation with aldehydes 37
privet, Egyptian 32
producers 25
propanil 144, 211
proteases 209
proteins 33
 -, breakdown 33
 -, series 34
Protozoa 25, 27
psamments 73
Pseudomonas 28
Pulex iritans 167 (*see* human flees)
purine 34
PVA (polyvinyl alcohol) 105

pyrethroids, natural 142
pyridine 144
 -, as a solvent for lignin 38
pyrimidine 34
pyrite 150
 -, oxidation 4
 -, raspberry shaped 24
pyrogallol 32
pyroxene 17
 -, examples 18
 -, oxidation 4

Q

Q-fever 171
quartz 23
quinones 32
 -, structure 33

R

rabbits 25
radioactive
 -, emissions, natural 129
 -, materials 280
 -, nuclides 134
 -, particles, depositional patterns 134
radionuclides 148
 -, uptake 131
rain forest 171
raindrop impact 107
realgar (AS_4S_4) 127
red tide 153
reducing type 203
reductive dechlorination 271
reductones 37
regolith 11
regosols 89
Reims 154
remedial techniques
 -, chemical 269
 -, oxidation 269
remediation, technologies 267
remote sensing methods 237
renal dysfunction 117
rendolls 79
resorcinol 32
retention 175
 -, curves 49
 -, nonadsorptive 187
Rhine
 -, alluvium 126
 -, pollution 125
rhizofiltration 277
rhodonite (Mn_3Si3O_9) 17
R-horizon 11
ribonucleic acid (RNA) 34, 168
ribose, structure 35
ricin 172
Rickettsiae 171
rift valley fever 170
ringed worms 27

risk level 267
RNA *see* ribonucleic acid
road cuts 150
rodents 166
  –, role in soil formation 10

S

salids 70
saline soils 112
  –, classification 112
salinity
  –, amendment 114
  –, indicators 113
salinization 112
sampling 228
  –, foliage 243
  –, litterfall 243
  –, procedures 229
  –, soil
    –, air 233
    –, solution 230
  –, solid soil matter 229
sand 11
Sandoz works 125
saprists 75
sarin 156
  –, structure 157
saxitoxin 160, 173
schedule
  –, 1 155
  –, 2 155, 161
  –, 3 155, 161
Schiff's base 37
sealing 108
seals 108
seawater, magnesium-rich 22
sedimentary
  –, cycle
    –, main 6
    –, pedogenic processes 7
  –, rocks 6, 70, 80, 86
selenium 280
series 63
sesquioxides 23
sewage sludge, heavy metals 152
sheet 19
  –, pores 193
  –, silicates 16, 18
silesia 154
silica
  –, colloidal 4, 13, 18–19, 23, 44, 49, 70, 89, 186
  –, leaching during eluviation 13
silicate 15
  –, tetrahedron 15
silicon 15
siloxane sheets 18
silt 11
silver 120
simazine 144

site characterisation 226
skeletal
  –, crystals 194
  –, fluorosis 127
Skierniewice 153
skins, on coarser grains 6
sludge 118, 151
slugs 27
smallpox 164, 168
smectite, 20
snails 27
sodic soils 112
  –, classification 112
sodicity, indicators 113
sodification 112
sodium 22, 51
  –, chloride 112
  –, sulphate (Na_2SO_4) 112
soil
  –, acidity 50
  –, avoidance and alleviation 111
  –, air 44–45, 106, 151, 190, 234, 272
    (*see* gaseous phase)
  –, major constituents 45
  –, average composition 15
  –, bulk density 96
  –, classification, American system 57
  –, colour 69
  –, compaction 105
  –, conditioners 105
  –, definition 6
  –, degradation 95
    –, human induced 98
  –, Turkey 99
  –, erosion
    –, control 104
    –, measurement 101
    –, modelling 103
  –, fertility
    –, exhaustion 99
  –, flushing 273
  –, formation 15, 100
  –, loss tolerance 104
  –, mantle 6
  –, matrix 185
  –, moisture regimes, in pedons 61
  –, monitoring, steps 227
  –, morphology 11
  –, nitrate 98
  –, on recent volcanic deposits 23
  –, organic matter, decomposition of 28
  –, orders of taxonomy 60
  –, organisms 23–25
  –, physical properties 47
  –, pollution, biologically controlled 175
  –, profile 12
  –, quality 95
    –, assessment 97
    –, biological indicators 95
    –, chemical indicators 98

–, definition 95
–, impact of acidity 111
–, indicators 95
–, physical indicators 96
–, respiration rates 96
–, rock dominated *see* desert soils
–, semi arid regions 61
–, solution, investigation 235
–, taxonomy
   –, 12 orders 64
   –, criteria of classification 57
   –, morphological criteria 57
   –, system 57
–, temperature regimes 60
–, tropical 61
–, types and classification 57
–, washing 272
–, water 42
   –, composition 44
   –, dominated 51
   –, principal constituents 45
   –, zones 6
solid phase, activation 4
solidification, bitumen-based 280
solonetz 89
solum 11, 58
soman (GD) 156
   –, structure 157
Soviet Union 156, 164
SO_x 148
Spain 126
speciation 123
   –, radionuclides 129
spherical bacteria 27
spodosols 82–84
   –, formation 83
springtails 25
stacked triplets 20
stacks 19
staphylococcal enterotoxin b 173
sterols 35
   –, in humin 42
storage 122
straining 187
Strecker aldehyde 37
   –, formation 38
structure
   –, essential types 48
   –, primary particles 47
subgroup 63
suborders 63
sucrose, hydroysis 31
Sudan 86, 171
sugar-amine condensation, theory 36
sulphates 148
sulphides 148
sulphur 114
   –, bacteria 220
   –, mustards 157
   –, oxidation 6

sulphuric acid (H_2SO_4) 4, 111
SURPL 43
Swedish Institute of Defence Research 156
Switzerland 125–126

T

Tanzania 142
tars 148
tea plant 127
tectosilicates 23
teratogenic effects 124
termintaria 27
termites 25
terra rossa 6
   –, colour 4
   –, derivation from lime-stone 6
texture 96
thallium 126
The Soil Loss Estimator for Southern Africa 104
Thelaspi caerulescens 276
thermal treatment 280
thermic 61
Thiobacillus
   –, *denitrificans* 220
   –, *ferrooxidans* 28
   –, action of 28
Third World 163
thread-shaped bacteriia *see* Chlamydobacteria
ticks 171
tillage pans 107
tin 117, 151
tinfoil 118
titanium 120
topaz, $Al_2(FOH)_2SiO_4$ 17
torrands 68
torrerts 87
torric 61 (*see* aridic regime)
torrox 81
toxic
   –, chemicals 155
   –, sludge 126
toxicity 123
toxins 172
transacylases 211
transaminases 212
transferases 211
transformation, bacterial action 219
transmethylases 212
transphosphorylases 211
transport
   –, effect of pore geometry 190
   –, of heavy metals 121
   –, routes 6
trapping 187
tremolite, $Ca_2Mg_5(OH)_2Si_8O_{22}$ 18
triazine 144
trichlorethylene 269
trigonal ring (Si_3O_9) 17

2,4,6-trinitrotoluene (TNT) 278
triplet 18, 20
 -, structure 18
tularemia 168
turbels 73
Turkey 98, 151
typhus 171

U

udands 68
udepts 79
uderts 87
udic moisture regime 43, 62
udults 86
ultisols 84–86
 -, formation 86
umbrisols 89
Umweltbundesamt 131, 163
United States 120, 131, 151
 -, Air Force 154
 -, Army 168
 -, Congress 167
 -, Department of Agriculture 59
 -, Soil Conservation Service 57
UO_x 148
uptake, passive 122
urbanisation 98
ustalfs 66
ustands 68
usteps 79
usterts 87
ustic
 -, moisture regime 62
 -, regime 43
ustolls 80
ustults 86

V

Van der Waal's forces, distribution on clay surfaces 54
vapour extraction, soil (SVE) 271
variola 170
Venezuelan equine encephalitis 173
vertebrates 25
vertisols 86–88
 -, formation 88
V-gas 156
Vibrio cholera 168
Vietnam 154
viruses 164, 168
 -, fate in soils 169
 -, transport in soils 169
viscosity 190
vitrification 280
volcanic soils 110
VX 156
 -, structure 157

W

Waksman (1932), lignin theory 38–39
Wales 121
war planners 164
Warsaw 153
waste, municipal 151
water
 -, -balance equation, Thorntwait 43
 -, colours 120
 -, holding capacity 96
 -, in hydration, envelope around cations 4
 -, infiltration rates 97
 -, logging 99
waxes, animals and plant 35
weapon tests 129, 131
weathering
 -, biological 3
 -, chemical 3
 -, mechanical 3
 -, physical 3
white arsenic (As_2O_3) 127
wilting point 43
wind erosion 100
 -, prediction equation 104
wiring 119
wollastonite 17
wood lice 25
World
 -, Food Summit 99
 -, Reference Base for Soil Resources 88
 -, War
  -, I 153
  -, II 154

X

xeralfs 66
xerands 68
xerepts 79
xeric
 -, moisture regime 62
 -, regime 43
xerolls 79
xerults 86

Y

Yersinia pestis 167
Ypres 153, 157

Z

Zaire 171
zinc 120, 151
 -, mobilisation 111
zircon 17

Printed by Printforce, the Netherlands